KINGSLEY AMIS

KINGSLEY AMIS

A Biography

ERIC JACOBS

St. Martin's Press

New York

A THOMAS DUNNE BOOK.
An imprint of St. Martin's Press.

ISBN 0-312-18602-9

First published in Great Britain by Hodder and Stoughton, a division of Hodder
Headline PLC

First U.S. Edition: July 1998

10 9 8 7 6 5 4 3 2 1

To Gillon Aitken
Emma and Daniel

Contents

List of Illustrations

Robert Conquest. *(Courtesy of the Daily Mail.)*
Anthony Powell. *(Courtesy of the Daily Mail.)*
Elizabeth Jane Howard.
Amis and Jane flanked by Mary and Mike Keeley in Greece in the mid-1960s.
Amis and Jane outside Lemmons in the 1970s.
Sally Amis.
Martin Amis. *(Courtesy of the Daily Mail.)*
Philip Amis. *(Courtesy of Richard Mildenhall.)*
With washing in Regent's Park Road in the 1980s.
Amis with moustache and broken leg after an accident in 1982.
Home and away: the Welsh party and the London team.
At St John's College, Oxford, 1991.
With Sarah Snow the cat.

(All illustrations courtesy of Lady Kilmarnock except where otherwise stated.)

Preface

Over the two and a half years it has taken to write this biography, I have usually met Kingsley Amis three or four times a week, in the Garrick Club and the Queen's public house or at his home in Primrose Hill. We have also visited some of the scenes of his earlier life: Norbury, Berkhamsted, Oxford, Swansea, Barnet, Hampstead. During all these excursions a good deal of eating and a great deal more drinking has gone on. So I have seen and heard Amis in the broad spectrum of his habitats and moods, from the deeply downcast to the hilariously elated. But, particularly in the sessions at his home, he has also answered my questions; sometimes reluctantly, sometimes with difficulty – memory being an imperfect instrument of recall – but always, I believe, honestly.

This is an authorised biography and Amis has therefore been free to go through the text, correcting as he went. But he has used his authority lightly. Most of the changes he has insisted on have been small points of fact or punctuation – he was particularly stern about my negligent use of the hyphen – and they have led to definite improvements in accuracy rather than plastic-surgical alterations designed to render his appearance more pleasing. I genuinely do not know whether he likes or dislikes what I have written, what matters of substance he might have preferred me to put differently or to add or leave out altogether. If he would rather his authorised biography presented a different version of himself to the world, he has not said so. And he has not asked me to remove anything at all that might seem to show him in a less than shining light.

The issues on which we have had the most argument have not

involved any event in his life or side of his character but rather the extent to which he himself is to be found in his work. Amis utterly rejects the idea of a novelist as someone who writes up his own life and calls it fiction. In the Preface to his *Memoirs* he insisted that his own novels were 'firmly unautobiographical'. Yet in the same paragraph he also declared that he had 'written an account of myself in twenty or more volumes, most of them called novels', in which every word 'inevitably says something about the kind of person I am'. So Amis is after all to be found in his fiction. But where? It is temptingly easy to match him to the hero of his first novel, *Lucky Jim*, because both were lecturers at provincial universities. As a human being rather than a job-holder, though, Jim Dixon is less like Amis than is Maurice Allington, the hero of *The Green Man*, whom few people identify with Amis because Allington runs an inn and Amis has never done anything of the sort. Amis steered me through such easily made misreadings and many more complex ones too. Even so, I thought I had found traces of him where he did not think they existed. We argued a good deal about this. But we agreed when Amis summed up our differences by saying that I probably saw more of him in his fiction, and he probably saw less of himself, than was there. If the resulting compromise between what I saw and he saw is in any way inaccurate, then I can only say it would have been terrifyingly much more so if I had been left to interpret his novels on my own.

I have of course done as much as possible of the other things a biographer is expected to do: talk to colleagues, relatives, friends; unearth whatever letters and other documents can be found; watch videos and listen to recordings. But it is on a close association with Amis that this biography is above all based. Partly for this reason, I have kept the scholarly apparatus to a minimum. To have loaded these pages with footnotes and references would have given a misleading impression of the balance and accessibility of my sources, which have predominantly been Amis himself and his published work, the first wholly inaccessible except with his consent, the second to be freely looked up in any bookshop or library. Besides these, my other principal source has been Amis's correspondence with his friends Philip Larkin, Robert Conquest and Anthony Powell, which is also unpublished and to which it is therefore pointless to provide references. In any case an elaborate apparatus would raise the costs and therefore the price of the book, which seems a bad idea, while making the work of Amis's next biographer easier, which doesn't seem much better.

I have tried as far as possible to identify the source of quotations in the text so that readers can know what they are reading without going to the trouble of looking to the foot of the page or the back of the book. In the Bibliography there is a full list of Amis's works which will help those who wish to explore further, and there are brief chapter notes beginning on page 361. Where unsourced phrases appear in quotation marks it is simply to indicate that the words used are Amis's, spoken or written, not mine.

This is not a literary biography in the critical sense. I have not attempted to evaluate Amis's novels and other writings against each other or measure them against other writers' work. Nor have I tried to describe the whole range of Amis's output, which a glance at the Bibliography will show is formidably large and varied. That list is itself an important part of the biography. For it demonstrates what is too often forgotten: that a writer, certainly one as prolific as Amis, spends most of his time alone at his desk grinding out words, not doing exciting things to liven up his biography, like getting drunk, making love or fighting wars. If I have referred to some of Amis's works and not others it is because I think some reveal the man and others do not. No judgement of quality, but only of biographical relevance, is intended.

It will be enough indication of my own feelings about his writing if I say that one of my main motives for embarking on this biography was the realisation that Amis was the only writer whose latest novel I still customarily bought on publication at the full hardback price – the kind of critical appreciation Amis himself most values, and rightly so. I should add that working on this book has been the most tremendous fun for me and he has had the good grace to tolerate my often irritating enquiries, ejecting me from his house, not usually, I think, from exasperation, but so as to watch *The Bill* on television without interruption. As I hope he knows, I am very grateful to him indeed for all the pleasure and instruction I have enjoyed in his company.

Acknowledgements

Many people and institutions have been most kind and generous with their help. They are listed and thanked below. But one institution has been consistently obtuse and obstructive and deserves a special mention: the Bodleian Library at Oxford. No campus novelist would be thought other than riotously implausible if he based a plot on my dealings with this extraordinary establishment. With Amis's consent, I asked permission to look at his letters to his friend Philip Larkin which are deposited at the Bodleian along with other papers of Larkin's. The library said no. Expanding on its reasons, it declared that it was Bodleian policy not to open up for inspection or use the letters of someone still alive, on the grounds that other living people might be hurt by what was said in them. I pointed out that the fact of a writer's death need make no difference to the hurtful effects of his letters. Larkin himself was dead; his letters had been published; and they were undoubtedly hurtful to some still alive, including Amis himself. But the library was unmoved. It remained unmoved even when the Chancellor of Oxford, Lord Jenkins of Hillhead, and the then Secretary for Education, John Patten, pitched in on my side. Nothing, it seemed, would budge the Bodleian from the policy it had arbitrarily imposed on itself. So I finished a version of the biography without the letters.

Meanwhile, however, Amis had had a bright idea. Why should he not write to the library and ask for copies of his own letters, the cost of making which he would pay? Lo and behold, the library agreed. Days after my first version of this book reached the publisher, copies of Amis's letters to Larkin began to arrive at his home and I began to rewrite large sections of my book. Even then, the library moved at glacial speed. It pleaded that copying was a time-consuming business, but all offers to pay a bit extra or send someone to do the copying were ignored. It took the Bodleian all of five months to finish the job of copying and posting 500 or 600 letters.

For a library to be so rigid in its policy, so unaccountable to anybody except its own Iron Guard of curators, so ready to fall for a simple ruse to get round to its own rules and then so ponderously slow in executing a simple task suggests a stunning combination of institutional fatuity, credulity and inefficiency. I can only hope that this brief account of its performance will act as the verbal equivalent of a small charge of dynamite placed beneath its comprehensive ineptitude – although I suspect that only a long overdue change to its statutes will do that trick.

I should like to record my thanks to the following organisations for giving me the freedom of their shelves and files: Victor Gollancz Ltd, the *Daily Mail*, the Huntington Library, the BBC Archives at Caversham, the Imperial War Museum, the London Library, the Public Records Office and St John's College, Oxford. My gratitude to the BBC is qualified by the fact that it is the only person or institution I have consulted to make a charge – £40 a day – for the privilege of looking in their records. By happy contrast, I am extremely grateful to the British Academy and The Huntington for funding between them an exchange scholarship which enabled me to study the Amis archive at the Huntington Library in California.

Of the many people who have helped bring this book to birth, I owe particular gratitude to the enthusiasm, encouragement and imaginative energies brought to bear by the team at Hodder: Richard Cohen, Angela Herlihy and Roland Philipps. I should also like to thank Linden Stafford, whose care, skill and subtlety as an editor are beyond compare. So delicate and sensitive were her improvements that I thought I must have made them myself, which may sound unpleasantly egotistical but will I hope be recognised by Linden for what it is meant to be: the highest form of praise that could be offered by a member of that egotistical species, the writer. And I must not leave out my daughter Emma just because she is my daughter. Her researches have been invaluable.

I am also grateful to the following individuals for the kindness and help they have shown me in any number of ways, from digging into their memories to lending me an Oxfordshire cottage in which to write. Some, of course, have been more kind and helpful than others; but it would be invidious as well as tedious to give full details of the scale and nature of all their contributions to this book. In thanking them equally they should know that I deeply appreciate what each has done personally. It would have been good clean malicious fun to give a second list, of those who declined to help, so that readers could entertain themselves by wondering why. After a mild struggle, though, I have decided to resist that temptation, on the grounds that the list would be short and achieve no more than to stir up unnecessary confusion and avoidable bad feeling. I apologise for any oversight which may have led to unintended omissions from the list that follows:

Brian Aldiss, John, Martin, Philip and Sally Amis, Sir Peter Baldwin, Sir George Blunden, Sir Mervyn Brown, Brie Burkeman, Esmond and Jean Cleary, Jonathan Clowes, Robert Conquest, Maurice Cowling, Simon Curtis, Sam Dawson, E.W. Denham, Annis Nicholson Desvergnes, L.R. Dowsett, John Doxat, Dr James Durham, Sir Edward du Cann, Richard Eyre, Professor Antony Flew, Russell Fraser, Paul Fussell, Clive Gammon, Davina Gammon, Richard Perceval Graves, William Hallett, Dr William Hayes, Richard Hough, Colin Howard, Elizabeth Jane Howard, Noel Hughes, Norman Iles, Paula Jabloner, Edmund Keeley, Delphine Kelly, Professor and Mrs E.J. Kenney, Alastair and 'Hilly' Boyd (Lord and Lady Kilmarnock), Richard Law, J.E.B. Marsh, John McDermott, Cyril Metliss, James Michie, Colonel Eric Milner, Dame Iris Murdoch, Mavis Nicholson, Anthony Powell, L.F. Richenberg, Hilary Rubinstein, Virginia and Michael Rush, Dale Salwak, James Silberman, Joyce Sims, John Terraine, Eve and Stuart Thomas, Donald Trelford, Dr Malcolm Vale, C. N. Vokins, John Wain, Brian Wormald and Sir Peregrine Worsthorne.

KINGSLEY AMIS

Introduction

Portrait of the Artist in Age

Kingsley Amis sits at a table upon which rests a glass of whisky. It is a large Macallan malt with a splash of water, his most frequent drink. He leaves the glass untouched for a few moments, now and then glancing at it expectantly. The delay is hardly needed to sharpen his appetite for this first drink of the day. But it gains time and so helps make sure his whisky ration will last the hour or more to lunch, which it otherwise might not do, even though the ration includes two more large malts or just possibly two and a bit or maybe even three.

The time of this little ritual is half-past twelve if the location for it is the Queen's public house, just along the road from Amis's home, or soon after one o'clock if it is the Garrick Club, which are the two places where Amis is most regularly to be found at this time of day. It is with his first whisky that he brings to an invariably satisfying close his not always satisfying morning, which began when he rose at 7.47 – on days, that is, when sleeplessness has not roused him an hour or two earlier.

At whatever time he wakes, doing so is often not much fun, since Amis is apt to find himself in a state of mind not far short of utter dread about life and complete despair at the novel he is currently writing. For at this early hour both his life and his novel can look very bleak indeed, scarcely worth going on with in either case.

So Amis rises in a mood of depression from his narrow, uncompromisingly single, bed in his ground-floor bedroom. He puts on dressing-gown and slippers and climbs upstairs to the kitchen on the floor above to make himself a cup of tea. Then it's downstairs again to start on the wearisome business of morning

1

ablutions in his bathroom next to the bedroom: shit, shave, shower and dandruff-destroying shampoo, always in that order and always finishing with his hair carefully Brylcreemed and parted on the right side, as it has been for decades. Next it's into the bedroom and on with socks, underpants, shirt and trousers, supported these days by braces, a belt being no longer a reliable device for keeping trousers in place against the downward-thrusting weight of the belly. For like Peter Thomas in his novel *The Old Devils* the Amis figure has undergone a 'historic escalation', which has made it necessary for Amis to buy most of his outer clothing at an outsize shop.

These 'getting-up procedures', as Peter Thomas calls them, have stopped being 'what you hurried heedlessly through before you did anything of interest'. They have become a major event in their own right, though conducted in strict privacy, as is only right for an oldster, whether Thomas or Amis. The shit, for instance, once taken in Amis's stride, has become outstandingly important because success or failure in this department can colour the mood of the entire day. Shaving, too, now involves intricate tracking through the folds and wattles of a jowly face, which once presented a smooth ride for a safety-razor. And then there are surgical stockings to be grappled with, shoes to be got on – the list of once non-existent problems seems endless, though they are made a little easier to bear by the track or two of jazz Amis often listens to on his cassette player. Faced with a similar early-morning obstacle course, another character in *The Old Devils*, Malcolm Cellan-Davies, contemplates giving in and spending the whole day in pyjamas and dressing-gown. But this would not do for Amis. You cannot go to the club or the pub in pyjamas and if you wish to live anything like a normal life you must go through the procedures of normality, even if minor steps along the way have taken on proportions of monstrous size and farcical complexity.

Washed and dressed, Amis is ready to climb the stairs again to the dining-room next to the kitchen where his breakfast has been laid for him by the person he sometimes likes to call his first ex-wife, Hilly. He doesn't much enjoy what he eats at this meal – grapefruit and cereal with milk, unsweetened, and a piece of banana, plus more tea – but he has been told that it will do his liver good and, being a fervently committed hypochondriac, he dutifully gets it down. While eating and drinking he studies the *Daily Mail* and *The Times* and while he is doing that Alastair Boyd, otherwise Lord Kilmarnock, Hilly's third husband, nips upstairs from the basement flat and makes Amis's bed for him. To hear about at a distance, this seems a curious ménage: a chap living in the same house as his ex-wife and her current husband. Like some bizarre domestic arrangement in a novel by the old friend he first met at Oxford, Iris Murdoch, Amis wryly concedes. But strange though it may be it suits him and them very well.

By 9.30 or a little later breakfast and the newspapers have both been consumed and he goes downstairs again to the study at the front of the house through the door from his bedroom. This study is no gracious literary man's parlour but a nonsense-free writer's workshop. The walls are fitted with plain, sturdy bookshelves and the floor is scattered with books and papers, often to a density that can make navigation around the room a matter of some difficulty. There is a photocopier, a desk and two cluttered tables, on one of which, beneath the glare of ceiling strip-lights, sits a typewriter, a heavy Adler office manual which Amis bought in the 1960s from a shop near the BBC's Broadcasting House because he thought it looked impressive, and which he has stuck with ever since. A stranger might think Amis's study remarkably undistinguished for a writer so distinguished as to be a Sir and a Booker prizewinner, but Amis does not notice. He is genuinely indifferent to his surroundings. So long as the essentials are there – drinks, typewriter, books, television, bed, bathroom, and so forth – he is happy to ignore the rest.

Amis justifies to himself the time spent on his breakfast scrutiny of the newspapers with the argument that it is an important part of the writer's job to keep up with what is going on in the world. There is something in this but not a lot, as he well knows. He does like to keep up, if only to keep himself in raw material for conversation. But reading newspapers also puts off the awful moment when he must get down to work. Could today be the morning when he runs out of drive at last? Like death, that day must surely come. The typewriter must therefore be approached in propitiatory mode, like a trainer advancing towards the lion's cage or a priest approaching the altar of an unpredictable god. He may panic when he reads through what he sees on the sheet of paper left overnight in the typewriter. 'This is no good,' he tells himself, 'no good at all!' But he calms down, taps out ten or so words and perhaps goes off to have a pee before returning to settle in for the morning.

If, though, things have ended up last night the way he always intends they should, Amis will find that he left off writing in the middle of a passage which he has a fairly clear idea how to continue. Under these circumstances the day's work will get off to a reasonably easy start. And as the morning goes by, the grey clouds of the early hours will disperse to let the sun peep, even blaze, through. A few hundred new words on paper, on top of bathing, dressing, breakfast, the newspapers and, with luck, a really decent shit, have somehow magically banished all that black depression along with any last remaining trace of the night before's drinking. Amis can be heard laughing out loud at what he is writing. No longer does the novel in hand seem utterly and completely worthless as it did when he first came to. Now it seems not bad, rather good – in fact, very well worth going on with.

The morning's progress will depend on all sorts of things but it will be easier, as it always has been for Amis, if he has dialogue to write, much less easy if it is narrative or description. Writing of whatever kind, though, always comes desperately hard, even now that Amis has written more than twenty novels and much more besides. T.S. Eliot, he thinks, got it absolutely right when he wrote, in 'East Coker', that 'every attempt / Is a wholly new start'. Amis brims with confidence when he lunches with his publisher. But alone, in front of the typewriter, he is a beginner again and all that accumulated experience seems to count for nothing. The process of writing a novel he sees as like making a journey from London to Edinburgh. You may know the first ten miles or so pretty well but thereafter, apart from the names of a few places scattered along the way, you know nothing and must work out every mile of the route to your destination as you go along. This is what he doggedly does each morning: a deskbound traveller groping his way through the unexplored fictional territory that always stretches foggily ahead.

Amis concentrates closely on the nuts and bolts – the mechanics – of his art. What preoccupies him are such practical matters as how a character can best be got out of the pub and into a taxi. Conrad, whom Amis doesn't much admire, was at least good at that, at getting a chap from ship to shore. The 'how' of novel-writing is what interests Amis, not the 'why' or the 'what for'. Talk of artistic aims or any other literary abstraction makes him uneasy. For one thing, it tends to phoniness and posturing, which he absolutely detests in any form, whether in art or in life. For another, he already knows perfectly well what his own artistic aim is. It is what it always has been: to write a good story about recognisable human beings in readable – which most definitely does not mean experimental – prose, the upshot of which will be a novel that he himself would enjoy reading.

The canon of English fiction and poetry that Amis personally endorses is small. Many sacred cows are unsacred to him. If it were possible to imagine Amis on a protest demonstration in the 1990s – difficult, since he has not taken to the streets since Suez in 1956 – and if he were to set up his banners outside some university Eng. Lit. faculty, the legends on them would probably read 'Down with Dickens! E.M. Forster! D.H. Lawrence! James Joyce!' Not to mention a fistful of more recent writers from Anthony Burgess to William Golding. Oh, yes, and the author of *Beowulf* – especially down with that Old English bore! Each of these Amis believes to be overrated, mainly because he finds them tedious or false, although he will allow that Chaucer and even the author of *Beowulf* have something to be said for them if you compare what they wrote with the work of their even more unreadable contemporaries.

Amis is as concerned with his reputation as any writer, and he would certainly like his own novels to be included in the popular English canon – but as novels people genuinely want to read for their own pleasure and enjoyment, not as monuments coldly admired because that is what the ringmasters of the literary circus have declared to be admirable. Amis is a free-marketeer by conviction and nowhere more so than in his own field. He believes the writer must satisfy his customers or go to the wall, just like the widget-maker.

To him there is no literary problem more worth bothering about than that of actually writing a novel or a poem. Worrying about the place of fiction in the late twentieth century and other such airy theoretical issues only puzzles the mind and puts the writer off his real business, which is writing. Doubts of this kind, he has long suspected, helped put his best friend, the poet Philip Larkin, off writing any more novels after he had written two in his early twenties, and Amis is determined not to let himself be snared in the same trap of self-doubt and impotence.

The present typescript being a first draft, Amis 'x-es out' words and phrases as he goes along and inserts new ones above the line, their place in the text marked by an oblique slash, in the time-honoured style of writers who use manual typewriters. Every ten pages or so he makes photocopies of what he has written and sends them to his agent for safety – just in case the house burns down or some other act of God fouls things up. Amis has never been a religious believer but he is by no means immune to superstitions of luck and chance. Handwritten changes to the text will come later, followed by a second and final draft. Sometimes, though not very often, Amis can get stuck for an hour or longer on a shortish passage. Can this piece of necessary business be got through in one sentence or will it take two? And if he has to go the longer way round how can he do so without the risk of being boring?

Boredom! For Amis, this is absolutely central, the crux of it all. Being a bore is the worst offence in fiction and not much less than the worst offence in life. He himself registers boredom in the way that a dog picks up frequencies inaudible to others. So, as he works, he is constantly aware that if what he writes bores him then it will surely bore the reader too, and that must be avoided above all other possible shortcomings.

The target for this day as for all days is a minimum of 500 words, the very number Philip Larkin set for his mythical novelist, 'the shit in the shuttered chateau / Who does his five hundred words / Then parts out the rest of the day / Between bathing and booze and birds'. Larkin fantasised about this Maugham-by-the-Riviera lifestyle but never came near to achieving it and probably wouldn't have much enjoyed it if he had. Amis has achieved the 500-word output and the wealth that could buy a pretty adequate version of the lifestyle, but he would not for one minute dream of living on the Riviera or anywhere else abroad – a large and dreadful place, strenuously to be avoided. What's wrong with London, he asks, and rarely stirs beyond it. What about Larkin's 'birds', then? Well once, yes indeed, that side of the writer's reward would certainly have appealed to Amis. But nowadays women as 'birds' do not interest him and he is content to settle for the booze part of the Larkin deal, beginning with some Macallans before lunch and proceeding at intervals throughout the day. Should someone be so injudicious as to ask if he 'needs' another drink he will very likely reply in the words of King Lear when asked if he really needs any knights, attendants or followers: 'Oh, reason not the need: our basest beggars / Are in the poorest things superfluous.'

Whatever the tally of words turns out to be, and whether cheered or cast down by the quality of the morning's output, Amis has usually quit the typewriter by about 12.30. Having rescued his false teeth from the glass in which they have been cleaning themselves since soon after breakfast and glued them firmly in place, he puts on tie and jacket and gets ready for the taxi which will take him to the pub or the club or, once a week or so, to a restaurant.

Seen fully clothed for the public world beyond his own house in Primrose Hill near Regent's Park, Amis does not cut the dash of a particularly snappy dresser. Although he has been heard to express admiration for those who are well dressed and occasionally to voice the desire to be well dressed himself, he never seems to get round to doing anything about it. He did once rather enjoy wearing boldish socks and ties but anything much more would require a lot of shopping and that is one of the many things that bore him rigidly stiff. His second ex-wife, Jane, has described how glum he was for several hours after she dragged him into a shoe shop in Rome when she discovered the pair on his feet were rather worn. Now Hilly does the shopping, and today Amis is quite likely wearing the pair of shoes she has recently managed to purchase for him at the astonishingly low price of £6. He shows them off with a touch of smug satisfaction, as if to prove that, cheap though they may be, they go as well as the next chap's with the rest of his ensemble of standard-issue men's gear – jacket, shirt, trousers, tie, and so forth – and that anyone who shops for better is only throwing away good money and time.

The whisky in the glass Amis at last raises to his mouth is one reward for his morning's work. Conversation is another. Amis loves to talk and listen – company is essential to him. He dislikes being alone for long except when sleeping or writing or reading at night. To live alone would be unimaginable. Even to make a train journey without a companion is unpleasant, so he never makes one. A solitary taxi drive as far as the Garrick is tolerable but fortunately a solo car journey is out of the question because he does not hold a licence and has not driven a vehicle of any kind since he left the army fifty years ago.

What Amis needs is the company of family or friends, of people he can talk to, the kind of people who belong to a club like the Garrick – or the kind of people who should belong there. But even here, in what he has described as the best place in the world, absolute security against the onslaught of bores cannot be guaranteed. Bores, in Amis's ample definition, are necessarily a numerous species and almost as likely to find their way into clubs as pubs or anywhere else. What's more, in a club like the Garrick which encourages its all-male membership to mingle and mix and never drink or eat alone, repelling bores is difficult. 'Why can't you just tell them to fuck off?' Amis is prone to wonder. But of course you can't, and Amis exposes himself doubly to the danger of being joined by them through his habit of sitting in the window of the club's first-floor bar where several empty chairs round his invite promiscuous occupation by whoever may choose to sit in them. The 'adhesive doctor', for instance.

At the approach of a bore Amis's face droops in despair. He is very bad at concealing his feelings. 'Oh, Christ,' he murmurs, not very quietly. If his unwelcoming appearance does not deflect the bore, Amis adopts other tactics. He pointedly does not address any remarks to the bore but keeps the conversation going over and round his head as if he were not there. If it is his turn to buy a round, Amis often does not offer to include the bore in it, which is an eloquent message from someone as rigorous as Amis in the etiquette of buying rounds. In really bad cases, Amis may heave himself out of his usual seat to go and find someone else to talk to at the bar, although he hates standing. And for top-rank, world-class bores he has been known to break completely with his own settled tradition by truncating the whisky session and going down to lunch before the watershed hour of two o'clock, concocting a tale about having an early-afternoon appointment as his excuse and sometimes embroiling his Primrose Hill neighbour and friend, the writer Richard Hough, in his strategy of deception.

In the pub a woman approaches. 'Oh, Christ,' murmurs Amis. She hesitates, flustered, perhaps having caught the look on his face. But Amis manages to half rise from his seat and assemble something like a smile, since he contrives to maintain an old-fashioned courtesy towards women whatever he may be thinking of them – and what he is thinking they will probably be more comfortable not knowing. 'We are seeing you tonight, aren't we?' he says. 'Yes. We'll be round at half-past six.' She withdraws, more flustered still. 'Of course, she'll be late,' Amis says when she's gone. 'Just to make herself seem more interesting than she is. An hour late, I should think.' And, indeed, an hour late is what she later turns out to be.

A young man, seeking Amis's help with a book he was writing, once plucked up his courage and plumped himself down alongside Amis's regular group in the pub. Amis successfully froze him out of the main conversation by the simple expedient of ignoring him. The young man eventually screwed up his last resources of boldness and addressed Amis. 'I know it would be a frightful bore . . .' 'Ye-e-e-ah,' interrupted Amis at the word 'bore', loudly and pointedly. The young man stuttered out the rest of his sentence somehow and disappeared, seen off for good with a bit of luck.

The Amis bore can be defined as someone with nothing interesting, intelligent or amusing to say. 'No harm in the chap,' Amis says, 'just happens to be a bore.' And if by chance he (or she) does have something to say that might be worth listening to he (or she) goes about saying it all wrong. The bore holds the floor. He (or she) gives the long version of every story instead of the short. 'He can't mention somebody without telling you all about his aunt and uncle too,' Amis complains. The bore does not understand that conversation is like a jazz piece with passages played in sequence on different instruments and never ever a one-man performance. An Amis bore is, in short, much like anybody else's, the difference being that the Amis rules of boredom are drawn so tightly that a person who might seem no more than mildly tedious to most people appears as a raging monster of boringness to him, fit to be repelled by any tactic short of outright physical violence or too overwhelmingly obvious rudeness. In the Amis view, there is nothing to be said for giving a bore a second chance. 'Nice enough chap, but if you let him talk to you now he'll talk for an hour and a half next time.'

Today, however, the club is mercifully bore-free. Amis sits at the head of the members' table with two or three or four agreeable chaps on either side of him. That is enough company because, at any rate until the dining-room empties a bit and the noise dies down, his increasing deafness will make conversation with a larger group hard to keep going without a lot of intrusive 'eh?'s 'what?'s and 'say that again?'s.

The wine-waitress, Muriel, appears alongside Amis, smiles, and asks, 'The usual?' 'Yes, please.' The usual is a bottle of Gewürztraminer, an Alsatian white wine Amis has adopted more or less permanently as an acceptable way to ensure that he makes his contribution to the drinks side of the meal, leaving the others in the party plenty of time to choose something red to follow.

Amis consults the menu and makes a rapid choice of dishes. These are usually modest in scale, for his appearance is misleading. In spite of his great girth and his intermittent restaurant columns, Amis is an unambitious eater. His tastes are traditional, and one reason why his column has so often visited hotel restaurants is his conviction that hotels have to serve up food that its guests will actually want to eat, unlike the odiously pretentious modern cooking so many restaurants go in for. The only place where he is guaranteed to turn greedy is a curry restaurant. He is also very fond of the sandwiches Hilly makes for him to consume after the Saturday pub session – the one lunch he eats alone every week, with only a newspaper and *Private Eye* for company. The sandwiches contain cheese, onion, Branston pickle and much more besides. Amis has not fathomed the full mystery of their contents but he likes their taste so much he begins to look forward to eating them soon after breakfast on Friday. At the Garrick, however, two starters will often do, unless Amis decides later that a piece of cheese or an ice cream will round them off nicely. It is with relief that he gets the ordering over. Food may be necessary to a lunch but it is drink and conversation that make the occasion worth getting fully dressed and into a taxi for.

Perhaps the talk is slow to get started, or something has jogged Amis's memory. 'Did I ever tell you the one about "Little Sophie"?' he asks, already laughing. 'Bob Conquest's, of course.' Robert Conquest is an old friend, best known to the world as the author of *The Great Terror*, in which he chronicled the murdering persecutions of Soviet Communism, but with a prodigious talent on the side for light, often obscene, verse, much of which survives only because Amis has a matching talent for memorising it. Some of Larkin's light and scabrous verses survive for the same reason.

Amis recites:

'The first man to fuck Little Sophie
Has just won the Krafft-Ebing Trophy,
Plus ten thousand quid,
Which, for what the chap did,
Will be widely denounced as a low fee.'

Laughter follows. And alternative readings. Didn't it go something-onus and something-bonus instead of quid and did? This limerick has been in circulation for a long time but no definitive version seems to exist. Why doesn't Kingsley write one down and the rest of the Conquest *œuvre* he has memorised while he's at it? But he brushes aside the suggestion. Reciting one Conquest limerick, however, invariably reminds Amis of another, which he also cannot resist reciting. This one is a condensed version of Shakespeare's Seven Ages of Man and is much admired for its economy and wit:

'Seven ages: first puking and mewling.
Then getting pissed off with one's schooling.
Then fucks and then fights,
Then judging chaps' rights
Then sitting in slippers, then drooling.'

Now the name of a journalist crops up because in his weekend column he has come out of the closet and declared himself to be . . . a republican. Amis makes short work of him. 'Of course, he used to be a Marxist. Now he can't be that any more, obviously, but he still wants to fuck things up. He realises the royal family is a force for unity in the country, so of course he wants to fuck that up too.'

Amis was once a Communist himself but that was many years ago and long forgotten. Now his politics are conservative, simple and blunt. 'What I think about things now is – hang Saddam Hussein, don't go into Bosnia, go into Rwanda but only to feed the people, not to stay.'

That takes care of politics. Amis is not much concerned with them now, truth to tell. With the departure of his heroine, Lady Thatcher, he seems to have lost interest in politics as such, though not by any means in issues like hanging or education. Once, not so long ago, he announced in the club bar with a very contented smile, 'It's happened at last. Mrs Thatcher and I.' 'Oh yes, what was that?' He paused for effect. 'We declared our love last night, in my dream.'

Now, unusually for him, Amis strays from his own code so far as to become just a bit boring himself. Someone remembers that the polymathic Anthony Burgess once took him to task for committing a solecism. When Amis had written that stopping a record interrupted Frank Sinatra 'in mid-phoneme' he was wrong, according to Burgess. What Amis should have said was 'in mid-allophone'. This linguistic reprimand has been quoted in a newspaper review and it gets mentioned in passing, which launches Amis on a robust account of why he happens to be right and Burgess wrong. Amis's defence is long, intricate and, well, not only unintelligible to the non-experts around him but not very interesting to them either. He lets himself be carried away for several minutes until the others begin to look almost as disconsolate as he does when confronted by a bore and even start to whisper among themselves. Amis pulls up short.

Talk of Burgess reminds someone of another writer who is having

11

a difficult time. His wife has left him, his last book has flopped, he is short of money. 'Old so-and-so is in a bad way,' the company is told. 'Not bad ENOUGH,' Amis booms back. This has become something of a catchphrase of his, 'not' and 'ENOUGH' being added on either side of any adjective indicating distress to give extra emphasis to Amis's disapproval of the person described. Thus, when it is said that 'Old X's book has had very poor reviews', Amis retorts, 'Not poor ENOUGH', or, when 'Old Y has dropped off the twig terribly young', Amis will have it that Old Y was 'not young ENOUGH'. Very likely the offending persons are fools or 'FUCKING FOOLS' – like bores, a pretty numerous species in Amis's human taxonomy. FUCKING FOOLS are not necessarily stupid, just wrong. Lord Longford, the prison-visiting, Catholic, former Labour Minister, is for Amis a particularly splendid specimen of this breed on any number of grounds.

Time for another story. Amis remembers one told him by his friend Stuart Thomas, a Swansea solicitor he has known since he taught at the university college there more than thirty years earlier. Thomas was in the army in India during the war and heard of a soldier in a train who resisted all efforts by an Indian to enter his carriage. 'But', complains the Indian, 'I am a judge of the Indian Supreme Court. You cannot refuse me.' 'Never mind that,' replies the soldier. 'Hop it, Cherry Blossom.' Amis does the accents to perfection – mimicry, with faces to match, has been one of his talents since childhood – and chortles deeply. What he likes about this story is not its racism, which he does not like, but the essential rightness of the details for its time and place – the imperturbable superiority of the ordinary British soldier faced with an Indian of no matter how exalted a rank and the soldier's choice of insult. Cherry Blossom! The name of a polish the soldier probably shined his own boots with! Perfect.

It may be the Swansea connection that later reminds Amis of another story. A Swansea man and woman have sexual intercourse in an alley after an evening's drinking, are caught in the act and are charged with obscene behaviour. In court, the judge is perplexed: 'You say they were standing up. But she is . . . what? . . . six foot six and he's only, well, four foot and a bit. I don't quite see how . . .'

'I understand a bucket was used, my lord.'

'Oh, I see. He placed the bucket on the ground and stood on it and she sort of . . .'

'No, my lord. He inverted the bucket over her head and swung from the handle.'

More laughter all round. But now Amis has finished his food and drunk his coffee and if there is one thing he does not like to do after lunch or dinner it is to stay sitting at the table where he has eaten. This is what his parents insisted he do when he was a small boy and he has hated doing it ever since. So it is time to leave the dining-room, but not yet time to go home. There is something for the road still to come, to be taken in the little lounge 'under the stairs' in the Garrick. Two Beaumes-de-Venise or one large Grand Marnier or a glass of claret or burgundy, or perhaps two. Then it is out into the street for a taxi. By now it is some time between, say, 3.15 and 3.45.

It is only when Amis gets to his feet and heads for the club's front door or the pub bar that his age really begins to show. On his feet he is suddenly an old man or, in his own preferred word, an oldster. Talking and drinking, he is as lively and engaged as if he were three or four decades younger than his seventy-odd years. Nor is there much sign of great age in his features. His hair is grey and thinning but it still effectively covers his scalp. The face is round, jowly, expressive, pinkish and not much lined, partly because of the 'historic escalation' of a decade earlier. Being fat has its uses. It irons out wrinkles as air inflates and smoothes the crinkled skin of a balloon.

Sitting down with drink and friends, Amis soars free as a bird. But when he has to walk the bird is grounded – clumsy and awkward. Nowadays he shuffles along in short steps, and not very far if he can possibly avoid it. If nobody can give him a lift home from the pub he summons a taxi, even though it is barely 300 yards from door to door. He suffers from claudication, or intermittent claudication in full, a term – he notes this with the detached interest of a lifelong enthusiast for words and dictionaries – that may owe its origins to the Emperor Claudius (Latin *claudus*, lame). This lameness is certainly a nuisance. It turns once insignificant events, like the ascent of the Garrick Club staircase, into expeditions which must be carefully planned and mounted. But it is not a serious handicap. Amis has not done much physical sport since he left school and has hardly bothered with exercise either. He may worry about his health but he takes no interest in his fitness. The modern enthusiasm for exercise regimes has quite passed him by. Not now being able to walk very far gives him an excuse to go in taxis, or not to go at all. Except for the nimbleness that comes in handy when evading bores, the inefficiency of his legs makes no great difference to the pattern of his life. And, slight deafness aside, he is otherwise in excellent condition. There have been occasional alarms, but Amis never seems actually to catch anything. After a recent examination his doctor pronounced the Amisian heart, liver and lungs to be no less than Churchillian.

By which the doctor no doubt meant that they were capable of handling the large quantities of alcohol with which Amis, like Churchill before him, assaults his body every day without fail. Drink is a subject on which Amis has strong views. He will admit to being a heavy drinker but rejects with contempt the word 'alcoholic', which he regards more as a term of abuse than as a clinical definition. 'The definition of an alcoholic', he says his doctor once told him, 'is a man who drinks more than his doctor.'

To Amis the point about drink is that everyone should know his own capacity and adapt his habits to match it. This is what he has done, to his own satisfaction at least, if not to everybody else's. When his second ex-wife Jane buggered off, as he puts it, she made his giving up drink completely and for ever the prime condition for her return. He refused. He has drunk at regular intervals throughout the day for the thirty years he has been a full-time writer, and who is to say that he would have done better as writer or husband without it? Might he not have done worse? At any rate, he has no intention of giving it up now or ever.

The daily routine continues after Amis leaves the club and finds a taxi in the street to take him home. He may try *The Times* crossword puzzle, though he does not linger long over this, wrestling with two or three clues before falling asleep, fully clothed, in his high-backed armchair in the first-floor sitting-room. The ensuing snooze will last half an hour or so, during which the lunchtime booze intake is slept off and the decks thus cleared for a second session at the typewriter. Work restarts at five or thereabouts and lasts for at least an hour and often nearer two.

It is with some reluctance that Amis finishes his stint. Not that he is filled with unstoppable inspiration; he simply knows from years of experience what an appalling prospect it will be to start up again in the morning. When he finally decides he has reached a point where he can plausibly quit for the day he goes back upstairs to the dining-room where he keeps a bottle or two of Macallan and some plastic bottles of Evian mineral water. The dining-room is one half of a room that runs from the back to the front of his house. The other half is the sitting-room. Under the sitting-room window, which looks out on to the back garden, is a large television set, and placed directly in front of the set is the chair in which Amis enjoys his after-lunch snooze. Here he settles down with his evening's ration of Macallan to watch his favourite television programmes.

His absolutely top favourites these days are the long-running soap *Coronation Street* and the police series *The Bill*, both of which run three nights every week. He watches these with the sound at or close to maximum volume, in the style of a football fan, cheering the best bits ('Bloody marvellous – you wouldn't want to change a word of that!') and booing the characters he doesn't like ('Get her off – I can't stand the sight of her!').

Were anyone to suggest it was a little odd that a man with claims as good as Amis's to be England's leading novelist should spend his early evenings watching this kind of (let's face it) lowbrow stuff, Amis's answer might likely take the form of a two-fingered salute. He likes and always has liked popular work – films, television series, thrillers, science fiction – as long as it is good of its kind. Quality is what matters – quality, that is, as defined by somebody who knows what he is talking about, which means somebody like Amis who thinks for himself and not one of the chorus of self-appointed pundits and trend-hounds who puzzle the public mind with their crappy opinions. By his own firm standards there is rather more good stuff around in popular television or fiction than there is in what passes for literature. On this point Amis will be difficult to persuade that he is wrong.

Then it is time for supper, cooked by Hilly or by his daughter Sally, who comes round from her flat close by, but never by Amis himself. Philip, the oldest offspring, may also join them. (The younger son Martin turns up on Sundays with his two children or meets his father for dinner during the week.) With supper there is beer and after that there may be more television, depending on what's available on the terrestrial or cable channels, or a video. At eleven Amis switches off, goes downstairs, gets into pyjamas and dressing-gown and reads. He chooses from the heap of books beside his bedroom chair which includes poetry, novels, thrillers – anything so long as it is not both literary and new. The only writer he still cares to read

15

sometimes on publication is George Macdonald Fraser, journalist, screenwriter and author of the Flashman novels. Otherwise he will stick to old reliables: Evelyn Waugh, Christopher Isherwood, C.S. Forester, Anthony Powell, Elizabeth Taylor.

The last new literary novel Amis read – more or less – was *Time's Arrow* by Martin Amis, whose career he watches with a combination of fatherly pride and barely suppressed irritation that readers under the age of, say, forty may now in literary chat be more likely to associate the surname Amis with the Christian name Martin than with Kingsley. It isn't that Amis is jealous in any ordinary sense. He is extremely fond of his children and does not resent other writers' success so long as it is deserved. The trouble is, it so often isn't. But you can't really say that to your son. Nor can you tell him to his face that he has in place all the essential elements of a FUCKING FOOL. Martin, of course, knows pretty much what his father thinks of him. But how can he tell his dad that the novel has moved on since his time, that the son's brand of postmodernism has taken over from the father's brand of realism? To avoid unwanted rifts they tend to steer clear of such topics, or at any rate steer round them. What makes it all the more annoying is Kingsley's suspicion that Martin may be having a subversive influence on his own work. Martin never seems to write anything as straightforward as 'They finished their drinks and left the pub.' His sentences always have to be busier, fizzier. And now he's got his father wondering occasionally whether his own sentences are as hard-working as they should be – carry as much punch word for word as his son's. It is extremely vexing.

Amis pours the last Macallan of the day around eleven. It lasts him until half-past twelve or so when, fortified by some sleeping-pills, he gets into bed in the expectation of being fast asleep a few minutes later. Should he wake up in the middle of the night there will be no need to go to the bathroom. The bucket by the bed takes care of emergencies.

Most of Amis's days are like this. He rarely goes out in the evening except to have supper with neighbours or with Martin. He only leaves London when the Garrick closes at holiday times, leading him to seek other company among his Swansea friends. Apart from random visits by interviewers, agents, producers and other unavoidable adjuncts of his trade he does his best never to break the daily pattern. His secretary comes once a month and that is all the time he gives to business matters. If he forgets to dictate a letter when his secretary visits, then too bad, the letter will have to wait another month. Nor does he involve himself in domestic affairs. He pours whisky into a glass but leaves the glass to be washed up by somebody else.

For many years now Amis has lived a routine like this, varying in detail but always built around writing, reading, drinking and the company of family and friends. Close or within easy reach he must have all the essential ingredients, from typewriter to Garrick Club. To keep going he must write but to do that he must satisfy his other needs too. Each is important but each is precarious. Writing, for one, is a perilous business. The knack or gift or whatever it is may disappear. This has happened to more than one of Amis's dried-up contemporaries, and he himself seems to have exhausted his own poetic powers. Poems simply do not suggest themselves to him any more. Could other writing stop too? Fearful thought! And might other parts of his life dry up – his lameness make the Garrick Club inaccessible, for example, or his surviving friends drop dead? Just as fearful a prospect!

Amis has been for many years beset by anxieties of every kind. He is hypochondriacal, afraid of being left alone overnight, filled with dread by any number of things from the thought of flying to driving a car or being trapped in a lift. At times he has been afraid of going mad. At other times he has wished that he had chosen to be something less solitary and grindingly hard than a writer of fiction and poetry – a journalist, perhaps.

His routine can appear grossly self-indulgent, the visible bits of it contributing powerfully to his late-life reputation as supreme clubman, boozer and blimp. There is truth in this image, which Amis quietly relishes, partly because he likes to present a challengingly clear-cut version of himself to the world and partly because he is confident there is more to him than the image allows. For his routine – the bits of it people don't get to see, like the hours he spends every day in his study – is also a form of self-discipline, as hard or harder than any nine-to-fiver's day at the factory or the office. Those who are in employment have much of their work prescribed for them, but imaginative writers have only their own resources to depend on. This has meant, for Amis, maintaining an endlessly tricky balance between conflicting sides of his nature, such as weighing the absolute necessity for solitude against his intense desire for the company of others. The routine through which he resolves these contradictions turns mere habit into something more, a pattern of behaviour which may seem weird but nevertheless contrives to keep Amis productive and reasonably content by subduing the fears of disintegration and failure which always gnaw at his heels and threaten to consume him.

Norbury

'It is a sad fate', Amis has complained, to be, like him, 'the child of the urban or suburban middle classes.' The middle stratum is 'drab and glum'. Its heroes are denied 'a brisk introduction to sex behind the coal tips' on the one hand and 'a fructifyingly bad time at Eton' on the other. Though the middle classes are beset by 'constant anxieties about decorum', alongside those of the working class their fears show up as 'neurotic, unreal and self-regarding'. To be born in the middle stratum is thus, it seems, to start life multiply handicapped: not well enough off to enjoy the freedom and security of the rich but too well off for one's anxieties to be taken seriously and yet far too insecure about maintaining status in the world to seek solace in any kind of working-class hedonism.

By this measure of Amis's own, few writers have started out more fully equipped with life's disadvantages than he. His early years were lived wholly within the rules and limitations of the middle stratum, and the lower and more stifling reaches of that stratum too.

Kingsley William Amis was born on 16 April 1922, in a nursing home on the edge of Clapham Common. 'William' was his father's name while 'Kingsley' came from a cousin of his mother, one of a pair of twins, the second of whom was called Queenie, which would probably have been Kingsley's name if he had turned out a girl. Kingsley was a large baby and his birth was difficult – so he was told later. His parents were both in their thirties. Whether it was because of their age or the pain of Kingsley's birth or some other reason like infertility in one parent or the other is unknown, but Kingsley was to be their only child.

William Robert Amis described himself on his son's birth certificate as a 'mustard manufacturer's clerk' – he was employed by J. & J. Colman at its Cannon Street office in the City of London. The word 'clerk' suggests some pinched and stiff-collared wretch perched high on a stool to scribble in a ledger, a latter-day Bob Cratchit. But in the 1920s and 1930s it covered a far wider job definition than that. In a City bank, a clerk might be anybody below the rank of partner. William Amis had a secretary and was required to learn Spanish in order to deal with South American customers to whom he sold Colman's Savora brand, a mustard concocted for the international market, blander than the home product. Today he would certainly have a title more important-sounding than clerk, perhaps export executive.

But if William Amis was not at the very bottom of the white-collar ladder he was not many rungs up it. The home to which Kingsley returned from Clapham with his mother was at 16 Buckingham Gardens in Norbury, a little semi-detached box of a house dropped in a place he came to think of as not really a place at all.

Norbury is the ancient name – a contraction of North Borough – now attached to a minor political entity, a ward of the London borough of Croydon. Once Norbury was a sub-manor within the manor of Croydon, which was at the time of the Norman Conquest a possession of the Archbishop of Canterbury. In 1337 the sub-manor was granted to Nicholas de Carreu and it remained in the Carew family for 500 years – with a brief gap between the execution of Sir Nicholas for treason against Henry VIII and the return of the property to his son, Sir Francis, by Queen Mary. In 1849 the last manorial lord, Charles Carew, inherited, only to sell up ten years later to pay off his debts.

According to a local history, 'Early man lived in the Croydon area but avoided Norbury.' Modern man, for a very long time, steered clear of Norbury in much the same way as his Stone Age ancestors. As late as 1866 Norbury contained only six houses. In 1893 Arthur Conan Doyle had Sherlock Holmes, in 'The Case of the Yellow Face', visit Norbury and walk 'from the high road into a narrow lane, deeply rutted, with hedges on either side', on his way to see a hop merchant. Holmes had reached this still-rural scene by means of a train, getting off at Norbury station, and the opening of that station in 1878 was the transforming event in Norbury's long and – a minor battle in 1264 aside – uneventful history.

Development followed the railways. Roads were laid out and houses built along them. On the tail of the speculators came the London County Council, which between 1906 and 1910 built 498 small houses on 17½ Norbury acres. Croydon's electric tram system reached Norbury in 1901, and in 1926 Croydon's trams were connected to the LCC's and so linked through to central London. Commuting into the city was encouraged by cuts in rail fares and more frequent services. Two private schools were started, both of which Kingsley Amis was to attend: St Hilda's, primarily for girls, and Norbury College, solely for boys.

Cinemas, churches, the King Edward shopping parade, a police station, a public library and primary and secondary schools all came in their turn. So did more housing, private and public. Until the First World War, Norbury could be described as 'healthy', at least by estate agents, because it was still a country suburb. But continued development ate away at its open spaces. Two golf clubs disappeared, the Norbury, which was built over, and the North Surrey, turned into allotments during the Second World War. In 1921, the year before Kingsley Amis was born, Norbury's population had reached 9,413, and ten years later it had grown to 15,538, where it stayed.

So, bit by bit, the minimal amenities of a twentieth-century suburb were stitched in among the roads and houses built around a railway station whose tracks connected it to the heart of London, and Norbury was drawn from its ancient isolation in the tranquil Surrey countryside towards the loose embrace of the vast, amorphous city. For a while the name Norbury Garden Suburb seems to have been attached to this new entity, but it did not last. The countryside had been too fully expunged, the houses were too crabbed and mean, the streets and terraces not so much laid out as haphazardly scrawled in the empty spaces between railway lines, streams, playing fields, old highways. Unlike some London suburbs, Norbury has hardly advanced up the scale of social desirability in the seven decades since Kingsley Amis arrived there, not having the necessary ingredients of space or charm to tempt the aspiring middle classes of the post-1945 years to invade and improve it.

'Cramped, thrown up on the cheap and never finished off, needing a lick of paint, half empty and everywhere soiled' is how Amis the novelist was later to have one of his characters, Stanley Duke of *Stanley and the Women* (1984), describe the Norbury type of South London landscape. 'Half the parts south of the river were never proper places at all, just collections of assorted buildings filling up gaps and named after railway stations and bus garages.' Stanley Duke contrasts his roots with others': they came from 'a place . . . I got out of an area.' Amis became as apt to identify his childhood home with its postal district, SW16, as with its name, Norbury. Not that he felt any great sense of loss or resentment at the lack of roots he shared with Stanley Duke. It spared him the sentimental baggage of a childhood environment richer in beauty or historical associations than the place, or non-place, where he grew up – Norbury, London, SW16.

The Amises, originally from Norfolk, were by the time of Kingsley's birth well settled across South London. Kingsley's paternal grandfather, Joseph James – Pater or Dadda – lived further out, half an hour by bus from Norbury, in Purley, Surrey, where he had moved from Denmark Hill in Camberwell. There were two live-in maids in his house, which was called Barchester, but maids were no more unusual at that time than would be a dishwasher or a CD player in the same house today; Kingsley's own first home for a while boasted a live-in maid too. Joseph James was a merchant dealing in domestic glassware, from which he prospered until his trade began to be undercut by mass-produced goods sold through chain stores like Woolworth's. He had married Julia Spinks and they had three sons, William, James and Leslie and one daughter, Gladys. James, like William, worked in the City, for the bank of Seligmann Brothers, while Leslie, a bachelor, stayed with the glass business.

Lucas was the family name of Kingsley's mother, Rosa Annie. She was called after two aunts, but since Aunt Rosa and Aunt Annie both died of drink she preferred to be known as Peggy. Her father, George, who married Jemima Sweetland, was a tailor's assistant in a gentleman's outfitter off Brixton Hill. He also played the organ at the Denmark Place Baptist chapel. It was there, according to family legend, that William and Peggy met as children and it was certainly there that they were married in June 1915 when Able Seaman Amis was on wartime leave from the Royal Naval Air Service at East Fortune in Scotland, where he helped maintain airships. Peggy had a brother and a sister – George, who was a postman and, most likely for that reason, rarely seen, and Dora, whose behaviour was spectacularly odd and who was to spend some years in a mental hospital because of it.

These grandparents, uncles and aunts – along with two cousins,

John and Joan, the children of James Amis and his wife Florence, or Poppy – were the family Kingsley knew. There was also an outer fringe of American relatives who came to stay with Joseph James from time to time – aunts, uncles and cousins. One uncle was Uncle Tom, Kingsley remembers, a cousin was called Uretta. But they made no deep impression on him and he made no effort to keep in touch with them.

Although his inner circle of relatives were not widely scattered, he knew them only at something of a distance. As there was no car in his own part of the family until Kingsley bought one for himself and another for his father in the 1950s, visits to relatives were small expeditions. It took two buses to reach the Lucas household in Lowth Road, Camberwell, while for John Amis to walk over from SE27 to see his cousin in SW16 was a one-hour hike. (Kingsley, John recalls, never hiked in the reverse direction to visit him.) Peggy took Kingsley to see her mother – his Gran – once a week when he was small, but family get-togethers were more likely to be prompted by dates such as birthdays or Christmas. The impression his relatives left on Amis tended to be vivid rather than profound – they were remembered for their eccentricities rather than their influence.

To Kingsley, as to many children, his grandparents seemed caricature grotesques. In his *Memoirs* he summons up a recollection of Joseph Amis as 'a small fat red-faced fellow with starting moist eyes and a straggly moustache'. His nose had strong purple tints with several hairs an inch or two long sticking out of it. When he laughed he blared but he did not often smile. Nor did he speak directly to his grandson except when greeting him. Joseph's wife Julia, known as Mater, had a reputation for meanness. She was said to collect grocery bags and other pieces of leftover paper for use in the lavatory, leading her son James to complain once of cutting his bottom on an acid drop. Kingsley's cousin John remembers Pater and Mater for much the same reasons. Kissing both of them was awful because both had moustaches, Pater's bristly and tinged with booze. John's version of Mater's meanness is a story which has her inviting six guests to lunch while providing only five chops, compelling two people to divide a single chop between them. Kingsley came to loathe and fear her as he never did Pater.

On the Lucas side, his Gran was repellent in her own way, like 'one of those horrible shrunken little old women dressed in black who used to sit on walls or outside shops on the Continent', and

23

kissing her was horrible too. But he liked her husband, his Grandad, who collected books, 'real books', of poetry. When Grandad read out favourite passages to his wife she would make faces at him which he could not see and, since Kingsley's sympathies were with Grandad, this helped him along the road to hating her very much.

If his grandparents had no lasting effect on Kingsley beyond leaving him with a collection of mostly unpleasant memories, their impact on their own children was little short of devastating. Kingsley's Aunt Gladys was barely allowed to escape from her parents even into respectable marriage. When she became engaged to a Harvard professor, Ralph Foster, Pater and, more enthusiastically still, Mater tried to put a stop to the wedding at the very last moment, though Gladys was over twenty-one and free to do as she pleased, as was Foster. Kingsley's parents joined the argument on Gladys's side, rather to Kingsley's surprise, and the wedding went ahead. 'What was interesting was her [Mater's] reason for being against the American,' Kingsley wrote in a version of this incident that appeared in his 1986 novel, *The Old Devils*. 'He was an American.'

Uncle Leslie found it even harder to escape. He stayed in the family glass business most of his life and remained at home with Mater when Pater died. One day he announced to his brother, Kingsley's father, that he liked men and wanted to go to bed with them. By this time Leslie Amis was well into middle age, a late date at which to make such a notable discovery. His brother advised him to see a doctor. But such a visit turned out not to be necessary. Leslie somehow found out that he liked women after all, though he did not seem to have fully realised this until after the death of his mother, aged almost ninety. Then he cashed in what capital he had and set out on a round-the-world cruise during which he was reported to have 'fucked every female in sight', something he continued to do for the short remainder of his life.

Aunt Dora, on the Lucas side of the family, also had to wait on the death of her mother for her emancipation. Dora was trained as a singer, being chosen over her sister Peggy for this privilege, even though Peggy was said to have the better voice, and it was Dora rather than Peggy who got the treats and the pretty frocks. This created an abiding resentment in Peggy, amounting, Kingsley believed, to near-hatred. Dora had professional singing engagements at public functions like dinners and was married to Arthur Mackness, who sang with a touring concert party called The Roosters and was later promoted as a 'well-known BBC tenor'. What went wrong with the marriage is not clear, though Kingsley's father was convinced that Dora's mother, missing her help around the house, lured her home with tales of what Arthur might be up to with the women he met on tour. At any rate Dora left Arthur to return home and look after her mother.

As Kingsley knew her there, Dora was very strange – indeed, so strange that she became his 'knockout relative', 'the star of Lowth Road'. Dora was forever on the lookout for waste to pick up – a fallen petal, a single hair, a scrap of fluff. If Peggy dropped a dead match in an ashtray, Dora would be sure to snatch it up and run it under the tap to make quite certain it was extinguished. There were other oddities too, and eventually Dora was put in a mental hospital where she recovered her wits sufficiently to become a trusty in the kitchen. Then her mother died and suddenly all Dora's symptoms of insanity vanished. She was now fiftyish with nowhere to go. But the hospital recognised her skills in the kitchen and gave her a job on the staff, where she flourished, providing Kingsley with a handy joke to tell his friends about 'having the only mad aunt who had done so well in the asylum that they offered her a fellowship when she graduated'. She was on the point of taking on the top catering job in one of the major London hospitals when she died – like Leslie, cut off cruelly soon after gaining her freedom.

These sad incidents may have had no effect on Kingsley beyond providing him with tales of family melodrama to be humorously recounted in his *Memoirs*. But they suggest something of the costs exacted by life in the middle-class stratum into which he was born. There was Peggy's resentment of Dora, the consequence of Dora being chosen over her as the sister to benefit from the family's limited affluence. There was the near-invisibility of Kingsley's Uncle George, the postman, virtually excluded from the family circle perhaps because he had committed the fearful crime of falling through the net of middle-class respectability into the pit of working-class employment.

More insidious and more painful than these was the way in which parents, no doubt themselves driven by fears over money or status, tormented their own children. There is no reason to think Kingsley's grandparents deliberately malicious, or that they believed what they did was anything but the right and best thing for their children. But a sense of righteousness may only have intensified their determination to bend their children to their own designs and so make the consequences that much worse – worse in missed opportunities, distorted sexuality and even lost sanity. Echoes of such benevolently high-minded parental tyranny were passed down the generations to be felt, though in milder form, by Kingsley too.

'They [Mum and Dad] didn't say it was going to be like this,' Amis remembered whimpering to himself on his first day at school as more children than he had ever seen before pelted bewilderingly around him in the playground. St Hilda's, not far from Amis's Norbury home, was an establishment for girls which took boys in the lower forms. It was in its playground that he found himself that first morning when he was about five years old, the confusion created by the chaos around him not being much eased when, appealing desperately for someone less wild than the rest to be his chum, he was instead bitten on the arm by a boy called John Skelton.

As he was to remember his brief time at St Hilda's, Amis was a solitary in his school hours and made no new friends. Before tea he was roped in to tie the laces on the outdoor shoes of children who could not or would not do it for themselves. Afterwards he would play with the children of people from his parents' circle of friends, including a Roman Catholic boy who 'traumatised' him by saying that God was always about and showing him his set of religious images. Beyond these, Amis's recollections of St Hilda's were to be few. They included one fellow pupil, an older girl called Freda Roberts, but only because 'she was once famously heard to fart', and two teachers. Miss Crampton taught French and would often say 'le shat, the cat' which, mysteriously to Amis, made the other pupils in the class want to laugh. Miss Barr taught English and was remembered by Amis as 'a tall, Eton-cropped figure of improbable elegance'. He fell in love with her and wrote in his *Memoirs* that it was 'here, perhaps, that we can date my first devotion to the glories of our literature' – and perhaps also trace the first connection between Amis, literature and sex.

Amis soon moved on to Norbury College – like St Hilda's, long obliterated – not far away from his first school along the London Road. Mr Howie, the headmaster, became an Amis family friend. Norbury College had been set up in a double-fronted suburban house in 1903 as a preparatory school for younger boys, catering for Norbury's burgeoning middle-class population. Since then it had expanded both its facilities and its age range until in 1925 it was recognised by the Board of Education as 'a properly-conducted secondary school'. The word college in its title reflected the school's superior aspirations. It operated the competitive 'house' system of the English public school, each boy being allocated to one of four houses, named after their first captains: Ward, West, Lingard and Chapman. Norbury's aim was to turn out that cherished English type, the all-rounder, a boy equally at home on the playing fields and in the classroom. As an incentive to this form of character building, a silver cup, presented by a Mr L.A. Aston, was awarded each year. It was inscribed 'Mens sana in corpore sano' and the winner had to prove himself in both school work and sports, half the points being scored in each. But the college aimed to be educationally advanced as well as socially traditional. The then headmaster, Mr T.A. Briggs, informed a local newspaper in 1932 that everything was done to foster boys' individual talents. 'The old idea of driving is gone,' he said. 'We try to make them feel we are going along the same road as they are.'

Amis did not stay long enough to be in contention for Mr Aston's silver cup, but if he had he might well have fallen short on the sporting side. He got no further in this than playing left-back for the Third XI football team, coming in last but one in the Under-Twelve 440 yards and, according to his own account, taking 40 minutes to score four not out in a pick-up game of cricket. He did no better in the workshop, neither his cardboard bus nor his plywood cigarette-box ever getting beyond the drawing-board stage.

Things were more promising in the schoolroom. Here a master called Waller got Amis interested in mathematics, especially algebra. So interested did he become that he thought for a short time he might devote himself to the subject. He abandoned that ambition when literature 'took [him] over' but his mathematical skills were to come in useful later, in army signals. A second master called Waller taught English but failed to stimulate Amis's interest. 'Big' Mr Waller, as he was distinguished from the mathematical 'Little' Mr Waller, was not much concerned with mundane aspects of English like parsing or sentence analysing; in fact he did little proper teaching at all. He preferred, as naturally did his pupils, to read aloud to the class books 'about the Great War ("I've copped it in the back, sir") or lethal espionage in Eastern Europe ("for God's sake shoot me and have done with it")'. But that was the sort of adventure Amis was reading for himself at home, and hearing it all over again at school did not encourage him to take literature seriously.

His interest in English as a subject for study was stimulated when 'Big' Mr Waller was replaced by a new teacher, Mr Ashley, who was sufficiently keen on his own speciality to introduce his boys to *The Merchant of Venice* and the Georgian poets, then still fairly modern. He made his pupils write essays on themes like 'Beauty' and try their hands at poetry too. On one occasion they were told to write in blank verse on the miracle of St Sophia. Amis turned in ninety-nine lines and Mr Ashley declared his effort the best. This helped make Amis begin to like English, as he put it later, 'for its own sake'. But Mr Ashley was 'properly sarcastic' about Amis's first published work, a story of 300 words called 'The Sacred Rhino of Uganda', which appeared in the school magazine. No copy has survived but in Amis's recollection it was about a Captain Hartly, a 'veteran hunter' who shot the rhino in ignorance of its sacredness and was forthwith set upon by its 'native worshippers', from whom he somehow escaped with his life. It was written in 'taut, impressionistic prose on which Mr Ashley was particularly hard: "He clutches at his side . . . pitches forward . . . unconscious . . ."' 'With this,' Amis wrote, 'the initial, experimental period of my writing came to an end.'

Although he went home for lunch and spent little of his school time outside the classroom, Amis saw enough of the playground to become aware of the 'vast free entertainment' that school life had to offer. He had fights and made a boy called Hall cry. He picked up a miscellany of playground lore about matters as diverse as sex, rhyming and rude words. And he developed his enthusiasms at the popular end of the cultural spectrum, discussing writers like Percy F. Westerman and Ian Hay and visiting the Streatham Astoria, where he saw film stars like Gary Cooper and Richard Barthelmess and stage shows featuring the Paramount Tiller Girls and Troise and his Mandoliers (though his parents, having seen the trailer of the Frederic March version of *Dr Jekyll and Mr Hyde*, took care to ban him from the Astoria the next week). The taste for popular art forms he acquired in these years was never to leave him.

City of London School

'For goodness' sake, Peter, stop ruining your eyes and get out into the fresh air. The first spot of sun we've had for weeks, and you have to sit poring over that tripe.' In these hectoring words Captain Furneaux addresses his fourteen-year-old son Peter at the start of Amis's 1973 detective story *The Riverside Villas Murder*, set in 1936. They are words that William Amis might well have used to his son Kingsley, who by no very extraordinary coincidence was also fourteen in 1936.

Captain Furneaux's outburst captures the style of paternal authority in the Amis household as it appeared to the young Kingsley. There is first the charge that reading is bad for a boy, particularly when a boy is reading a comic (*The Wizard*, in Peter's case). Along with this goes the notion that fresh air and exercise must somehow be better. There is a touch here of the English philistinism that puts outdoor activity above all others. But it isn't so much what Peter's father prefers. The real point is that, whatever a boy happens to be doing, his father will surely want him to do something else. Even if the boy can fit in both things, his father will try to stop him. That is what happens in Amis's 1994 novel, fittingly called *You Can't Do Both*. In this Robin Davies, also fourteen years old in the 1930s, is invited to tea on Saturday afternoon by a school friend, Wade. Robin and his parents have also been invited round at eight by the Carpenters, their neighbours. Robin tells his father he will be back from tea by 6.30, which will leave plenty of time to make it to the Carpenters. But his father is adamant. 'I'm sorry old boy. You can't do both . . .' From such a decision, however obtuse, there is no appeal.

Seemingly gratuitous parental bossiness afflicts every child. But it will be a heavier affliction where there is only one child to boss. Being an only child has its pluses and minuses. Parents will devote all their love and resources to one person instead of having to divide them among several. On the other hand, having only one child to love, they may spoil and over-protect to the point of smothering the child, while at the same time expecting it to fulfil too many of their own cherished hopes and ambitions. Kingsley benefited and suffered from all these contradictory pressures.

His mother doted on him. She called his fingers 'puddies', and Pud became a nickname in the family that Kingsley was only able to shake off in his teens. She was gentle, timid and pretty enough to attract what appeared to be a couple of admirers, one a mysterious Catholic priest called Steinmitz, who smelt of drink, and another, 'Uncle' Tommy, who made a habit of turning up as if by chance, even once when the Amises were on holiday in Norfolk. Kingsley, however, had no reason then or later to suspect there might be anything that was not perfectly proper in his mother's relations with either of these men.

'She was a jolly little woman for all her "nerves" and shortness of breath,' he recalled, 'fond of a giggle, a fag, a gin and tonic (no more than a couple).' His cousin, John Amis, remembers her as volatile – 'darting, like a bird, with a quick, hissing kind of laugh'. Besides her nerves, she was superstitious in all the usual ways with some unexpected superstitions thrown in, like having to sit down briefly if she had for some reason come back into the house soon after leaving it.

One of her preoccupations was Kingsley's health. Like many, if not most, mothers of the time, she was particularly anxious to make sure her family was regular in its bowel movements. This meant frequent use of laxatives, each one foul-tasting to Kingsley, so much so that in retrospect they seemed to him proof of the 'supposedly very English idea that, if something was really going to do you good, it must taste horrible: senna "tea", California syrup of figs, worst of all Gregory powders.'

Even more important to his mother was Kingsley's intake of food. In childhood, as later, he had a small appetite, but in his early days it showed. He was thin, so much so that his mother complained of being able to see his breastbone. The family GP, Dr Pringle, was summoned. He examined the boy's fingernails for white spots, supposed evidence of malnutrition, but pronounced Kingsley a 'well-nourished child'. His mother nevertheless continued to worry. 'She tried to beef me up with stuff like Parrish's Chemical Food,' Kingsley wrote in his *Memoirs*, 'a fearful reddish liquid which was supposed to turn your teeth black unless you bypassed them by sucking it through a straw'. Besides such patent remedies, his mother was determined to make sure he ate up his food at mealtimes. She would sit opposite him at the dinner table 'pushing forkful after forkful' into his mouth. (Perhaps, he was to speculate, this contributed to his later preference for liquid over solid refreshment.) At twelve or thirteen Kingsley was still being spoon-fed in this way. Then a new regime took over, though only slightly more adult. After some minutes toying with the food on his plate, Kingsley would say, 'Mum, would you sort it out for me?' Mum then divided what was left into two parts, the food that definitely must be eaten and the rest that Kingsley could leave if he wished.

Kingsley did not very much enjoy this insistence that he eat up, but it was too obvious that his mother had only his interests at heart for him to resent it. His mother's tactics contributed to his later reluctance to linger longer than he absolutely need at any dinner table. It had other consequences too. If there was a touch of bullying in his mother's spoon-feeding, there was more than a touch of cosseting as well. It was she who sorted out the food and bore it on the spoon to Kingsley's open mouth at an age when other boys had long got into the habit of doing that for themselves. All he had to do was chew and swallow, repugnant though these actions may sometimes have been.

Kingsley came to be spoon-fed in more ways than this. If, for instance, he wanted to invite a friend round to tea it was likely to be his mother, not he, who called the friend on the telephone, then still a novelty in the Amis household (and a rarity in the country at large). Peggy's care for her son was a source of great security. But it also gave him a lifelong unwillingness to do things for himself, to make his own arrangements, even to use the telephone. 'I dislike, in a sense, standing on my own feet,' he told an interviewer many years later when he was in his fifties. If his mother's support gave Kingsley the confidence to be independent in ways that helped him strike out on his own as a writer, she also left him heavily dependent on other people, especially women, to see him through the humdrum details of daily life.

Peggy can be credited with giving her son the first small shove along the path that made him a writer. She liked reading herself – 'a book was as much part of her accoutrements at home as handbag and knitting' – though she was no highbrow. Her taste ran to popular writers like Ursula Bloom, Norah C. James and Ann Bridge. Not the classics, but not 'slop' either. When Kingsley was at a loose end on wet afternoons with nothing to read she would suggest that he 'do a bit of writing'. It was surely not her intention to turn him towards a writing career but that is what she seems unknowingly to have done. 'Mum probably did as much as anybody to get me going' was the mature Amis's judgement.

In the Amis household, as in many others, Mum played the role of the soft cop, the nice partner in the parental team. It was her, not his father, from whom Kingsley wheedled the ninepence he needed to go to the cinema. Father played the tough cop, the not-so-nice guy. It was he who laid down most of the rules of behaviour and gave Kingsley a clip on the ear if they were broken. William Amis followed the Shavian maxim that a child should only be hit in anger, a rule Kingsley was to follow later with his own children. There was physical punishment in his father's house, but never any heavy-handed drama of the 'your father would like to see you in the study' variety.

William Amis was a man of medium height, with broad shoulders and, in his son's view, 'passable' good looks. These he kept all his life, never losing much hair and not much changing in appearance until his final illness. 'He took a lot of things seriously,' Kingsley remembers; 'he was not one for letting things slide. He thought you had to interfere with things.'

But if he had his stern aspects he had his funny side too, possessing a talent for physical clowning and mimicry that made him, in his day, one of the funniest men Kingsley ever knew. He told stories using 'the full deployment of facial, vocal and bodily resources'. 'It is in mid-story that I see my father most clearly,' Kingsley wrote in his *Memoirs*, 'quite a dapper figure in one of his grey or light-brown lounge suits (though he never could tie a tie properly) hobbling across the room in the style of some decrepit director of the firm, or forcing his face into lines of disquietingly silly uncouthness as he became the man next door'. Peggy had some of the same talent, so that in his teenage years Kingsley came to believe these humorous techniques were standard practice in telling anecdotes and for the rest of his life found something lacking when they were not used. Kingsley certainly employed the same techniques in his own conversation, as anyone who ever talked to him for more than five minutes was bound to discover. And this aptitude for fun and mimicry passed into his fiction, beginning with Lucky Jim's barely controllable habit of making faces, and informing the novelist's comic manner from then on.

Like his two brothers and in time his own son, William went to the City of London School on the north bank of the Thames near Blackfriars. But he did not proceed from there to university because, Kingsley believed, of Pater's meanness. Instead, he found a clerk's job with Elder's and Fyffe's, the banana importers, and, after his service in the First World War, joined J. & J. Colman, where he stayed until his retirement in the 1950s. Cricket and tennis were two of his passions and he played both well into middle age, captaining the local club side in his sixties. He encouraged Kingsley to play too, even paying for him to be coached, though he could not easily afford it. But Kingsley had a 'lazy' eye and lacked the enthusiasm that might have made up for it. Attempts to get him a trial for Surrey came to nothing, though he remained a knowledgeable spectator of the game for the rest of his life.

As with cricket, so with career. William Amis wanted his son to be like him, only better. He was not a businessman in the entrepreneurial mould. The importance to him of his job was the security it promised. He told Kingsley once that he regarded himself as a failure. But he was matter of fact, not bitter, about this and it did not mean he wanted his son to take off in some radically different direction. A job in the City – but a more senior, better-paid job – would be a decent ambition for Kingsley to aim at. William enjoyed City life well enough but nothing could have been further from Kingsley's taste. Long before he had formed any ideas about what career he might follow he experienced his first remembered taste of the savage boredom that became his lifetime's enemy from hearing his father talk about City matters. The distaste endured. Late in life, when the broadsheet Sunday newspapers had all spawned business supplements, he would extract those sections and pile them out of the way before he tackled the other sections. Mere mention of business – 'the word that concentrates the whole thing' – would bring on 'a reliable diffusion of boredom' throughout his life.

Yet William did not put pressure on his son to follow in his footsteps. 'He had learned never to try to influence people too much,' Kingsley recalls. 'He would have liked me to be something like himself, but he hoped I would do it voluntarily.' Nor did he discourage his son from writing poems, though he did warn, sensibly enough, that Kingsley could not expect to make a living out of any kind of writing.

In the early 1930s, as the slump intensified, Kingsley became aware that his father was disturbed at the prospect of losing his job. Later, too, when Colman's merged with Reckitt's, there was a danger that, if he was not fired, William would be shifted to Reckitt's headquarters in Norwich. Luckily, both disasters were avoided. William suffered nothing worse than having his office moved from the City to Bedford Square in Bloomsbury. He even prospered. The annual summer holiday, taken at Pevensey in Sussex and later East Runton in Norfolk in a cottage or digs, increased from two weeks to three. The Amises moved house twice, from Buckingham Gardens to Ena Road and later to Galpins Road, though they did not move far. All three homes were in Norbury, all within walking distance of each other. Each was slightly larger than the last, representing one small step up the social ladder. But none was significantly grander than the others and all conformed to the same basic pattern. Downstairs – drawing-room, dining-room and kitchen. On the floor above – two double bedrooms and a boxroom. And above that an attic, used for storage only. The style in which the rooms were furnished was spare and practical, like the rooms Peter Furneaux was used to, which had in them 'two kinds of objects, ones you used, like chairs, and ones you looked at, like his mother's china cats'.

Drawing-room and dining-room were rather grand names for what were in fact rather small rooms, separated by thin walls. Such physical circumstances made it difficult for Kingsley to find much privacy, even if his parents encouraged it – which they did not. His father's 'constant concern' was to stop Kingsley 'getting away from him'; 'reading in public was deemed rude, while reading in private was anti-social.' Kingsley was supposed to join in the family circle, not escape into solitude.

What Peter Furneaux, an only child too, felt about his family reflects closely what Kingsley felt about his. Peter is envious of Daphne, a young neighbour, 'of her having a brother and a sister . . . of what he imagined the life of the whole family to be like. Being an only child did not mean that you were by yourself too much; on the contrary, you got the whole of your parents reserved for just you instead of divided up into three, say. Peter liked his father, but would prefer on the whole to have him as an uncle, even one living in the same house. His mother he quite loved, but there again she talked to him about quite a lot of things he could not remember afterwards, even the next minute.'

Robin Davies in *You Can't Do Both* has some of the same feelings. In his teens he is effectively an only child, like Peter and Kingsley, because his older brother and sister have left home. Far from feeling unwanted, as a kind of afterthought to the family – he is the youngest by seven years – Robin feels the opposite and thinks longingly about what it must be like really to be unwanted, a condition he thinks he could willingly put up with six days out of every seven, leaving only Sundays for him to be wanted on. Like his fictional creations Peter and Robin, Kingsley found the domestic atmosphere in which he lived 'claustrophobic', 'constricting'. He could have 'settled for less love'. His parents added to his isolation within their closed little circle by ruthlessly restricting his choice of friends and his opportunities to meet them. They could not have done these 'more unflaggingly if there had been a long family history of male prostitution or juvenile dipsomania', Kingsley recalled in his *Memoirs*.

One of their reasons may simply have been the desire to keep their only child to themselves. But they also wanted to make sure that Kingsley did not fall into the wrong, that is into lower-class, company. When Robin Davies's father bans him from having tea with his school chum Wade, Robin wonders why. Does he think Wade is a bad influence: a smoker, drinker and gambler, perhaps even a sex pervert? But it is simpler than that. Mr Davies thinks Wade 'looks a bit of a hobbledehoy' and sounds even more of one over the telephone. For Mr Davies, as for William Amis, even a hint of hobbledehoyness in his son's friends sets alarm bells frantically ringing and he expects his son to accept that hint as sufficient reason for barring all contact out of school. The fear of falling from social grace had passed from William's parents to William and he expected his own son to keep to the same strict social rules. Kingsley might boil with frustration but there was no point in turning to his mother for help. There was nothing she could do, even if she wanted to.

Two incidents in Kingsley's life before he reached adolescence were to leave lasting marks. One was his first flight on an aeroplane, taken in 1932 when he was ten. It was a five-bob (25p) 'flip' around Croydon aerodrome, then London's major airport, which he made with three or four friends in a De Havilland Dragon Rapide, whose tapering wings and wheels with 'spats' Kingsley always afterwards remembered. Flying was a novelty at the time, not the routine means of transport it has since become, and this flip was supposed to be a treat. That, though, was not what it turned out to be. Looking through the window beside him, Kingsley suddenly saw the ground beyond his shoulder. This was the wrong place for it – the ground was supposed to be under your feet! The shock put Kingsley off flying for the rest of his life. He was tempted, but in the next six decades he never flew again. It became a matter for some regret that this inhibited his chances of travelling. If he had been willing to fly he could have visited, say, America more often. However, balancing such missed opportunities against the horror of finding the ground where it ought not to be, Kingsley never felt he had made the wrong decision.

The other experience with lasting consequences was cumulative rather than a single incident. When his parents went out at night (as they often did), even by the time Kingsley was old enough to be left on his own they sometimes got in a babysitter, but if he told them he could manage without one they did not. Yet, alone in the house, Kingsley felt abandoned. Perhaps someone would climb through the window and murder him. Sometimes he screamed. Throughout his life Kingsley was to be afflicted by anxieties of all kinds. His horrified reaction to flying and to being left alone were the first signs.

The fees at the City of London School were low, £30 a term, but would have consumed close to a fifth of the income of someone on a salary as modest as William Amis's £500 or so a year. In September 1934, aged twelve, Kingsley proceeded there from Norbury College all the same. William gambled that his son would get a scholarship in time for his second year and the gamble paid off. Kingsley was to boast ungratefully for a time that he had been a scholarship boy from the start; but he was not, and he acknowledged, to himself at least, that it was no small risk his father had taken in making a commitment to pay at least one year's school fees. For if Kingsley had failed to win a scholarship his father would have been faced with either the humiliation of withdrawing his son or the burden of paying £90 a year for seven years.

The City of London School stood at that time on the Victoria Embankment, just inside the City's western boundary near Blackfriars (hence the name of Peter Furneaux's school, Blackfriars Grammar). The site had been occupied since the Great Fire of London successively by a theatre and the offices of the New River Company and the City Gas Works. In 1883 the school had transferred there from elsewhere in the City to new premises purpose-built in Renaissance style, mainly of Portland stone and brick, which Kingsley found on first arrival 'rather oppressively dignified' with 'lots of corridors and a vast agoraphobic playground filled with self-possessed boys in black coats and striped trousers'.

The school has now moved again to a site along the riverfront east of Blackfriars but the buildings Kingsley knew remain. In the 1990s guides conducting boat-trips down the Thames are apt to point them out as having once housed the school Winston Churchill went to before going on to Harrow. This sets the tourists goggling but it is not true. The school can boast another prime minister among its old boys, however: Herbert Asquith, Earl of Oxford.

The City of London School is a relatively modern foundation, though it has ancient origins. A town clerk of the City called John Carpenter (also the executor of Dick Whittington's will and a Member of Parliament) left money when he died in 1442 for the education of four poor boys. His bequest was – after several centuries, protracted politicking in the City and an Act of Parliament – eventually incorporated into the funds that were used to launch the school in 1837. This was a time when progressive educators in London were challenging the monopolies of Oxford and Cambridge where tests of Anglican orthodoxy still governed admissions. In a spirit of rebellion two colleges were founded, University College and King's College (now parts of London University), which in turn opened their own schools. Religion was taught in these establishments, but it was not a condition of entry to any of them to be a subscribing Anglican. Nor was it at the City of London School, the Act of Parliament by which it was founded laying no religious restriction of any kind on masters or boys.

The result was a liberal tradition that endured to Amis's day and beyond. Catholics, Free Churchmen and Jews were pupils from the start. Low fees and scholarships for boys from London County Council and Middlesex primary schools ensured a wide social as well as religious mix. Kingsley was immediately sensitive to this spread of backgrounds, hearing accents both posh and common, varying 'from ones that discomforted me to ones that made me feel superior'. But he also learned something more important. Watching how the other boys behaved he realised that class did not count, here at least, whatever might be the case at other schools or in the wider world. Neither did religious difference. After announcements at morning assembly, but before prayers, 'the Head nodded as always to the senior Jewish boy, who led from the hall the others of his faith, the Roman Catholics and the Dissenters.' This was how Kingsley remembered the school's spirit of tolerance in the version of it he described in *You Can't Do Both*. He himself, though not a believer, stayed on for prayers.

From all his years at the school he could remember only one anti-Semitic incident. On parade with the Officer Training Corps he was standing in the front rank alongside his friend Leonard Richenberg when the regimental sergeant-major attached to the school corps from the regular army stopped in front of Richenberg and asked his name, which he knew perfectly well. The sergeant-major's second question was 'What's your religion?' to which Richenberg replied, 'Jewish, sir.' Something like 'an audible hush' fell on the parade, Kingsley wrote in his *Memoirs*, 'as if every Gentile in hearing turned white'. No training in good race relations had been necessary to produce this reaction – tolerance was part of the climate. The author of the school's history, A.E. Douglas-Smith, felt able to write in 1937 that 'the anti-Semitism of Fascist countries is incomprehensible to boys who have made their life's friendships in the School.' Kingsley himself wrote: 'Nobody at the school had told me not to behave in the sort of way [the sergeant-major] had; I just knew, like everyone else.'

Such attitudes to race and class stayed with Kingsley all his life. His later brand of right-wing conservatism did not include race or class prejudice, and the unreconstructed white racism which he found in the American South in the late 1960s was to make him deeply uncomfortable during the four months he spent teaching at Vanderbilt University in Tennessee, as he described at length in his *Memoirs*.

Kingsley might have turned out differently, might have become like his own Stanley Duke in *Stanley and the Women*, who admits to 'making the kind of mildly anti-semitic remarks that came naturally to someone like me born where and when I was', which is roughly when and where Kingsley was born. The City of London School steered him away from such attitudes. That did not stop him relishing and enjoying differences of class and race, a very different matter. He could always laugh at a joke about an Englishman, a Scotsman and a Jew or about an earl and a dustman, so long as it was a good one. But his enjoyment came from the comic possibilities of the differences between them. He would not find funny a joke based on the stereotypical notion that Jews or earls were funny or ridiculous, let alone contemptible, just for being who they were.

The social manners of the school were thus by the standards of the time tolerant, liberal, easygoing. But the school also kept a traditional framework of formality. Boys called masters 'Sir', masters called boys by their surnames, and boys addressed each other in the same way, unless good friends. Canings were administered by both masters and prefects, a maximum of six strokes on the backside. Masters, though, were easily accessible. A boy with a problem had only to put his head round the door of their common room to have one come outside and discuss it. There was even an element of democracy in the school's administration, with a Boys' Representative Assembly whose resolutions were submitted to the headmaster for approval and usually endorsed by him. One such resolution, passed the year after Kingsley arrived, called for the universal adoption of the uniform dress of black coat and striped trousers, the sight of which had helped so to intimidate him on his first morning. Strange, we might think, that boys should vote to impose such a funereal costume on themselves, but so they did – though in summer they sported the less sombre rig of blazer and boater.

The headmaster appointed three years before Amis's arrival, F.R. Dale, had made a number of reforms, among them the introduction of streaming. This involved allocating the cleverer boys in each year to an A form, the less clever to a B. According to the school history, this system was intended to let the cleverer pupils develop quickly while not depressing the less advanced with over-strong competition. It was thought to be a progressive innovation at the time and would now be thought exactly the opposite. Nor did such naked intellectual segregation stop at the shuffling of boys into A and B packs. A master might make no bones about further discriminating between the bright and the dim, even within a class that had already been streamed.

C.N. 'Biff' Vokins, who taught in Kingsley's time, remembers how one of his colleagues, the Rev. C.J. Ellingham, approached his A-streamers:

Ellingham handled every class with vigour and originality. In translation from the classics, for example, he had greyhounds and tortoises. He spent only half an hour of each lesson on the routine of translation: a tortoise started, but was allowed only limited time. 'Right, now we'll have a greyhound: off you go, Amis, and you tortoises be ready with your pens and notebooks.' The method suited Amis and he forged ahead.

Kingsley confirms that this teaching strategy was employed, though remembers Ellingham's terminology of discrimination as greyhounds and duds, not tortoises, surely even more crushing to the less bright and, whether tortoise or dud, certain to cause riotous protest in a classroom of the 1990s.

Teaching was divided vertically as well as horizontally, by subject matter as well as by ability. Boys could opt for one of three 'sides' – classical, modern or science – and there were A and B streams in each. An A-streamer in the 'classical side' would learn Greek and Latin, while a B-streamer learned only Latin together with some extra science. The 'modern side' learned modern languages – French and German or Spanish, with Latin thrown in – while the 'science side' did physics, chemistry and biology. (The school was also divided in yet another way, into houses, though this was done almost entirely to promote competition in sport. Kingsley's house was Hale.)

The curriculum was, however, not so rigidly structured as all this might suggest. All boys learned at least some science and could specialise in history, maths or economics by the time they reached the sixth form. Classics, though, were still the fashionable subject as they had been in English education for decades, indeed centuries. One reason for their survival, besides the commonly held view that classics helped produce the best sort of English all-rounder, was the large number of classics scholarships that had accumulated over the years at Oxford and Cambridge, naturally a tempting target to boys of modest background. Another may have been that the headmaster, Dale, was himself a classicist, thus setting the school an example from the top. Kingsley remembers him as a scholar 'in the best old style'. In his *Memoirs* he wrote: 'To hear him read Greek verse, observing tonic accent, metrical ictus and the run of the meaning all at once, was to be given a distant view of some ideal beauty as well as to marvel at a virtuoso . . . He was human too. If ever a kind of man vanished for good, his did.'

Dale was evidently a remarkable person. Kingsley held him in 'terrified veneration'. He was variously described as tall and dignified, friendly but austere. At Cambridge he had taken a double first in classics and during the First World War won the DSO and the MC, becoming second-in-command of a battalion of the Royal Welsh Fusiliers. His war experience gave him 'perfect sympathy' with ordinary people, according to one of Kingsley's contemporaries, who adds that Dale 'dominated' the school. Another contemporary goes further and says Dale was 'worshipped'. The school history, in the manner of school histories, finds it difficult to decide whether Dale was more the Roman for his 'unresting sense of duty' or more the Greek for his 'Sophoclean harmony and serenity of outlook'. The boy who was captain of the school on Dale's arrival, T.B. Williamson, summed up his impact more pithily: 'Here was a man, we felt, who would stand no nonsense. Within a few months we had revised our opinion on this point. Under this Head there would be little if any nonsense to stand.' Whatever his exact combination of qualities, now difficult to recapture from the awed memories of his pupils and the respectful official record, Dale clearly played an important part in creating a school which is remembered with an uncommon combination of real pleasure and genuine gratitude by its old boys fifty years on.

Kingsley signed up with the classical side at first but later began to question it with his friend Richenberg. They recognised that it was a good education in its way, but what was the point of it? Nobody explained. It was simply taken for granted that classics were best, except among those who had decided for themselves that science or languages were better. Richenberg hankered to switch to maths, though he stayed with classics. Kingsley began to hanker after English but there was no English class for him to join. Eventually one was created and he became its first pupil.

It was the influence of Ellingham that did most to draw Amis towards English literature, after the uncertain leads in that direction given by Miss Barr and Mr Ashley. Ellingham was also a remarkable man, 'the best teacher I have ever met', Richenberg remembers. In his enthusiasm for high standards Ellingham wrote a book for thirteen- to sixteen-year-olds called *Essay Writing, Bad and Good*. It begins by saying that to write a good essay requires a good English style which is hard to acquire – 'for bad style is catching, and we are surrounded by examples of it.' A hundred pages of lucid and robust advice follow, under headings like 'Conciseness', 'Vigour', 'The Sentence', 'The Paragraph'. The book finishes with a blunt warning: while some people may be clever enough to write *tours de force* proving that 'matricide is a virtue, and chronic drunkenness the height of bliss' it is 'unlikely that you are one of them.' From which it is not difficult to deduce that Ellingham would have thought little of the late twentieth-century notion that young people are full of innate creativity just waiting to be poured out on to the page, if only it is not stifled by tedious lessons in grammar and construction.

But Ellingham was no mere pedant. The school history describes him as 'imaginative and unconventional'. His habit of letting grey-hounds set the pace in class was part of his strategy for creating time for things that were not on the syllabus. He thought the two years' work which went into preparation for the School Certificate exams could be done in one year, so leaving half the available time for other interests. He brought a gramophone into the classroom to introduce his boys to classical music and taught them about painting from a National Gallery catalogue. It was from him that Kingsley first learned about Picasso, Braque and Matisse – 'those horrible people', as he would later describe them. His interest in music was further developed after school hours in the appreciation classes given by the music master, Mr Taylor, who explained Brahms's First Symphony alternately on the blackboard and the piano, and did it so well that it remained the piece of its kind Kingsley understood best for the rest of his life.

Kingsley continued his education outside the classroom, as he had begun to do at Norbury College. From other boys he now learned about a new diversity of possible interests – Fats Waller, Charles Morgan, air-pistols, dating girls, Borodin, Rimsky-Korsakov – and from H.G. Williams-Ashman he learned something of upper-middle-class life as represented by 'the possession of actual pictures (not reproductions) by living artists, the voluntary reading of *The Oxford Book of French Verse* and the notion of foreign travel'.

He began, too, to discover literature for himself, notably the writers G.K. Chesterton and A.E. Housman, 'the first two grown-up writers I ever read outside school', he was later to recall. Chesterton's *The Man Who Was Thursday* became a particular favourite. Kingsley read it so many times he claimed that, if shown one sentence, he could quote the next from memory. Housman was the modern poet he came to like best of all, topping the list of his twentieth-century favourites: John Betjeman, W.H. Auden, Philip Larkin, Wilfred Owen and Robert Graves. One day Mr Ellingham revealed that Housman was his favourite poet too. This was important as confirmation of Amis's own judgement. But it was something more. Mr Ellingham was a convinced Christian while Housman was an atheist. So poetry was a form of truth even if it did not tell the whole truth and it was possible to admire poets with whose beliefs one did not agree. 'I saw for ever that a poem is not a statement and the poet "affirmeth nothing",' Amis wrote in his *Memoirs*.

3

Evacuation to Marlborough

As Kingsley grew up he began to talk back to his father. The family's hard cop found the suspect standing up for himself. The soft cop, Peggy, was not much involved, except when required to smooth things over. This was a contest between males, a joust between father, settled in his ideas, and son, groping to discover what his ideas were. The result was a violent quarrel 'at least every week or two for years'.

There was no shortage of issues for the two to fight over. One was politics. As the 1930s drew on, Kingsley found they had made a socialist of him – indeed, like many bright youngsters of the time, a supporter of Communism and of Stalin. He was sometimes pious though not humourless or single-minded about his beliefs. These were vague, based on his reading of writers like Auden and Orwell, rather than on first-hand acquaintance with social evils of which, in Norbury, there were few conspicuous examples. Kingsley was 'against capitalism' rather than in favour of any very precise alternative. William Amis, the City man, was of course all for capitalism. He had been a Lloyd George Liberal, after the First World War switching his allegiance to the Conservatives and becoming active in the party and the local ratepayers' association.

In retrospect, Kingsley came to cringe at the ease with which he was then willing to swallow beliefs like the one contained in the question and answer, 'What would happen if somebody voted against Stalin?' – 'But who would ever want to?' Swallow them he did, though, and such attitudes could hardly have been more contrary to his father's. This, of course, was a large part of the reason why

45

Kingsley held them, though he did not realise that at the time. His beliefs seemed solid and convincing when he argued them and it did not occur to him that it might be his father he was struggling against more than it was capitalism.

Art, broadly defined, was another fruitful source of dispute between father and son. William's tastes were middle- to lowbrow. He enjoyed now forgotten detective-story writers like Freeman Willis Croft, R. Austin Freeman, H.C. Bailey and Francis Grierson, the masculine equivalents of Peggy's feminine romances. The theatre to William meant West End successes, especially musical comedies, while in music his preference was for Gilbert and Sullivan and Edwardian ballads with, as Kingsley remembers them, mostly one-word titles like 'Because' and 'Until'. William took Kingsley to see *The Pirates of Penzance* and *The Yeomen of the Guard* but they only bored his son, who 'meanly exaggerated' his boredom to discomfort his father. Kingsley preferred promenade concerts at the Queen's Hall or John Gielgud playing Hamlet at the Streatham Hill Theatre.

Ballads were sung at the piano during musical evenings when William and Peggy entertained their friends. But why, Kingsley would caustically wonder, was there no Schubert or Wolf among the sheet music beneath the lid of the piano stool? He would make this deficiency an excuse to accuse his father of not really liking music, which was untrue. William did like music, only not Kingsley's sort. And he had a swift comeback to his son's disdain, to the effect that attitudes like Kingsley's would come better from someone who could actually play what he liked on the piano, as he could and Kingsley could not. There was no easy answer to that, in fact no answer at all. It was one of Kingsley's lifelong regrets that his father, willing enough to boss his son around in other causes, failed to bully him into learning the piano.

Musical evenings were a dying form of family entertainment, as Kingsley was aware. The wireless and the gramophone were replacing them. Battle was therefore joined over control of the switch on the wireless set and the needle on the record. Kingsley might wish to hear, say, Brahms's Second Symphony when it got one of its rare performances on the sparse radio music programmes of the period. But his father, after a day at the office and a long journey home, would not want to hear Brahms 'blaring through the house all the hours there were and upsetting' his wife. It was bad enough, in William's view, for Kingsley to waste his own time over books or art galleries but far worse when he wanted to waste other people's time by flooding the whole house with music. And it wasn't just Brahms that William anathematised. 'My father's catholic distaste ranged from Haydn to Troise and his Mandoliers, from Benny Goodman to Borodin,' Kingsley wrote in his *Memoirs*. But his father had the last word.

The growling mood of low-intensity domestic conflict is captured in this passage from *The Riverside Villas Murder*. Having come downstairs after doing his homework, Peter Furneaux 'ate supper (liver and bacon and cold apple pie) with his parents, and then he was allowed to listen for a time to *Romance in Rhythm*, with Geraldo and his Orchestra – not his Gaucho Tango Orchestra, just his Orchestra, and featuring Olive Groves and the Romantic Young Ladies, so that when objection to all that nigger ragtime stuff was voiced, Peter noted the phrase, to pass on to [a friend] and instantly switched off.' First round to father. (The period details, incidentally, are accurate. Kingsley put in more research on what had been going on in his own youth than he had done for any previous novel, reading through a month's back numbers of *The Times*, *Punch* and the *Radio Times* to get his facts straight.)

But Peter comes back fighting. Having consulted the *Radio Times*, he suggests they switch on again later 'and listen to the B.B.C. Orchestra Section G, conducted by Leslie Heward and with Heddle Nash (tenor), performing stuff called things like K.205.' This puts Captain Furneaux on the spot. He can hardly admit that the reason he doesn't want to listen to this obviously worthy music is that he doesn't like it, any more than he doesn't like the more obviously unworthy 'nigger ragtime'. He therefore switches the argument to 'it was getting late and there was school in the morning'. This is pretty lame stuff and Peter senses that he is ahead, making this 'the perfect moment for asking permission to have a bath, since refusal could only look like retaliation, which could only look like acknowledgment of having taken a knock'. Permission is granted. So Peter has won the bout. He is pleased about this, although it does occur to him wearily to wonder 'if everybody who lived with anybody kept carrying on in that sort of way'. It was as though the Marxist vision of class war and endless crisis was being played out in miniature between Kingsley (the proletariat) and William (the ruling class) on a daily basis.

Like Peter Furneaux, Kingsley used his growing intellectual interests as evidence of superiority over his father and as useful weapons to deploy in generational skirmishing. But at least in rows over politics or art he could say exactly what he wanted without fear of anything worse than being argued back at. Sex was a different matter. Here intellectual airs and superior attitudes were no help to Kingsley at all.

Sex was never willingly discussed or referred to in the Amis household. Kingsley was given no instruction in the basic facts of the matter by either parent. By the age of seven or thereabouts he had picked up a rough working knowledge – with plenty of omissions and guesswork – from his schoolmates, who had presumably picked up what they knew in the same way, through the tribal network that tells children things about the adult world their elders are reluctant for whatever reason to pass on. The playground, rather than themselves, seems to have been the source from which parents of the time took it for granted that children would learn the facts of life. Some ten years after Kingsley had pieced the fundamentals together, his mother remarked that she supposed he knew all there was to know 'about marriage and so on', without thinking it necessary to ask how he had found out. Kingsley replied that he thought he did know, forbearing on his side to add that she could have made his life a good deal easier if she had explained it all to him years earlier.

But this would have been too much to hope from parents for whom sex was too embarrassing to be mentioned even obliquely in the presence of their son. Kingsley remembers his mother silencing the woman next door with a warning head-shake when she dared to mention in front of them both something as sexually unalarming as it is possible to imagine, like the fact of somebody going on their honeymoon. Standard middle-class reticence aside, it may be that William and Peggy had their own reasons for silence. Perhaps Kingsley's painful birth put an end to their sex lives out of fear that Peggy would have to go through it all again. But this is speculation. The only thing about his parents' attitude to sex that Kingsley ever knew for sure was their unwillingness to talk about it in his presence.

It was not his parents' failure to tell him the basic facts Kingsley minded. He was content at having picked these up in the casual way he did. It was their refusal to acknowledge that sex existed at all. What he missed from his parents, and what he made sure to give his own two sons when their time came, was not the information that would have been revealed in a frank talk about sex – he knew enough about that already, as he was correctly to assume his sons did later – but the 'permit' such a sex talk would have given him to acknowledge the sexual world's existence. Why go around carrying one's sexual awareness like a guilty secret when everyone shared the same knowledge and feelings? 'In no sphere is it truer that it is necessary to say what it is unnecessary to say' was how he put the case for candour in his *Memoirs*.

Robin Davies suffers the same frustration in *You Can't Do Both*. His mother and father discuss the impending birth of his sister's first child in what they take to be a series of impenetrable exchanges between them, but which Robin understands perfectly well. Why must he behave like 'a prodigy of ignorance and a monster of incuriosity', he wonders. His parents should treat him like 'an average boy of nearly fifteen' or they should refrain from all mention of the subject. He wants to bawl at them 'a demand to be told what the hell was being talked about'. But he refrains, knowing that his father would only tell him 'to be a good chap and not poke his nose into matters which were none of his business'. And then it might embarrass his mother. Sensing what his son may be thinking, Mr Davies changes the subject. 'Have you done your bigs this morning?' So Robin is put in his place by being reminded of his 'only-just-out-of-nappies status'.

If sex in general was a threat, it was masturbation, not girls, that boys needed to be protected from, at least at the start of their adolescence. Fears about what this could do to a growing lad's health and sanity had declined since the high Victorian years when experts could claim it led to blindness and other extremes of debility. But fears were still strong. A friend of Kingsley's was told of a mythical being who was warned in apparent seriousness that if he kept on masturbating his thigh would eventually 'snap like a rotten stick'.

Kingsley began to masturbate when he was fourteen and was insufficiently vigilant to prevent his father finding out what he was up to. His father harangued him along the lines of how this hateful practice was a danger to his general health, warned him he would 'go barmy' if he carried on and urged him to stop before it was too late. Somewhat ashamed at this onslaught, in which his mother joined, Kingsley duly promised, though with undeclared reservations. He thought it was just like his father to be 'horrified by what everybody else took for granted' but to appease him he would enter into a sort of bargain: admit a little and be forgiven the rest on the promise of not doing it again – then be careful. 'Thereafter,' Kingsley wrote, 'I went my own way under a pact of silence and dissimulation.'

William kept a watchful eye on his son's behaviour and now and then repeated his warnings. A sort of hide-and-seek or low-key witchhunt followed, the atmosphere of which is evoked by Peter Furneaux's experience. 'As part of his undeclared, unremitting and quite unavailing war on Peter's baser nature, Captain Furneaux had decreed that he should never lock [the bathroom] door "in case somebody wanted something" (the shaving-mirror? the soap rack?), and had been known to come stealing upstairs at times like this to check that his wish was observed.' On one occasion, determined on a session of self-abuse, Peter gets ready to 'lock the door, not spin things out too long, and be prepared, should his father show a sudden desire to carry off the tooth-glass, to plead thoughtlessness, habit, sieve-like memory.'

Kingsley, too, was not supposed to lock the door when he had a bath. But his father did not pursue him with quite the missionary or police-state zeal of Captain Furneaux. Kingsley sensed that he was genuinely shy about sex and would prefer not to have to deal with it at all. William had done what his conscience told him to do by delivering the required warnings against self-abuse but, having done his own duty, he was largely content to leave Kingsley to do his, at any rate so long as nothing awkward or unpleasant came to light.

Unlike Peter Furneaux, Kingsley never went so far as mutual masturbation with other boys. Nor did he take the further step into sexual experience that the plot of *The Riverside Villas Murder* required Peter to take. Peter loses his virginity when he is seduced by the much older wife of a neighbour. Real life was not so kind to Kingsley, who had to wait several more years for his first chance to make love to a real woman.

In sexual matters William Amis not only toed the standard middle-class line, he made it his own deeply felt personal attitude too. But in matters of religion he was much more his own man. He attended the occasional church service at Christmas or on Armistice Day but he discarded his own Baptist upbringing – which Kingsley describes as 'a fairly gruelling nonconformist nurture' – and completely rejected the system of beliefs that the services he went to celebrated. In this he regarded himself as an apostate, an emancipator, even a rebel, which Kingsley thought laughable at the time, since he could not bring himself to see his father as any sort of rebel against anything. The consequence for Kingsley in one respect at least was the same in religion as sex, though the intention was utterly different: he received no instruction from his father in either subject. If William was altogether silent about sex – silent even about why he was silent – he was more forthcoming about religion. He resolved that the 'chapel stuff' he had had to put up with when he was a boy would not be rammed down Kingsley's throat.

Whether or not Kingsley took his cue from his father, he never felt the slightest temptation to believe in the central Christian idea, the existence of God. He came to endorse the proposition that 'the traditional God of Christianity [is] very wicked' and wrote that he had no belief in God, 'not the first beginning of one, not a shred, and never have had as far back as I can remember – not no belief in him as all-wise, all-loving, all-powerful, difficult as these might be to acquire, just no belief in him as an eternal supreme being'. Apart from a few vaguely remembered 'moments of doubt' when he was around the age of eleven, Kingsley never deviated from this utter rejection. Though aspects of Christianity and of Christ continuously attracted him, the first necessary step in belief – faith in God – he came to think of as a gift, and one he did not have.

This did not remove him from the influence of the religion he could not accept. There were daily prayers at school and a weekly Scripture lesson, in which subject he even got a credit in his School Certificate (though less through study than through the luck of happening to read some notes about the authorship of the St James Epistle on the morning of the exam, which then turned up on the question paper). He studied the Greek Testament in the sixth form and joined the school chapel choir. In the army, he went to Church of England services. And then, of course, there was English literature – 'from beginning to end English poetry as a whole has been shaped by religion, constantly reflects it, looks back to it often when it seems furthest away', as he later wrote.

Kingsley studied Scripture because he had no choice, joined the school's chapel choir because of the singing and attended services in the army because it was the easy option. But one way or another he absorbed a great deal of Christian lore and acquired much respect and affection for it. When he came to reread the Gospels after more than twenty years he was surprised at how much he still recognised, and 'recognised with some intimacy'. Later still he could describe himself

51

as an 'unwilling unbeliever' and wrote a novel, *Russian Hide-and-Seek* (1980), about the pitiful plight of an England in which both art and religion had been obliterated by a Russian Communist takeover. He even became a stout defender of the traditional English church against what he saw as the depradations of its modernisers, declaring himself to have much more in common with 'a believer of my own generation than either of us shares with the modern Christian'. If he had only been able to accept God, it is possible to imagine him a thoroughgoing Christian – cranky, anti-modern and opinionated, no doubt, but devout in his own way too.

Christianity not only haunted Kingsley's imagination; it percolated through into his character by way of his father. William may have thought he had got rid of all the baggage of religion when he stopped believing in it, but traces remained all the same. He objected to the use of Christ as a swear-word and would even so far forget himself as to put the blame for Kingsley's misdeeds down to his son's lack of religion. (This was something that was to puzzle Robin Davies in *You Can't Do Both*. His father, like Kingsley's, has been inculcated with a lot of 'mumbo-jumbo stuff' in his childhood and is all against inculcating Robin with the same. 'Then why,' wonders Robin, 'when he's really fed up with me, why does he tell me the trouble with me is I've got no religion?') It was in his morals, though, that William's Christian upbringing most truly survived. Conscientiousness, patience, frugality, as Kingsley listed them, the morals of a 'pious Protestant'. And in so far as Kingsley felt able to claim such virtues as honesty, responsibility, thrift and industry for himself he believed he would have less claim to them if he had been brought up 'quite outside the shadow of the chapel'.

Meanwhile Kingsley tried his hand at writing poetry. One of his early efforts was a description of the journey to the City of London School playing-field at Grove Park, written in imitation of Chaucer's *Prologue*. 'It was so good', says the master for whom Kingsley wrote it, J.E.B. (Boggy) Marsh, 'that I kept it but it disappeared in the confusion of 1939.' Later pieces were published in the school magazine. The first of these was called 'Prelude'. 'Thus I set out,' it begins, and 'Thus I return,' it finishes. In fifty-eight unrhymed lines it describes the passage of a year and the stirring of vague adolescent moods and longings.

As for happiness
What is happiness? I am too young to know.

Kingsley later described it as 'a kind of suburbanite's Waste Land tizzied up with bits of Wilde'. More pieces followed which the mature Amis judged to be 'of appalling pretentiousness and affectation'. One was called 'Prometheus', after the mythological figure who stole fire to restore it to the human race, and deals in fantastic images, such as of the world becoming

A roaring holocaust, and I alone
Untouched, stood gazing as if turned to stone.
For in that furnace was eternity,
And God threw back its doors that night, for me.

Next came a prose poem, 'He Curses his Insensibility', which 'drew Mr Ellingham's proper contempt'.

Shadows, trees, sky: this is the holier trinity. But its divine power fails to transfigure a reluctant heart, and in vain the wind of love fans the cheek of the libertine.

The last to be published in the school magazine was 'He Reads Immortality in the Eyes of a Virgin':

Be sure now, my soul, that this is the moment for which you pine,
This the split second which you have ever desired to fashion
Into eternity; treasure this when all else has no worth;
This is your one happiness before death; this is the hour of your birth.

Not much sign in these of blooming literary talent – rather of adolescence trying to express itself in borrowed, ill-digested forms. It was not until after he left school and other influences had time to make themselves felt that Kingsley began to find his feet as a poet.

He became one of the school magazine's three editors and contributed reports on the proceedings of the school parliament. The Easter edition of 1941 covers three of the parliament's sessions and catches Kingsley in the act of moving politically leftwards. In the first session we find him playing the moderate part of 'Deputy Leader of the Labour Opposition'. But by the third session he has become the only Communist in the House, in which guise, he wrote of himself, he 'rose and fiercely attacked the Government with a speech of which three-quarters was unacknowledged quotation from Lenin'. He might have become a Communist, but in reporting his own plagiarism Kingsley showed he had not lost his sense of humour in the process.

By then the Second World War was well under way. The coming of war made Kingsley more politically aware than ever before and opened up a new phase in his school life. Although he had been sufficiently conscious of international tensions to join the school branch of the League of Nations Union and to attend some of the meetings it held jointly with the City of London Girls' School – girls, of course, being an extra incentive – the outbreak of hostilities still came to him as a 'great surprise'. The school authorities had been more alert to the dangers of the times. They had first thought up a scheme of evacuation to Wales but abandoned it in preference to a deal with Marlborough School in Wiltshire, struck in 1938, a year ahead of need. Under this arrangement the City of London School was to share kitchens, dining-room, classrooms and sports fields with Marlborough. But Marlborough could not offer spare beds, so the City of London School boys were to be billeted in whatever accommodation could be found nearby.

In the last days of August 1939 boys and masters carried suitcases to school and bought daily train tickets, not daring to buy new season tickets in case they did not return home that night. Labels were issued and the school paraded in columns to rehearse the older pupils in not marching so fast that they left the younger ones behind. Gas-mask drills were held and boys persuaded to give up large suitcases for small ones. Eventually, on Friday 1 September, two days before war was declared, the column of boys and masters marched the few hundred yards from the school's front steps to Blackfriars station, where normal train services were alternating with fast specials carrying schoolchildren out of East London. At 12.30 the school started across the city to Ealing, where a special train for Marlborough was expected. But the only train at the station was a Taunton express, into which zealous evacuation officials proceeded to herd the rear end of the school column. Protest was of little use, since, by a typical piece of baffling wartime bureaucracy, even to know where the school was headed was some sort of offence against security. A compromise was eventually negotiated and the Taunton train made an unscheduled stop at Savernake, where the school arrived two hours earlier than it would have done if the right train had shown up.

The new confusions of war continued when the local evacuation authorities found they had on their hands not only a whole school two hours early but a trainload of mothers and babies instead of the primary school children they had been expecting. An unplanned-for wave of soldiers also needing billets soon followed. Kingsley slept for the first few days in an army-type bed in a barn, living mostly on tins of condensed milk issued as part of the school's iron rations. For several days stragglers continued to arrive, masters and boys, some cycling all the way from London.

Soon, Kingsley was allocated with four other boys to quarters in a small farm labourer's cottage on the Bath Road which had been empty for some time. Here Mr Ellingham and his wife Edna were in charge. Amenities were sparse. The room Kingsley shared with Leonard Richenberg was tiny, perhaps six foot by seven, with truckle beds and no electric light, only candles. Baths were taken in the kitchen in a portable tub filled with water from a single cold tap and heated on the stove. There was only an earth closet which had to be cleared and Kingsley earned the reputation of being reluctant to do his bit on this and other tasks. He came to suspect that Mr Ellingham 'did not greatly care for' him and wondered whether that might not be due among other causes to his not having pulled his weight in the cottage.

Things got worse as autumn gave way to winter, which that year was freakishly cold. The only fire was in the living-room and Kingsley remembered it as the coldest building he ever lived in, 'far worse than anything in the army'. Conditions one day were such as Richenberg never saw before or afterwards. Trees, he recalls, were coated with ice so that they looked like 'glacé fruits'; 'every twig on every tree [was] encased in its own translucent jacket of ice, sometimes inches thick, and twenty-foot boughs came crashing to the ground like matchsticks beneath the weight of accumulated ice.' The temperature fell below 20 degrees Fahrenheit and on at least one night the urine in Richenberg's and Kingsley's chamber pot froze. The ground became too slippery to use bicycles, and even walking was dangerous. A master, Mr J.H. Wheeler, remembered boys arriving at school in the morning 'looking like bedraggled Shire horses, with the feet wrapped in sacking or old socks'. Not surprisingly, perhaps, an epidemic of flu broke out, with at one time more than a hundred boys laid up in the gymnasium, the college sanatorium being already full up with Marlborough boys. Many years later Kingsley recalled that winter's day vividly in a poem, in which he wrote that, walking to school,

> I saw that every twig,
> Every leaf in the vicar's privet hedge
> And every stalk and stem was covered in
> A thin layer of ice as clear as glass
> Because the rain had frozen as it landed.
> The sun shone and the trees and shrubs shone back
> Like pale flames with orange and green sparkles.
> Freak weather conditions, people said.

Conditions were not improved, in Amis's view, by the 'outrageously stand-offish' attitude of Marlborough College. Contact between the two schools was not simple, since timetables had to be organised so that when one school was at, say, lunch the other was very much not. Without such 'Box-and-Cox' arrangements, as Mr Dale described them, coexistence would have been impossible. Still, there was plenty of time for what Kingsley calls 'host-like gestures'. None was made. The school captains discussed rules and the like, some of the younger City of London School boys were loaned to the Marlborough choir to 'dilute' its baritone strength with their treble voices, and a handful of sportsmen clashed on the squash courts and elsewhere. Otherwise there was scarcely any contact at all. During his five terms at Marlborough, Kingsley reckons to have spent no more than two minutes in conversation with Marlborough boys, a fact that more than fifty years later he still finds 'disgraceful'.

The atmosphere of the school changed too with the shift from day school to boarding. In London, CLS boys went home at night and became the responsibility of their parents. In Marlborough, the school was responsible for them twenty-four hours a day. But no institutionalised system of discipline existed and the masters were not accustomed to exercise the kind of authority necessary to keep several hundred boys out of mischief round the clock. Away from parental authority too, some of the boys ran wild. There was fighting with catapults and airguns, Richenberg recalls. One boy nearly lost an eye and came close even to being killed. Local shopkeepers suspected them of shoplifting. Much of the job of sorting this out fell on the prefects, of whom Kingsley became one, while his room-mate Richenberg was made captain of the school. Richenberg remembers this duty taking up much of his time and finding that first year in Marlborough 'very distasteful'. There were signs, too, that CLS boys might acquire some of the homosexual tendencies which were endemic in boys' boarding schools, much less intense in day schools. But this mostly amounted to little more than some 'silly talk' among the older boys who had developed, or thought they had, a fancy for one or another of the better-looking younger ones.

Compulsory games were introduced, perhaps as a means of keeping CLS boys' minds off such dubious distractions, certainly as an extra form of discipline. In London, Kingsley had mostly avoided games in spite of being harangued by Mr Copping, his junior housemaster. He pleaded that games would interrupt his studies and besides the route between the school sports ground and his home was a 'horrible dog-leg' train journey, an excuse that did not carry much weight with Mr Ellingham. Now that Kingsley had no choice he found he enjoyed sport after all. He turned out for Hale house at cricket, rugger, tennis and athletics. He also found a place in the school's Second XV and was even promoted for three matches – not enough to secure his school colours, as it happened, since one match was for the A team, not the First, and so did not count. All in all, though, his time at Marlborough must be counted the most physically energetic of Kingsley's life, an outburst of regular sporting activity never to be voluntarily repeated in any form of exercise much more demanding than a stroll to the pub.

By leaving London, Kingsley lost the small excitements of a great city, like eating lunch in the Lyons teashop near Blackfriars Bridge or scouring the secondhand bookshops and record stores around the Charing Cross Road with his friend Cyril Metliss. But there were advantages to exile. One was the pleasure of the Wiltshire countryside which gave 'a kind of zest to [Kingsley's] adolescent melancholy'. For years afterwards whenever he tried to 'visualise a generic country scene' it was always some image of Marlborough that appeared to his mind. He joined the chapel choir and found that singing in four-part harmony was 'the apex of non-sensual pleasures', though listening to a choir he was not part of, he later discovered, offended him: 'sonic wool roughened with sibilants'.

He took part in further, non-political, debates, once winning a competition against sixteen rivals by defeating a motion that 'half a loaf is better than none'. This he did with schoolboy ingenuity by redefining 'loaf' in the Cockney sense to mean brain and arguing the dangers of a half-developed intellect. In the dramatic society he read the part of the conjuror in Chesterton's play *Magic* 'thoughtfully . . . but without any really bitter feeling'. He claimed that he understood Auden and Isherwood's play *The Ascent of F6* even if nobody else did and explained that 'the main theme was the "Demon-Mother" who, too ambitious for her son, led him on to final destruction'. He read the part of M.F. and 'succeeded in conveying his doubt in the whole journey'. Kingsley became secretary of the music society too. He would give a short lecture at its meetings, then play records of a symphony. One boy, Len Dowsett, owed a lifelong interest in classical music to being introduced to it by Kingsley.

All boys had to join the Officers' Training Corps at the age of fourteen, unless they claimed to be pacifists or conscientious objectors, in which case they joined the Boy Scouts and were mildly despised by their fellows. Boys put on uniforms like those of the Rifle Brigade twice a week. As cadets, they drilled, did weapons training, went on field days and spent an annual two weeks in summer camp. They learned to shoot, and Kingsley became a first-class shot and a sergeant. While the school was still in London he was drilled at Wellington Barracks and was once one of the school contingent in the annual Lord Mayor's Show. A boy called Moss discovered a bookshop near the Temple which sold War Office publications, including a book about 'section leadership' in the cavalry. Kingsley joined the corps's mounted section and took riding lessons every week from a Colonel Lawrence. He hated being on a horse when it cantered but liked wearing breeches instead of trousers and tying his puttees in a way that marked him off from his schoolmates.

Like others of his age, Kingsley was brought up on legends of soldierly heroism. Storytellers like Rudyard Kipling, G.A. Henty, John Buchan, Anthony Hope and Percy F. Westerman described the adventures of brave officers or civilians with service life not far away in their backgrounds. Many such tales were about the First World War. One, called *The Dispatch Riders*, had two English schoolboys joining the Belgian army and driving off a German patrol with soda siphons. This appealed to Kingsley not only for its bravery but because, as an only child, being a soldier implied membership of an organisation in which 'imaginary comrades were shadow brothers'. He developed a taste, too, for poems with military themes by writers like Macaulay, Southey, Browning, Kipling and Housman (some of which he was to include in his 1978 collection of stirringly declaimable verse, *The Faber Popular Reciter*).

The army was to hold a long fascination for Kingsley. For a while he even thought he might become a soldier if he did not become a writer. Romantic enthusiasm for the military life was, however, tempered by caution and a couple of dashes of cowardice and laziness. At Marlborough, Kingsley found a new niche in the signals group, which was given instruction in Morse code and signalling by flag. Its advantages included not having to wear uniform or do foot-drill and to prepare him to join the Royal Corps of Signals when he reached the army proper.

In spite of its drawbacks and inconveniences, Kingsley found his terms at Marlborough more liberating than otherwise. There were no parents to boss him around and he had time to cultivate his relations with boys who in London might go home after school to the other side of town. Friendships flourished with Richenberg, Metliss and Saul Rose, another resident of the Bath Road cottage.

Early on in his career at the City of London School, Kingsley had learned that neither class nor race, intellect nor prowess at sport were the keys to schoolboy popularity. 'To be accepted you had only to be amiable; to be liked you needed pre-eminently to be able to raise the occasional laugh' – for which no great resources of wit were necessary. One particularly effective way of doing this was to be able to mimic members of the staff. Kingsley, already adept at mimicry from experience at home, developed his version of Mr Dale: 'Get it right, not wrong. Black, not white. Cat, not dog.' This sort of thing, he found, easily earned him the esteem of his classmates.

Like most, perhaps all, schoolchildren, Kingsley saw his teachers not so much as real people but as caricatures whose eccentricities stayed in the memory for life. He was ever afterwards able to remember Mr Marsh 'sucking the earpiece of his glasses', Mr Penn 'accusing us all of having eaten his biscuits', Mr Carruthers remaining imperturbable when Kingsley and another boy, Rumsey, deliberately dropped a suitcase full of broken glass on the floor to make an ear-shattering crash, and Mr Copping, who spoke 'with an Attic Greek accent of the fifth century'. Masters make this indelible impression on boys in their earlier teens because schoolboys see the adult world, according to Kingsley, 'with the delighted, faintly hostile astonishment of the tourist'. The schoolboy 'is entertained to the limits of endurance by its quaint tribal customs, its grotesque ritual dances, its capering, scowling, gesticulating witch-doctors'. For most boys, this way of seeing the world passes when he enters the adult world himself. But the novelist must retain something of the 'adolescent's coldly wondering stare'. Or so Kingsley came to believe. In his observations of his schoolmasters he can perhaps be seen unconsciously rehearsing his novelist's gift for appropriating useful parts of the characters around him.

If his masters were making a lasting impression on Kingsley, he was making the same on his schoolboy contemporaries. He had a knack for getting on with people, for making friends. 'There was a great affinity between us,' Richenberg remembers. 'We were on the same wavelength.' Cyril Metliss says, 'It was an intense period – we sparked off each other.' According to Sir (as he later became) Peter Baldwin, 'He was a really good friend – very kind. He had an acerbic tongue, but about things rather than people.'

Besides his private friendships Amis had public dash. He was bright, gregarious, good fun, a great maker of faces and pricker of pomposity. In the judgement of Sir (as he also became) George Blunden, Kingsley in his last year 'was clearly the intellectual leader of the school'. Metliss remembers him at that time as 'a great mimic and full of fun'. To Baldwin, 'irreverence was a very great factor of his character . . . his charm helped him get away with things . . . we expected him to be different, an iconoclastic chap . . . he was great fun in class – more of an undergraduate than a schoolboy.' Richenberg says, 'the pastiches he did of French, German and Russian generals were childish but very funny.' Blunden recalls that in his last year Kingsley 'set the standards of cultural and intellectual activity out of class. He revelled in being the intellectual star.'

When his first novel, *Lucky Jim*, was published thirteen years after he left school those same people and others who knew him in youth were quick to identify the Kingsley they had known with the Jim Dixon of his novel. 'In that last year,' Blunden recollects, 'he turned into Lucky Jim.' 'When I first read *Lucky Jim*,' says Richenberg, 'I thought, that's Kingsley.' When he read *Lucky Jim*, Dowsett recalls, 'I thought – that was Kingsley at school.' 'Lucky Jim', says Metliss simply, 'was Kingsley.' Such straightforward identifications between Jim and himself – or indeed between any of his fictional characters and any real person – were to infuriate Kingsley for the rest of his life. Just because Jim 'drank rather a lot', for instance, why should it be assumed that his creator did too? But people do make such identifications between writer and character. And, whether rightly or wrongly, the unanimous verdict of his contemporaries was that Amis at eighteen already bore a striking likeness to the Lucky Jim he had not yet even begun to think of.

For Kingsley to go on to university a further scholarship was required, William being in no better position to pay for Oxford or Cambridge than he was for the City of London School. There were plenty of scholarships in classics to be had at both universities and Kingsley believed himself good enough to walk into one of those if he had wanted. But in his last school years he was already thinking of becoming a writer and so set his sights on reading English instead. This created two problems. There was no precedent at CLS and so no provision for teaching English to the necessary level. And there were far fewer scholarships in this discipline – so far as Kingsley knew, only two each at Oxford and Cambridge.

The school rallied round to accommodate him with the extra lessons he needed. Thus brought up to the mark, Kingsley tried for a scholarship at St Catharine's College, Cambridge, and failed. He then tried for an exhibition at St John's College, Oxford, and succeeded. Kingsley was to be always grateful to CLS for the special efforts it made to get him up to standard. Indeed he was to remain grateful for the whole seven years he spent at the school. If not the happiest years of his life, they were certainly as good as he could have hoped for. In 1958 he paid his old school a golden tribute, writing:

I have never in my life known a community where factions of any kind were less in evidence, where differences of class, upbringing, income group and religion counted for so little. In particular, although perhaps fifteen per cent of the boys were Jewish, not a single instance of even the mildest anti-semitism [apart from the incident involving the regimental sergeant-major and Richenberg] came to my attention in the seven years I was a pupil there. The academic teaching was of a standard not easily to be surpassed, but more important still was that lesson about how to regard one's fellows, a lesson not delivered but enacted. Thanks indeed for that.

During Kingsley's last summer holiday from school in 1940, the Amis family suddenly quit Norbury for good. Peggy became frightened at the closeness of the German bomber flight-path overhead – though Norbury was to escape largely unharmed, only some fourteen bombs falling on it throughout the war – and the family headed north for the commutable town of Berkhamsted in Hertfordshire, beyond the bombers' usual targets. They stayed for a while at the flat of William's secretary, Miss Woolcock ('Woolly'), until one house was found – a sort of workman's cottage, as Kingsley remembers it – and then another, in Shrublands Avenue, which was to become the final Amis family home.

The move made Kingsley 'a bit lonely, isolated'. Going to school in central London and later at Marlborough, on top of his parents' ruthless exclusion policy towards potential chums, had kept him short of friends out of school hours. But against the odds he had succeeded in making a few in Norbury and now he was cut off from them again. Oxford offered rich potential for making up this deficiency, for ending what he thought of as the 'underpopulation' of his childhood and youth. One consequence of this was to make it difficult for him to judge his own early years, whether or not they might have been better or happier. Having no brother or sister, living a circumscribed suburban existence, he had few standards by which to measure other young lives against his own. When he looks back, his judgement is ambiguous. A cousin, say, who happened in on his home might think he had 'never seen a worse case of a poor lad being terrorised' or, equally, that 'his parents were a model of compassion and tenderness'. Both views could be valid, depending on the temper of the moment the cousin chose to visit. In retrospect, though, Kingsley felt no reason to complain, no grounds for believing that his relations with his parents were better or worse than anybody else's.

Paradoxically, the years of maximum conflict between father and son had also been the years of greatest intimacy. Kingsley and William were never so close as when they were at each other's throats, though perhaps this is not as much of a contradiction as it seems. The closeness weakened when Kingsley began to live away from home, first at Marlborough, then at Oxford and in the army. As their lives grew apart, Kingsley's relations with his father changed from conflict to indifference to boredom. Until he married in 1948, the 'chief magnetic influence' that continued to draw him home was lack of money with which to stay away. When his parents visited him after his marriage the time dragged. William and Peggy never gave up their habit of trying to do everything together as a family.

Instead of meeting at mealtimes, say, and going their own way for the rest of the day they had to be around all the time. Observing this, Kingsley's friend Philip Larkin commiserated with him: 'Your parents never leave you alone.'

When Peggy died of a stroke in 1957, Kingsley brought his father to stay with him in Swansea and took him along for his sabbatical year in America in 1958–9. William enjoyed America but found it difficult to fill up his days in Swansea and returned to London, finding a new job with a company which dealt in domestic cleaning materials. There, he perked up enough from his desolation at Peggy's death to have 'a couple of lady friends', about whom he was unusually forthcoming to his son, though still diffident, as always, in matters of sex. William died of cancer on 18 April 1963, leaving such possessions as he had to one of his ladies, who insisted on reclaiming even the watch and wallet that had been passed on to Kingsley's sons, Philip and Martin. The only memento of his father Kingsley had left to keep was his last letter, thanking his son for getting him into a nursing home at Cambridge, where Kingsley was then teaching. Kingsley had nothing of his mother's.

After his parents' deaths, Kingsley's feelings changed to regret. Of his mother he wrote in his *Memoirs* that she was 'the first of the appallingly long line of figures in my life whom I have come to value altogether more highly, to appreciate the uniqueness of, now they are gone'. To his father's memory he wrote a poem, 'In Memoriam W.R.A.', in which he asked William to forgive him that it should have taken so long for his pride and love to become clear and to arouse Kingsley's love in return:

I'm sorry you had to die
To make me sorry
You're not here now.

It is possible to see Kingsley's adult life from Oxford onwards as a kind of gigantic spree in perpetual celebration of his good luck in escaping the restrictions of his family home, but heavily modified by an equally endless attempt to live up to his father's expectations, or at least not let him down too badly. There is a Kingsley Amis who wants nothing more than to have a good time and another Kingsley Amis who, with his father's ghost at his shoulder, desperately wants to achieve.

This may be fanciful, no more than a piece of suggestive imagery. What is certain is that Kingsley left home for Oxford already firmly moulded. He had developed a large appetite for sociability and for sex, the first not adequately satisfied, the second not satisfied at all; he had learned to work hard; his interest in literature and other forms, major and minor, from music to films, had been intensely aroused; he had formed the intention of being a writer, even if that was more a vague ambition to 'be' something than a concrete plan actually to write something; his skills as a mimic, face-puller and entertainer of his chums – all to be important in his writing and his social life – were well advanced. He was a committed non-Christian who was nevertheless attracted to and influenced by the beliefs he did not hold. He was also a Communist, though that turned out to be a phase, unlike his lifelong non-Christianity.

The young Kingsley had also formed habits that the mature Amis kept up all his life: for example, dislike of the telephone, of being bored, of domestic chores, of looking after himself in practical ways, whether cooking, shopping or keeping his accounts. (Even at Oxford, William continued to look after his son's finances, paying his bills out of his exhibition and allotting Kingsley 10 shillings (50p) a week pocket-money.) Kingsley's darker side, his anxieties and phobias – fear of flying, of being alone, of being trapped in an enclosed space or in the dark – were in place too. Apart from sex, almost the only thing he had yet to be properly introduced to was drink. Oxford got him started on both.

∞ 4 ∞

Wartime Oxford

Amis went up to Oxford for the first time in April 1941, arriving from Berkhamsted in the family butcher's old Morris car in what he called in his *Memoirs* 'impeccably proletarian style'. In normal times he would have started, with all the new undergraduates of his year, in the autumn term but this was wartime and now undergraduates arrived from school and departed for the forces in every one of the academic year's three terms.

War had changed the whole mood and tone of the university as it had the rest of the country. The Oxford of aesthete, aristocrat and wealthy playboy – of Max Beerbohm's *Zuleika Dobson* or Evelyn Waugh's *Brideshead Revisited* – had all but vanished. A young man with money was at no great advantage over others without, since the mobilisation of the nation's resources for the war effort saw to it that there were few luxuries left for anybody to spend money on. Night-time blackout was only the most obvious expression of the drab and sombre mood that settled over Oxford while the war lasted.

'Life in college was austere,' Philip Larkin, who was Amis's contemporary at St John's, remembered. 'Everyone paid the same fees (in our case 12s a day) and ate the same meals. Because of Ministry of Food regulations, the town could offer little in the way of luxurious eating and drinking, and college festivities, such as commemoration balls, had been suspended for the duration. Because of petrol rationing, nobody ran a car. Because of clothes rationing, it was difficult to dress stylishly. There was still coal in the bunkers outside our rooms, but fuel rationing was soon to remove it. It became a routine after ordering one's books in [the Bodleian Library] after breakfast to go and look for a cake or cigarette queue.'

To this list of deprivations Amis's memory adds that there were 'no lunch parties . . . food was rationed, liquor was hard to get, beer was short – everything was cut back. Everybody was spending hours in trains waiting for things to happen, or hours queueing for food – not essential food, but for any luxury . . . there was enforced pub-crawling because one pub would sell beer early in the evening, and another later on, so there would be groups of us just drifting around.' (Pubs were supposed to be out of bounds to undergraduates but in this respect at least war made no difference to their habit of defying the authorities by ignoring the ban.)

There were other abnormalities in the pattern of Oxford life. Younger dons were subject to war service like everybody else and were drafted into the forces or the newly inflated Whitehall bureaucracies, leaving behind dons who were elderly and somewhat remote, partly at least because of the administrative burden created by the ceaseless turnover of undergraduates. St John's did its bit for the war effort by housing sections of the Ministry of Agriculture which dealt with white fish and potatoes, so earning the college the temporary nickname of Fish and Chips. Consequent shortage of space meant that undergraduates often had to double up in rooms normally occupied by a single man – there were then, of course, no women at any of the Oxford colleges other than those set up especially for them – although Amis was allocated 'a nasty little pair of rooms in the top corner of the front quad' to himself.

Only those undergraduates studying subjects with some useful connection to the war effort, such as medicine or science, were allowed to stay the full length of their courses. The rest could expect no more than a few terms before being called up, leaving their studies to be completed later if they survived the war to return. Amis managed to secure the maximum grace of four terms by taking advantage of a regulation that gave extra time to those who said they were willing to become schoolmasters after they graduated; and he turned up at the vice-chancellor's office to promise faithfully he would do so. Amis's promise may have been made on tactical grounds but it was not wholly insincere. A schoolmaster was what Amis then thought he would probably become on his way to becoming a writer.

In the meantime, though, he was obliged to put on a uniform and turn out a day and a half each week for the Senior Training Corps. He was reluctant but knew that compliance meant a shorter spell in the ranks and a smooth path to officer status when he finally joined the army, while refusal would have earned him an instant call-up as a private soldier with no certainty of becoming an officer at all. Corps service involved weapons training and map reading but mainly meant drilling, which was 'economical of space, equipment, instructors and preparation' and at which Amis, with the experience of the school corps behind him, found himself to be 'quite good'.

Amis did signals training too. Once he was faced with the certainty that he would have to join the army during a real war with real guns firing, the military life no longer seemed quite so romantically appealing. The choice, as he saw it, was between signing up with either 'the Polovtsian Lancers or the Second Royal Fusiliers depot regiment', or between heroism and safety. Now that the moment of decision approached at full speed, Amis was in no doubt. Friends who had been declared unfit for call-up seemed likely to survive. Amis was 'bloody well' determined he would not let his good health prevent him doing the same. One friend with similar life-saving intentions, Norman Manning, recommended the signals section as a step on the way to the Royal Corps of Signals when the time came. 'We're always in the back,' Manning told him. 'We don't go further forward than Brigade.' Amis saw the sense in that and joined up. The Signals were to serve him well as a vehicle of survival; but the unfortunate Manning succumbed to tropical disease within weeks of being sent to India.

Besides military training there was another wartime chore: fire-watching. Hitler had threatened to unleash 'Baedeker' (guidebook) raids on England which would ravish the country's historic towns as well as the usual military-industrial targets, and Oxford was considered a likely stop on the bombers' tourist route. There were stirrup pumps and ladders all over St John's, and fire-watching was done in two-hour shifts throughout the night, undergraduates taking their turn every ten days or so.

The atmosphere created by the war may have robbed Oxford of most of its glamour, but there were balancing compensations in the new regime of austerity. Traditional class distinctions were compressed if not entirely eliminated. Sergeant-Major Reid of the Scots Guards, reputed to have killed fifteen Germans with his bayonet at Dunkirk, bawled with equal ferocity at common folk like Amis and at 'Lord This of Magdalen or the Hon. That of Christ Church, "You're the worst bloody soldier I've ever seen, sir! Now get down to the hut and back at the double!"' There was for Amis 'a wealth of satisfaction' in hearing upper-class undergraduates berated in such language.

War was the overwhelming reality of those first four terms, but it was also curiously distant. It was 'something you read of in the paper and heard on the wireless', according to one of Amis's contemporaries at St John's, Nick Russel, 'about as actual as newsreel pictures'. Nor, contrary to abiding myths of a nation united in patriotic fervour, was there in Amis's recollection any great enthusiasm for actually joining up and fighting, at any rate among undergraduates. He and his contemporaries would have preferred to ignore the war altogether if they possibly could in the hope, perhaps, that it would go away if they did. Philip Larkin wrote in November 1941 to a friend that 'I just don't want to go into the Army. I want to pretend it isn't there: that there's no war on.' Elsewhere he wrote, 'I was fundamentally – like the rest of my friends – uninterested in the war.' Larkin managed to avoid service himself by being declared medically unfit because of his poor eyesight, as did others of Amis's undergraduate friends like Bruce Montgomery and John Wain, thereby incidentally increasing Amis's sense of grievance about the whole thing. (And, also incidentally, helping to resolve in his mind a familiar question: why did the Second World War produce so few good poets compared to the First? Answer: because so many who might have been good war poets avoided being called up.)

Nevertheless, the war, however distant or unwelcome, permeated everything and one of its now forgotten effects was to stop people thinking about the future because there could be no certainty of having one. To Oxford undergraduates, this meant not worrying about their careers, at the university or after it. Even more than class distinctions, careerism disappeared for the duration. Set beside the possibility of getting killed, such ambitions as becoming a rugger blue or an officer of the Oxford Union or a star of the social scene were bound to seem frivolous and unimportant; while fretting over the relative merits of the law, the Civil Service or teaching as career choices was utterly pointless. All this went into what Larkin called the 'distinctive quality' of Oxford at that time. 'A lack of *douceur* was balanced by a lack of *bêtises*,' he wrote, 'whether of college ceremonial or undergraduate extravagance . . . I think our perspectives were truer as a result. At an age when self-importance would have been normal, events cut us ruthlessly down to size.'

When he first arrived, Amis's only friends in Oxford were fewer than a dozen former schoolmates from CLS, all of them at other colleges. They threw a party for him in Balliol at which he got seriously drunk for the first time in his life on half a bottle of sherry. Having staggered back the short distance to his own college and settled himself in an armchair with a chamber pot on his lap, his sufferings were interrupted by two representatives of the university Conservative Association hoping to recruit him. But on seeing his plight they politely withdrew, offering to come back at a more convenient time.

There did, however, turn out to be one young man in St John's whom Amis had met before. While in search of a scholarship at Cambridge, he had shared digs with a Bristol schoolboy, Norman Iles, who was also in search of a scholarship, also failed to get one from St Catharine's, Cambridge, and also turned up at St John's, reaching it, in fact, ahead of Amis and becoming Philip Larkin's partner in tutorials. It was through Iles that Amis met Larkin and so began the most important friendship of his life.

In Larkin's recollection, Iles spotted Amis's name in a list of new arrivals on the noticeboard in the college lodge and immediately recommended him as a 'hell of a good man' because 'he shoots guns'. The significance of this odd talent puzzled Larkin until, later the same afternoon, Amis appeared in the front quadrangle as he and Iles were crossing it. Without a word, Iles raised and pointed his right hand and fired an imaginary pistol shot at Amis. 'The young man's reaction was immediate,' Larkin wrote years later. 'Clutching his chest in a rictus of agony, he threw one arm up against the archway and began slowly crumpling downwards, fingers scoring the stonework. Just as he was about to collapse . . ., however . . ., he righted himself and trotted over to us.' Introductions were made and Amis treated the two to variations to the pistol routine on which he had been working: a pistol fired in a ravine, then the same thing again but with ricochet noises added. Iles laughed while Larkin stood silent. 'For the first time,' he recalled, 'I felt myself in the presence of a talent greater than my own.'

Amis has no memory of this incident at all, though there is no reason to believe it is untrue on that account. His own first memory of Larkin has a warmth of a different sort. Larkin generously, if uncharacteristically, offered him a cigarette, 'the equivalent in those days of a glass of rare malt whisky'.

Iles was for a while the centre of a circle that included Amis and Larkin, who were drawn to him by the force of his eccentricities, if not much else. Amis describes him as 'large, large-faced, out of condition, with an air of half-serious hostility to the world in general, to any received idea, but unaffectionate tolerance towards individuals of his own standing'. It was hard, Amis thought, to say what Iles believed he was doing at Oxford at all, since he did 'what he could to undermine the academic outlook by representing the university as a place where charlatans lorded it over ambitious or apathetic noodles'. Larkin confirms this version of Iles as square-peg-in-round-hole. Iles would roar like the Metro-Goldwyn-Mayer lion at anything which implied respect for 'qualities such as punctuality, prudence, thrift or respectability'; 'ostentatious courtesy produced a falsetto celestial-choir effect, ostentatious sensibility the recommendation to "write a poem about it"'. But Iles treated everyone equally in the same uncharitable way and had the further saving grace of keeping his 'most hilarious mockeries for himself'. His total rejection of the Oxford environment, Larkin thought, at first increased his influence over his friends, but later tended to cut him off from them. Much to Amis's surprise, Iles passed both his exams and his certificate for officer training, though he was later sent to an outfit for unemployable officers and at the end of the war became part of a Friends' relief unit in Poland where at last, according to Larkin, he seemed to be doing what he wanted.

With Iles and Larkin, Amis was drawn into a loose group of St John's men who called themselves 'The Seven'. Other members were Jimmy Willcox, Philip Brown, Nick Russel and David Williams. It began with one formal meeting but quickly turned into something more convivial, supper parties once a week or simply drinking beer in someone's rooms. Now and then The Seven would become serious enough to pass a poem around, and they held at least some ideas or attitudes in common: literature was to be taken seriously though it was not to detach them from drinking and swearing; dons were not to be respected but not challenged either; they were to be different from their parents' generation but that did not mean abandoning all traditions. Such attitudes were congenial to Amis then and for the rest of his life. But The Seven did not constitute anything so coherent or purposeful as a literary movement, more a source of annoyance to the college for their rowdy behaviour, playing records too loudly and the like. Much of what they got up to was simply childish. Larkin and Amis were not above pissing in the common-room fire after an evening on the beer. At various times they were summoned by the college dean to account for offences like these, an experience retold in Larkin's poem 'Dockery and Son', which recalls how,

> Black-gowned, unbreakfasted, and still half-tight
> We used to stand before that desk, to give
> 'Our version' of 'these incidents last night'.

Amis became best known at Oxford for, in Larkin's words, his 'genius for imaginative mimicry'. Over the years, Amis was to develop a whole repertoire of set-pieces. There was President Roosevelt broadcasting across the Atlantic through crackling air-waves; a Japanese soldier mutilating a corpse, which started with Amis's head appearing from behind a sofa; a bomb exploding among a crowd of hearties drinking at a table, with what one witness called a 'wonderfully prolonged' detonation; a variety of scenes from gangster movies with parts improbably assigned to well-known Oxford literary figures like Lord David Cecil (the gang leader) and C.S. Lewis (the squealer). His masterpiece, according to Larkin, 'involved three subalterns, a Glaswegian driver, and a jeep breaking down and refusing to restart somewhere in Germany'. This one was so demanding that Larkin only heard Amis do it twice. So realistic were Amis's impressions that once, beaming his motor-bike-failing-to-start noise in the direction of a motor-cyclist who had just got off his machine, he caused the man to return to it, puzzled at why it was trying to restart its engine without him. John Wain called him the best raconteur in Oxford: 'I've seen him surrounded by people begging him to stop, tears of laughter, choking from all his funny stories.'

At the beginning, though, Larkin found Amis's wit to be not so much a matter of set-pieces like these as 'the quickest way of convincing you that something was horrible or boring or absurd – the local comrade ("Eesa poincher see . . . assa poincher see"), the Irish tenor ("the sarn wass dee-cli-neeng"), the University [Company Sergeant-Major] ("Goo on, seh") . . .'

*   *   *

71

Amis naturally enjoyed being admired for his wit and his mimicry. But celebrity in this department had its drawbacks. It stopped people taking seriously his serious side, which in his first terms at Oxford primarily meant politics. As Larkin put it, 'in his efforts to prove he had a serious side he became insulting in his emphasis, which made him disliked by many [who] could not appreciate the humour that he breathed like air.' In the long run, however, it may have been just as well that Amis's humour was better remembered than his politics. The reputation of having once been a whole-hearted Communist is hardly one that Amis would have welcomed snapping at his heels for the rest of his days, as it did other undergraduates of the period such as Denis Healey, later a Labour Chancellor.

But a Communist is what Amis was and although some people thought this was just another of the faces he showed the world he was wholly sincere about it. Being a Communist put him at the far end of the prevailing spectrum of Oxford politics, which was heavily tilted to the left, influenced by the slump of the 1930s and the rise of Nazism, especially as these were interpreted to intelligent young people by the Left Book Club.

As often, the political left was divided. In 1940 the Oxford Labour Club had split in two, one faction toeing the Communist Party line, then determined by the Nazi–Soviet pact, which meant being against Finnish resistance to the Soviet Union and the war in the west. The non-Communist faction was in favour of both. Roy Jenkins and Tony Crosland, both later prominent members of Labour governments, founded and led the breakaway anti-Communist Democratic Socialist Club. Others, like the novelist Iris Murdoch and the military historian John Terraine, went with the Communists. (As treasurers of the rival factions, Jenkins and Murdoch had to negotiate the division of the Labour Club's assets, he addressing her in proper bourgeois fashion, 'Dear Miss Murdoch', she responding with Communist correctness, 'Dear Comrade Jenkins'.)

George Blunden, an old friend from CLS and by then an under-graduate at University College, remembers Amis calling round to ask him along to a Communist meeting which Amis wanted to attend but not by himself. They went to three meetings in all. Blunden found them full of boring people who talked in an extraordinary jargon and he did not join up. But Amis did. He went to meetings of the Berkhamsted branch of the party during vacations too. Amis tried to recruit other members of St John's to the cause and he also attended meetings of the People's Convention, a peace campaign with strong Communist links. He read Marx, Lenin and Plekhanov, went to Communist study groups and meetings, sometimes speaking himself, sang in the Labour Club choir and edited the Club's *Bulletin* for a term in 1942.

Amis was an 'open' member of the party, meaning that if he got up to speak at a meeting he was free to begin, 'As a Communist, I want to say this . . .' There were 'closed' members too, people who did not reveal their party affiliation. At one time, it was rumoured, three out of five members of the committee that ran the Oxford Conservative Party were actually such closet Communists, though this may have been no more than a piece of self-serving Communist disinformation. The 'unquenchable assiduity' with which the party carried out such penetration exercises into even non-political organisations like churches was the one useful lesson Amis believed he learned from his brief membership.

Edward du Cann, a contemporary at St John's and later a Tory cabinet minister, remembers thinking that Amis had 'political steel' in him. It seems hard to reconcile Amis's attachment to the dour dogmas of the party line with the life-and-soul-of-the-party reputation by which he was best known and best remembered, and it was often hard for his friends to reconcile the two Amises at the time. They sometimes laughed delightedly at his mimicry and abused him violently for his opinions in the same evening. Communism was an aberration in Amis's life, not to be repeated in any other dedicated political attachment, as it was in Terraine's and Murdoch's (she describes her two or three years in the party as 'first enthusiastic, then very unhappy').

Standard youthful rebellion against his father was one reason for Amis's Communism; but there were other reasons too. Reflecting on this episode in a radio talk twenty years later, Amis discovered a likeness between himself and Jimmy Porter, hero of John Osborne's *Look Back in Anger* (1956), in their common youthful desire to feel enthusiastic about something – about anything. As the Spanish Civil War and the Hitler menace had developed, Amis said, 'I got to the right age at the right time to feel the pull of the good, brave cause, the attractions of warning England of dangers insufficiently visible to the majority, the established, the reactionary, the complacent, the old.' But the fact that his causes were righteous enough 'was no more than a fortunate coincidence', the worthiness of a cause not being established simply by its ability to attract 'young, generous, compassionate, reasonably intelligent minds'. Other, 'less creditable motives, fears and aggressions' were also at work. He remembered 'how necessary it can feel to be one of an embattled minority'. If, he thought, 'your deep-seated need is always to be swimming against the stream, you can usually manage to find a stream that will do at a pinch.' In 1941 Amis did feel the need for such a stream. Twenty years later he had got over it and felt able to recommend an antidote for those who might still not have recovered, beginning with a wholly Amisian course 'of drink and *amour*' and proceeding to a more prosaic programme of resistance to 'the pressures of temperament and emotional need that constantly threaten to distort one's picture of reality'.

But Communism had its compensations. The orthodoxy of the left included permissiveness about sex, and left-wing girls were apt to be earnest in their duties, including their duty to sexual freedom. (This female earnestness could, however, have its dampening effects, one Communist girl being reputed to interrupt her embraces with a stern reminder to her partner: 'Remember, the party comes first.') It was through the Labour Club that Amis contrived to do what he had been wanting to do for several years – lose his virginity.

One female member of the club tipped Amis off that another girl would be willing to go to bed with him if he asked her, so he did. She agreed, but gave him an improving book to read first, a manual with a title along the lines of 'Happy Marriage' and an author with a Dutch-sounding name, both now forgotten. By way of further preparation, Amis needed to buy a packet of condoms, or French letters as they were always known in the 1940s. Feeling shy about this, he dropped in again on his friend George Blunden and asked if he would go with him to the chemist. Blunden agreed and the necessary French letters were purchased. Later he became Sir George and a deputy governor of the Bank of England, helping Amis lose his virginity being his one 'enormous' contribution to English letters.

Something remarkably similar happens to Robin Davies in *You Can't Do Both*. From being a schoolboy in South London he moves on to being a wartime Oxford undergraduate, like Amis. Robin talks to two girls at a meeting of the Classical Society, Patsy Cartland and Barbara Bates. Patsy tips him off that 'any advances he might make to Barbara would be well received.' This turns out to be true when he makes a trial run of 'limited advances'. Robin and Barbara arrange another encounter at which it is understood they will go the whole way. Barbara gives him a book to read first, a sex guide, Vanderdecken's 'Happier Love'. (There is no such book, by the way – Amis borrowed the name Vanderdecken from *The Flying Dutchman*.) They meet to have sex on two occasions in Robin's rooms. Neither encounter is a success. After the second, Barbara disappears for good, leaving Robin feeling ashamed and humiliated.

Where Robin goes wrong is in ignoring the advice of Vanderdecken. Admittedly, he is not a very easy guide, recommending FOREPLAY but warning that too much of that amounts to PERVERSION, so leaving the reader to guess how much is too much. But Vanderdecken's broader point is that men should understand women's different needs, and Robin fails to do so. His first attempt with Barbara is perhaps no worse than gauche but at his second he simply plunges ahead without any thought for her at all. He is rather proud of himself for having abandoned Vanderdecken in favour of the do-as-you-please style recommended by such writers as Blake and D.H. Lawrence and an American psychologist called Homer Lane. But in pleasing himself Robin has committed a near-rape on Barbara. No wonder she runs away.

The influence of Homer Lane had reached Oxford by way of a psychologist called John Layard, who gave a series of lectures at the university which Philip Larkin attended. Lane followed the progressive, which is to say anti-repressive, line in the treatment of juvenile delinquents and in psychoanalysis. He subscribed to the Lawrentian notion that God stood for physical desire; he thought, for example, that 'boys develop wisdom and character by free choice and self-expression' and that juvenile delinquency was a matter of 'positive virtues wrongly expressed'. Lane had scored a certain success in Britain during and after the First World War, attracting support for his projects of reform from such eminences as the Earl of Sandwich and Lord Lytton, Viceroy of India, who once summarised an idea of Lane's in the words 'no attempt must be made to inculcate moral standards; they must be allowed to grow'. Lane was eventually compelled to leave the country after allegations that he had taken advantage of girls in his care. Through John Layard, Lane's message had reached Christopher Isherwood and W.H. Auden in Berlin, and through them it reached Robin Davies in Oxford, and Amis too.

Amis did not in fact think equally highly of all these writers, especially Lawrence. Here he disagreed with Philip Larkin, who, in a letter to a friend, described Lawrence as 'a sort of touchstone against the false', the exact opposite of Amis's view. Amis did, though, think highly of Auden and Isherwood. But literary judgements and precise sources of influence are less important than the fact that a climate of permissiveness, reinforced by a certain amount of intellectual high-mindedness, was congenial to young men like Robin and Amis as providing justification for the sex they already had on their highly charged minds. The responsible Vanderdecken is the sort of writer who believes that 'lovemaking is unlikely to be successful or even pleasurable unless each party is able to contemplate with serenity the state of marriage with the other', which is exactly what Robin does not want to believe. Homer Lane, on the other hand, told you it was 'all right to do what you wanted to do', which was a slogan not without its problems, but had the merits not only of being what Robin did want to hear but of being precisely the opposite of how his father thought about everything. Robin's first sex with a girl from the Classical Society starts him off on a series of adventures with women in which his own selfish appetites are all the time at odds with his responsibilities, much as Vanderdecken's values are pitched against Lane's. Amis's first excursion with the girl from the Labour Club started him off in the same erratic direction.

Girls, however, did not play as large a part in Amis's first four Oxford terms as male friends, especially Philip Larkin. After they were first introduced by Norman Iles, Amis found Larkin 'instantly affable, to be seen as one who erected no barriers'. He thought Larkin's name a trifle comic and his clothes – tweed jacket, wine-coloured trousers, checked shirt, bow-tie – flashy. The eighteen-year-old Larkin's hair was already thinning and he was afflicted with a stammer (both hair and stammer were to disappear progressively over the years). He turned out to like films, cricket, science fiction, programmes on the wireless and other things congenial to Amis's taste, such as smoking, drinking and behaving badly.

Romantic expectation of such a pairing as Amis and Larkin – two budding literary talents encountering each other in the flush of youth among the lawns and spires of Oxford – would have them discoursing on high art and other lofty matters, as they lolled in the sun by the river or in their rooms late into the night. But alas for romantic expectation. That was not the nature of their relationship. For one thing, at Oxford they met mostly in the company of others – it was not until after Larkin had left the university that their relationship became more directly personal.

It is possible that Larkin would have liked more serious literary exchanges with Amis than they had. He enjoyed exchanges of this kind with others, particularly his art student friend Jim Sutton. But this was not Amis's way, at Oxford or ever, and Larkin learned to steer clear of it in his company. Literary matters as a subject of social intercourse always made Amis uneasy. Unless in the form of anecdotes, jokes, gossip, such talk was apt to be pretentious, fraught with potential for boredom, and risked distracting writers from their real business of writing, even filling them with doubts about their ability to do so, which was what Amis came to think went wrong with Larkin. On serious matters, too, Amis and Larkin were liable to have serious differences, as over Lawrence. Such dabblings in the unconscious as interest in Lawrence and Layard involved were very much to Larkin's taste but very much not to Amis's. So, when they were together, they put these matters to one side and got on with what they could agree about, which was mainly the things they both found funny or derisible.

Amis's lifelong objection to straying much further into literary country in conversation than quoting an obscene limerick, discussing a grammatical crux or arguing the definition of a word often gave the impression that he had few real literary interests – that he was even a bit of a philistine. If somebody happened to mention a writer admiringly, his way of dismissing that writer in peremptory style as 'No good' or, worse, 'No good AT ALL' reinforced that impression. But the impression was misleading. Amis always had strong opinions but he knew what he was talking about, even if he didn't talk much about what he knew. You might catch a glimpse of just how much he knew by challenging him with a quotation or a rare word, in which case you were likely to be overwhelmed by the comprehensiveness of his answer. But then the curtain fell and the conversation moved on.

Nor did Amis dislike every form of serious literary intercourse. He enjoyed it enormously provided it was conducted in the right context, which for him came to mean teaching students. With them, he felt, you could say, 'Jolly interesting, that thing Milton said in *Paradise Lost*. Book Six, isn't it? Let's look it up.' But you could not say that in the pub or the club or over the lunch table. At least Amis could not. It was one thing for him to take himself seriously – very seriously – as a writer, but it was quite wrong to inflict this seriousness on other people by behaving like a writer in public.

What he and Larkin actually did together was drink, gossip, tell jokes, swear, and in general behave like a couple of college hearties – without, of course, the hearties' appearances on the sports field.

It was no secret between them that they wrote poems, but so did plenty of others. Writing poetry was not what they often discussed and in so far as they did discuss literature it tended to be in disparaging terms, particularly with the writers they were both required to read for the English course, almost none of whom was excluded from their common dislike. Amis found that Larkin had written at the bottom of the college copy of Edmund Spenser's *Faerie Queene*: 'First I thought Troilus and Criseyde was the most boring poem in English. Then I thought Beowulf was. Then I thought Paradise Lost was. Now I know that The Faerie Queene is the dullest thing out. Blast it.' In describing some piece of Old English as 'ape's bumfodder' Larkin probably said no more than most of his fellow students of that language thought. But his dislikes went further, applying almost equally to Middle English, for writers in which other undergraduates could find at least something good to say: Chaucer, for example. But not even the Romantics escaped Larkin's criticism. Amis cannot remember his having a good word to say for any writer on the syllabus, with the possible exception of Shakespeare.

Much of this jeering was a pose for both of them, as much to do with dislike of having to read writers to pass exams as dislike of the writers themselves. It became something of a point of honour never to say anything good about any writer who had to be studied. For, unlike the glamorous butterflies who flitted through the pages of Oxford novels and never seemed to be troubled by such trivialities as exams, passing exams was ultimately what Oxford was all about for humbler folk like Amis and Larkin. Having been liberated from school to become men, they found themselves back on the old exam-passing treadmill after all. They resented the grind and took it out on the grist that fed the mill. Most people reading English, Amis wrote in his *Memoirs*, treated literature 'as a pure commodity, a matter for evasion and fraud, confidence trickery to filch a degree'. You read what you had to and went to lectures about it to prepare for 'the only significant event, the coming battle of wits with the examiners'.

The effect was paradoxical. Writers on the syllabus were resented for that reason – for being part of the whole tedious apparatus of exam-passing and required learning. Even if you liked what you read it was hard to put in a good word for the authors among friends similarly afflicted by the same set texts. Ploughing your way through, say, *The Fairie Queene* just because it was its turn to be read did not seem the ideal way of getting to grips with what was said to be a classic. On the other hand, would you read it at all unless you were compelled by the curriculum to do so? Very likely not. There seems no alternative to compulsion, even if compulsion creates resentment. For Amis at least, resentment was in time to be replaced by gratitude. He was glad he had read *The Faerie Queene*, even if only once and under duress, for he would have been the poorer if he had never read it at all.

And there was, perhaps, a hidden benefit in the aggressive attitude Amis and Larkin took towards the syllabus. If they treated virtually every single set text with suspicion – not to mention the commentators on those texts they had to read too – then they had to have their own agenda. They could hardly claim to be interested in literature if they wrote off the entire English canon as no good AT ALL. That might be how mere traders in exam results felt, even if they hardly dared say so out loud. But it was not enough for people aspiring to be real writers themselves. They had to make up their own minds what was good and what bad. In later life Amis always knew clearly what he liked and disliked, and when he was a university teacher he always tried to make his students judge set books for themselves and not simply recycle other critical opinions. It seems likely that the process of sorting out his ideas and priorities began when he found himself rubbing up against and challenging the orthodoxies of Oxford.

Meanwhile, other writers could be all right, even talked about, provided they were not on any reading-list. Larkin introduced Amis to or 'quickened his interest in' such modern writers as Auden, Isherwood, Betjeman, Anthony Powell, Montherlant and Henry Green. But they had the immense advantage of being alive and therefore in no danger of appearing on the syllabus, at any rate the compulsory part, which contained no writer who flourished much beyond 1830.

Amis arranged for two of Larkin's poems to be published in the Labour Club *Bulletin* when he was its editor – and was denounced for bourgeois obscurantism by the club committee. After his war service he also read the early pages of Larkin's novel *Jill* (1946) and even offered suggestions. Otherwise, the only literary co-operation between the two at Oxford was in writing pastiche. One of their joint efforts was 'The Bursar's Tale', done in Chaucerian style. Another was a set of obscene fairy-tales in an archaic, sixteenth-century sort of English. One Amis wrote was called 'The Jolly Prince and the Distempered Ghost', which included a farting contest between the prince and the (invisible) ghost. They also wrote a soft-porn lesbian story, taking turns to write a chapter, then sending it to the other to continue. The story featured two girls sharing digs. The girls were dreamy, they batted their eyelids, their hair fell into their eyes . . . In their room they had a cupboard full of jazz records which Amis and Larkin had heard about but which had not been issued in Britain. One record they both had was called 'I Would Do Anything for You', and that became the title of the book. Other records supplied chapter titles.

It was in jazz more than in literature that Larkin seriously influenced Amis in his first Oxford terms and it was to jazz that they both applied the zest for enthusing and discriminating that might have gone into literature had that not been ruled off-limits by virtue of its status as exam fodder. Jazz was a new enthusiasm for Oxford undergraduates. Those with a taste for it tended, like Amis and Larkin, to have picked it up indirectly at first, through hearing dance-band music on the wireless into which occasional 'hot numbers' would be infiltrated, a piece called 'Tiger Rag' being a leading early example. Jazz had the twin advantages of being disapproved of by parents and not being taught at school or Oxford. To Amis, liking jazz seemed a completely natural thing which he learned could be not only entertaining and enlivening but emotionally moving too, if not quite as much as parts of some classical works. But to Larkin it appeared altogether more thrilling. Jazz possessed, he wrote, 'that unique private excitement that youth seems to demand' which in another age might have been supplied by 'drink or drugs, religion or poetry'.

Larkin's expertise and passion had advanced further than Amis's. His parents had bought him an elementary drum kit on which he 'battered away contentedly' and he could play jazz piano with, Amis recalls, 'some proficiency in an unemphatic style'. Sometimes, at sessions in the Victoria Arms in Walton Street, Amis would sing, 'or rather bawl', accompanying lyrics, providing no strangers were present. Larkin brought with him to Oxford a small record collection he had built up with his friend Jim Sutton, who was a student of the Slade, then evacuated from London to the Ashmolean in Oxford. The Oxford University Rhythm Club, set up in 1941, held sessions in rooms above the George Hotel to which they invited players from London. George Shearing came twice. Oxford also produced a band of its own, the University Bandits. This had started out as a band for college dances, and perhaps a useful source of income for its members, but it could often not raise enough of these to fulfil its original purpose. At one time it had only two saxophonists and a rhythm section so was obliged to resort to jazz improvisations. One of the saxophonists was Mervyn Brown (later Sir Mervyn and High Commissioner to Nigeria), the only one who could read music among the circle of St John's jazz enthusiasts, which included most of The Seven and others like Edward du Cann and Graham Parkes.

Amis and his friends built up their collections as best they could, though this was not easy in wartime. Sometimes records out of print (known as deletions) could be found at two shops which later merged into one, Acott's and Russell's, and undergraduate enthusiasts were not above stealing them if they got the chance, though recollections about this are hazy, if not contradictory. Nick Russel recalls Larkin wearing a 'bulky brown-check overcoat' with a poacher's pocket big enough to conceal a ten-inch record, but insists that, though Larkin might pretend he was about to raid a shop, 'we were far too respectable, middle class, and nervous actually to shoplift.' In Amis's version, the coat with the poacher's pocket was a sports jacket belonging to Edward du Cann which was borrowed by his friends for exactly this purpose, Amis himself once walking out of a shop with a much envied Sidney Bechet disc stowed away.

Whichever version is the truth — and perhaps both are — mere contemplation of law-breaking is some indication of the excitement jazz aroused in those who had caught the bug. Having found jazz, Larkin wrote, 'we made it bear all the enthusiasm usually directed at the more established arts', and it was towards a hundred or so jazz records that he and his friends devoted their 'early anatomising passion', which more conventionally would have gone into books or perhaps classical music. Pee Wee Russell, clarinet and saxophone player, was 'our Swinburne and our Byron', Larkin wrote, and he sat around with the others enthusing over and analysing their favourites in a way they never did with writers. But then jazz music was not on the syllabus. It was a private joy, unofficial and unsanctioned, even treated with gratifying contempt by older generations.

Later, all that changed. The long-playing record replaced the shellac 78, so ending the 'concentration and concision' enforced by the three and a bit minutes of music which was about as long a bout of dancing as people could manage at the energetic pace it set and was also as much as one side of a 78 could carry. Concerts, critics, journals, university courses, all appeared in their turn to make jazz respectable to a larger audience and so diminish the subversive charm it held for its pioneer devotees. Far worse, from Amis's point of view, was the emergence of a new kind of jazz called bebop, a movement led by players like Charlie Parker and Dizzy Gillespie, which might have been more fun for them to play but certainly was not more fun for others to listen to – at any rate, not for Amis.

Although he could not know it at the time, in the very year, 1941, when Amis was developing to a high pitch his passion for jazz, Parker and the others were conspiring to destroy the music that enthused him. Amis, however, did not give up on it. For the rest of his life he remained faithful to the jazz heroes and heroines he had learned to adore at Oxford: 'the white Chicagoans, Count Basie's band, Bix Beiderbecke, Sidney Bechet, Henry Allen, Muggsy Spanier, Fats Waller, early Armstrong and early Ellington . . . Bessie Smith, Billie Holliday, Rosetta Howard . . . and Cleo Brown'. Still today he plays half a dozen tracks from among these favourites on his tape recorder every day. He might even play a tape or two of Artie Shaw or Benny Goodman, performers who in Oxford days were likely to be sniffily dismissed as too commercial, but now seem touched with the magic of a long-lost golden age.

∞ 5 ∞

In the Army

After four Oxford terms, the Second World War was still going inconveniently strong. Amis had no choice now but to join the army. Soon after he did so a fellow recruit asked Amis his first name. 'Kingsley,' Amis replied. Finding that a bit of a mouthful, the man asked if he had any others. 'William,' said Amis. And William, shortened to Bill, is what he became for the rest of his military service and the only period of his life – so, aptly enough, losing his familiar name along with his civilian identity for the duration.

Becoming a soldier meant for Amis venturing beyond the restricted circles of home, school and university to live for the first time among people from every social background and every part of Britain. This might be called entering the real world. As Amis quickly discovered, however, the army was an unreal world too, following its own arbitrary and bizarre rules, particularly in matters of order and discipline. He never did encounter such fabled absurdities as lumps of coal kept specially polished to provide for the eye of the barrack-room inspector a smooth surface to a burnished coal-bucket, but he was perfectly willing to believe such things could happen somewhere. The army's inscrutable workings had that effect on Amis, as on others – of making the mythical or absurd seem entirely possible and coming as no great surprise when they sometimes happened. He came to see the military life as resembling some strange society out of science fiction, 'a world much like our own in general appearance but with some of the rules changed or removed'.

In July 1942 Amis turned up wearing civilian clothes at Catterick

Camp in Yorkshire, was handed out denim fatigues and a few days later a proper private's battledress uniform. He was soon immersed in that mixture of bull, boredom and hectic activity which distinguished every recruit's basic training. Amis thought he looked 'a goon and a bleeding civvy in uniform' but he retained at least one skill he had acquired in the school and university corps – he was good at drill, an improbable accomplishment, perhaps, but one of which he was proud. Not only was he good at being drilled – at following orders to quick-march, turn-right, halt-stand-at-ease-stand-easy – he was good at drilling others too. He thought of this as a 'matter of timing and confidence, a kind of acting and showing off' rather than as evidence of physical prowess. He claims once to have amazed a drill sergeant by a virtuoso performance in which he put a squad through its paces in a shed no bigger than a Nissen hut, bringing his display to a brilliant climax with an immaculately timed 'right turn' when the men were just half a pace short of crashing into a wall.

Because of his corps experience Amis's basic training was cut to two months, too brief a period in the ranks, he believed, to give him as full an understanding of how the army worked as he might have got if he had spent more time with ordinary soldiers. When basic training was over he was posted to the Royal Signals unit for officer-cadets, 152 OCTU, also at Catterick, where twenty-six weeks' further training brought him a commission as a second lieutenant. A series of assignments followed – to Southern Command Signals at West Farnham near Salisbury, to High Wycombe, then back to Catterick for an advanced wireless course and on to Headington Hall in Oxford (later the home of the disgraced tycoon Robert Maxwell). At Headington in November 1943 he joined Second Army HQ Signals, a unit at full strength numbering some fifty officers and a thousand men, with which he was to spend the rest of the war.

A series of exercises across the south of England followed when Amis's unit, like dozens and hundreds of others, prepared for the invasion of Europe. The exercises – with codenames like Eagle, Mara and Blank – mainly involved setting up and dismantling communications networks. For one, codenamed Goldbraid, each soldier was issued with what were bluntly described as '2 Bags, Vomit' (meaning in Armyspeak not two bags of sick but two bags to be sick into). Another exercise, called Spartan, involved sending men out with no rations, though whether this was intended to teach them how to cope with hunger or how to live off the land is not clear. At any rate, Amis was lucky enough to miss it.

All this training and exercising was to expand Amis's sense of the absurdity of army life; for many of the things he was taught he never actually had to do – learning 'to fight as infantry', for example. What this meant he described twenty years later in a short story, 'I Spy Strangers', and repeated in his *Memoirs*. There were 'the cross-country runs, the musketry competitions, the three-day infantry-tactics schemes with smoke-bombs and a real barrage, the twelve-mile route-marches in respirators'. Such infantryman's exercises may have been a wise precaution in case the Allied armies were routed and had to fight to the last signalman, though they were just as likely motivated by 'the inferiority feelings common to all technical troops' because of the safety of their jobs compared with those of tank or infantry soldiers. But whatever the reason for it, the training was wasted. Amis's unit travelled across Europe in vehicles and did not go near the front line, so he finished the war 'without having had to walk a step or fire a shot'. His advanced wireless course turned out to be equally wasted. It involved him in experiments with early versions of telex transmitters and qualified him to operate a radio link as far as Moscow. Under the supervision of the prophetically named Major Dixon, he finished up with a Q2 rating, the military equivalent of a third-class university degree. But once he had completed the course the army kept him away from all wireless equipment.

If all this superfluous training was part of some larger plan into which the likes of Amis were magically slotted without their knowledge, then the mystery was never revealed. More probably they were the consequence of having to rough-hew in short order a vast military machine for an unpredictable invasion from such fit young men as were available. But, however inspired, the dead-end training added to the sense of unreality that had begun to form in the barrack-room and on the parade ground.

Still, there were compensations to be had in all this wasted energy. The circumstances of war which had plucked Amis out of Oxford had detached many others from their normal lives too, making this no bad time to be a sexually eager young man on the lookout for unexpectedly available women. At Catterick, Amis took out an ATS soldier called May but did not let things go too far, regarding her civilian clothes and especially her 'liquorice allsorts' hat as a statement of over-enthusiasm he very much didn't want to encourage. In a caravan in High Wycombe he had a fling with a South African girl, whom he borrowed, as it were, from Thomas Balogh, the Hungarian refugee and economist, later an adviser to Harold Wilson. More seriously, he began a two-year affair with a married woman, Elisabeth Simpson, of which he gave a version in his first serious adult try at fiction, 'Who Else is Rank'. This was intended to be a novel of more than 80,000 words written jointly with a brother officer, E. Frank Coles. They took it in turns to write sections, working after dinner in Belgium over a period of months in 1944–5. Amis appears as something like himself under the name of Francis Archer, and Coles appears as Stephen Lewis. The novel was never completed or published but several sections survive, mostly tales and sketches of army life. Two sections, called 'Rhapsody' and 'Ecstasy', give a version of Amis's affair with Elisabeth.

In 'Rhapsody', Archer goes on a course at Catterick. One evening he drops in at the King's Arms in Richmond with a friend. They meet two girls. One has 'dark brown hair and rather peculiar eyes a long way apart and rather long from corner to corner. Her nose turned up and she had rather a big mouth, with little teeth like a child's. Her name was Betty.' She is married. At first she is evasive about her husband – and evasive towards Archer too. She avoids his kisses, then relents, but when he puts his hand on her breast and tells her he likes her very much she says, 'You wouldn't have done that if you really liked me'. Betty tells Archer that she married a childhood friend and got pregnant, only to find that her husband, now in the Middle East, was seeing another woman. Archer tells her that he loves her – 'But I've no idea for how long; probably not for always.' Yet, Betty thinks, 'it may quite possibly be for always.'

'Ecstasy' describes Archer on leave, travelling from London to York to spend time with Betty. They book in at a hotel, eat, go to the cinema, look around the shops, make love. 'I warn you darling, I'm not very good at this,' says Archer. 'I'm not either, so it won't notice,' Betty replies. But the love-making goes all right – 'for the first time I was lost in another person and none of me could stand aside,' Archer tells himself. Betty tells him that she is going back to her husband for the sake of her daughter, and Archer admits that he has been unfaithful to her, thinking that she no longer wants to see him. 'Ecstasy' ends with their future in the air. Through the window of Betty's railway carriage, Archer says, 'I'm seeing you again', and Betty answers, 'Mind you do.' Then Archer watches her train steam off to Darlington.

In 'Ecstasy', Betty asks Archer what his father thinks about her and Archer says, 'He's a bit suspicious. I think he thinks you're a designing woman.' In real life, William Amis's reaction when he discovered his son was having an affair with Elisabeth Simpson was more than suspicious. It was explosive, his outburst of anger no doubt aggravated by the way he found out the truth, which was through 'a series of implausibilities and coincidences unacceptable in any kind of fiction or drama', as Amis wrote in his *Memoirs*.

What happened would certainly be denounced by the critics as ludicrous if it turned up in a soap opera. In 1943 Kingsley spent the first part of a leave with Elisabeth and the second part at home in Berkhamsted. One evening his father ran out of cigarettes and went rummaging for supplies in Kingsley's overcoat pocket. There he found what he was looking for, a packet of Player's, or so he thought. But when he opened the packet he discovered not cigarettes but French letters. Kingsley had left himself with a surplus, having over-estimated his needs. He thought his leftovers were well enough hidden in the cigarette packet but he was wrong. (His carelessness over details like this was to trip him up repeatedly throughout his married life.) William immediately realised that his son was fucking a girl, the truth of who she was soon emerged, and a blazing row followed.

William was genuinely shocked to discover that Kingsley was having an affair with a married woman. But the old methods of dispensing retribution no longer carried weight, as they did when punishment could be achieved by switching off the wireless. Kingsley could not be sent away because his mother wanted him at home. William could not stop his pocket-money since it was the army that paid him now. So he did the only thing he could think of. He wrote Kingsley a 500-word letter, carefully thought through and written out. Its tone was such that Kingsley was 'heartily glad' he did not keep it, though he remembered its message. This was part reproach to him for having let his father down and part self-reproach that father had not trained son in morals sufficiently to resist temptation. 'I'd ask you to give her up,' Dad said, 'if I thought you'd take the slightest notice.' Kingsley was 'daunted' by this but immediately felt better when his father went on to make a 'snobbish insinuation' about Elisabeth being the sort of girl who was no better than she should be. However justified William might have been in berating his son for adultery, Kingsley thought, he had no right to berate the adultress for being who she was.

The affair with Elisabeth lasted two years and as Amis remembers it now it took place 'mostly in the mind', since there was not enough money for it to take place in hotel bedrooms; 'all we knew was cold and never private,' he wrote in his poem, 'Letter to Elisabeth'. 'Where none would pause we found an hour to wait / And clung together when the streets were wet.' The circumstances of the affair may have been uncomfortable but it was as fulfilling an experience as Amis had yet had. 'At last, love, love has taught me to speak straight, / To make my body walk without a strut.' His connection with Elisabeth brought a new deterioration in his relations with his father. But it also inspired the first of his poems he thought good enough to collect.

If the invasion of Europe had gone strictly to plan, Amis would have arrived on D-Day plus fourteen, or 20 June 1944. As things worked out he got there on the 30th, driving undramatically up on to the Normandy beach in a jeep through an inch or two of water from a landing craft manned by indifferent Americans to whom crossing the Channel was already a routine. It was not only Amis's first approach to a battlefield; it was his first trip abroad as well.

The purpose of his unit was to provide comprehensive communications for the British Second Army headquarters under Lieutenant-General Dempsey – communications upwards to 21st Army Group commanded by General (later Field Marshal) Montgomery, downward to Second Army's own corps and divisions, and sideways to other formations, especially 83 Group of the RAF. Second Army HQ was divided into three parts – Tactical, Main and Rear headquarters – and the Signals unit was likewise divided to service each of those HQs. A handful of Signals men had landed on D-Day itself as part of the small and mobile Tactical Headquarters from which General Dempsey conducted his operations. Amis's arrival more than three weeks later, when Second Army's frontline troops were already shaping up for the battle for Caen, may be taken as some indication of the lowish priority of his role in the fight against Hitler.

On his unit's manpower returns, Amis was first described as '2 Coy HQ Sub' and, from November 1944, as 'OC 91 Tele-Op Sec'.

He was also promoted to lieutenant. Most of his time was spent as a signalmaster running an office on eight-hour shifts, first at Main Headquarters, later at Rear. Through the office came messages of all kinds which had to be dispatched and received by wireless, line or dispatch rider, according to the degree of urgency assigned to each one and the means of sending it available at the time. It was Amis's job to make sure that his shift of signalmen-clerks managed all the traffic with maximum efficiency. This was rather like being in charge of a medium-sized but very busy post office, with the significant difference that nearly all the customers were officers of superior rank to Amis, as a consequence of which he was 'bollocked all the time'.

Amis never came under direct fire but he got close enough to the action to feel scared on occasion. Once an anti-aircraft shell fired from an Allied gun fell quite close by and went off, causing a spatter of fragments to fall on Amis's tent. During the battle for Caen a Mosquito broke apart, its two halves falling on either side of the caravan containing the chief signals officer, Brigadier Steward, giving rise to a mild sense of grievance among Amis and his chums, who thought it might have done better to make a direct hit on the brigadier. In Normandy there was heavy air activity at night, though it created more noise than danger. Since it was thought bad form to seek cover in a slit trench, Amis lay in his one-man tent sleeplessly wondering whether he should use his tin hat to protect his head or his balls from stray pieces of shrapnel and inconclusively transferring his helmet from one to the other until he fell asleep. Mentioning this in the mess one morning, he was gratified to find that his fellow officers were all filled with similar doubts, though a majority seemed to favour protecting the head.

If Amis was never shot at, he saw something of the consequences for those less lucky. He was on the move soon enough after the battle for Falaise in mid-August to come across plenty of the dead, so recently made corpses there had been no time to bulldoze them from the road. He thought, as did most people, that they looked like 'life-sized dolls'. Animal corpses were strewn about too, for Normandy was farming country and the German army, in spite of its formidable technology, still depended heavily on horse transport. As Amis wrote in his *Memoirs*, the dead horses 'seemed almost more pitiful' than the dead men, 'rigid in the shafts with their upper lips drawn above their teeth as if in continuing pain. The dead cows smelt worst.'

The breakout from the beachheads, of which the Falaise battle was a part, set Second Army headquarters into rapid motion behind its frontline troops. Main Headquarters advanced leapfrog fashion, sending on an advance party, then letting the rear party catch up and take over, while Rear Headquarters tagged along ten miles or so behind. In September the pace of advance was such that headquarters' units stayed in the same place for as little as two days. It was all packing, moving, finding the new location, setting up communications, then doing it all over again, much of it in the dark. By November, the pace had slackened and Rear Headquarters settled in for five months at Sonnis in Belgium. *Ennui* had time to set in too, and Amis had the leisure to observe his fellow soldiers and even to do some writing.

When off-duty from his signals office, Amis entertained himself as best he could. He smoked – heavily, since the army saw to it that cigarettes cost almost nothing, so entrenching a habit it took him until the 1980s to give up. He got drunk when opportunity allowed, the first such being at a Normandy farm where, in a sitting-room looking out on an orchard, a white-haired farmer introduced him to Calvados. Another drinking bout with a grimmer feel to it he described in a letter to his co-writer, Frank Coles. It happened in Belgium at Christmas 1944. As a way of getting hold of Amis's jeep, three sergeants asked him to a Christmas Eve party. The promised girls did not appear, at the party or at a café they went on to. So they drove back to camp, where they shared two bottles of whisky and slumped into bed. After less than four hours' sleep Amis was roused at seven so that he could be part of the army's traditional Christmas ritual in which the officers serve tea to the men. When that was done he drank himself drunk again – 'painfully, slowly and sickly'. He served the men's Christmas dinner and then was on guard duty until two in the morning, after which he went to bed desolate and depressed and almost in tears.

Later, when the war in Europe was over and he was waiting to be demobbed, he studied works of mild pornography accumulated by the batmen of a holding unit. One item in their library, he remembers, was *Lady Chatterley's Lover*. Amis read fifty or so pages before deciding that that was enough of that – nor did he feel the need or desire to read more when he put his name to the campaign to have the ban on Lawrence's novel lifted fifteen years later. Amis never found much pleasure or consolation in pornography even if it came bearing literary credentials. He much preferred real women and in Belgium, Holland and Germany he searched them out whenever he could, unsuccessfully for the most part, though during a weekend leave in Brussels he contracted scabies from one of two females. One was a tart, the other a waitress, but he never knew which one infected him or whether, if it was the tart, he might regrettably have passed her infection on to the waitress.

Drinking and chasing women are things Amis would have been doing if had stayed at Oxford. The army offered other subjects of study. There can be few better places for an English novelist to get himself a grounding in questions of authority and class, all the more so in a war which had inflated the services with men from every imaginable non-military background.

Second Army's chief signals officer, Brigadier Steward, was a professional, an officer since 1916. But there were no regular soldiers among the officers in Amis's own unit apart from a couple of pre-war NCOs promoted from the ranks. Nor was technical skill necessarily a key to advancement, even though Signals was pre-eminently a technical service. Keeping line and wireless communications going was the whole point of the unit's existence, but that side of things was left to be got on with by engineers and technicians, headed by a major from the Post Office. Actual command of the unit – in the sense of keeping it fed, supplied, billeted and disciplined – was in the hands of men who were experienced neither as soldiers nor as communications experts. The seniority required for these positions was largely a matter of long service, which could only have been achieved by a civilian through pre-war membership of the Territorial Army. Amis's CO, Lieutenant-Colonel G.F.H. Walker, had reached his position by this route, as had the adjutant, Captain W.F. York. Men like these were less certain in the exercise of their authority than professional officers, to whom it was part of the job. But they relished exercising it none the less, no doubt because they had enjoyed nothing like it in civilian life. And they were not reluctant to exercise their power in arbitrary fashion, if only to show they possessed it.

So far as Amis was concerned, officers like these tended to suffer two further drawbacks. They 'seemed to share a low level of general culture', in other words were not interested in books and other things that interested him. And they were not only suburban and conservative but Conservative, hostile to Amis's brand of politics and wary of the rising tide of support for Labour among the troops which was to see Winston Churchill expelled from power at the 1945 general election. For all these reasons, Amis very much disliked his colonel and his adjutant, while they evidently disapproved of him.

In this atmosphere, Amis had to go carefully or risk retribution. Punishment need not be tailored to fit the crime – might, indeed, be doled out without any crime known to King's Regulations having been committed. The offence of 'dumb insolence', under which a soldier could be charged merely for the look on his face, has long been abolished. But it lingered on in spirit, since a punishment could usually be devised for someone judged to deserve one by his superior. Amis had the uneasy sense that all he had to do to earn, say, an unwelcome snap inspection of his section transport or find himself called upon to supply the duty jeep a week ahead of his turn on the roster was to speak up in the officers' mess for Joe Stalin or declare that the Warsaw Concerto was 'a piece of worthless commercialism'. No link between indiscretion and subsequent misfortune could be demonstrated. But could one following the other be simply a coincidence? He might be getting paranoid, Amis thought, but the 'atmosphere of that Mess was certainly not one to discourage paranoia'.

The higher the rank, of course, the more absolute the authority. Amis rarely tangled with the very top brass. General Dempsey, the Second Army commander, he knew about from a distance, mainly from his habit of first telling Signalman Reynolds, or one of Reynolds's colleagues, what he was going to say at his daily conference – sometimes taken as a sign of the general's easy touch with the lower ranks, though more probably a way of getting things clear in his own mind. Amis once tangled with Brigadier Steward. Visiting Amis in his signalmaster's office, he spotted a cigarette burning in an ashtray. When Amis admitted it was his, the brigadier said: 'Put it out – it's a disgusting habit.' (That Amis should have remembered this not very exciting incident for fifty years says something about the powerful impression senior officers could make on him.) But, among officers with red tabs on their lapels, Colonel the Lord Glenarthur was Amis's particular hate, 'the biggest shit on the entire staff', as he described him in his *Memoirs*.

As Brigadier the Lord Fawcett, Lord Glenarthur turns up in Amis's story 'My Enemy's Enemy', here more moderately described as 'the largest and sharpest thorn in the side of the entire Signals unit'. Both real and fictitious lords present the same problem to their respective signalmasters. They both want a special dispatch rider to make a trip of a hundred miles or more across Belgium to Brussels to pick things up, and they want him at once. SDRs were supposed to be reserved for items of military importance too urgent to wait for the regular dispatch-rider service. Tom Thurston, the signalmaster in the story, suspects that the real 'purpose of the trip was to take in the Brigadier's soiled laundry and bring back the clean stuff, plus any wines, spirits and cigars that the Brigadier's Brussels agent ... might have got together for him'. But what can Thurston do to stop it? He has already raised the brigadier's behaviour with the adjutant, 'only to be informed at length that the job of Signals was to give service to the Staff'. Rank rather than military necessity was what counted, so Lord Fawcett gets his SDR from Thurston, as did Lord Glenarthur from a reluctant Amis.

The wartime army offered a bewildering variety of opportunities to be punished. At one extreme, you might be shot for refusing to obey an order. At the other, you might suffer tiresome extra duties of the kind Amis believed he was landed with for being thought tiresome himself. The territory between was a minefield littered with a legion of other punishments. Amis was hauled up before a court of inquiry, in an incident which he later described in another short story, 'Court of Inquiry'.

Amis (reappearing here under his 'Who Else is Rank' name of Frank Archer) arrived at a new location to discover that a charging-engine was missing from his stores and sent a sergeant back to where he thought it must have been left. The sergeant returned hours later without it. The engine was no great loss. It did not work and never had. Besides, it was obsolete and there were no spare parts available. On top of which the quartermaster had another one in his store, an engine in full working order which he could perfectly well have issued as a replacement if one was needed. All this, while relevant to the efficient working of the unit as a means of transmitting signals, was irrelevant to its efficient operation as a system of discipline. Never mind whether the engine's disappearance mattered a hoot to the conduct of the war. The point was that the engine was 'on charge' to Archer, as to Amis; Archer, like Amis, had lost it; so Archer, like Amis, could be charged with negligence.

In both real life and fiction a court of inquiry was formally convened, although there seems to have been no proper legal ground for holding one – nothing more substantial on which to base a charge, in fact, than Major Raleigh's feeling (in 'Court of Inquiry') that he's 'had just about enough of young Archer'. A friendly member of the three-man court, Captain Jock Watson, tells Archer, 'It's a complete farce – just a bit of sabre-rattling.' But Archer decides to play safe. He admits inefficiency, negligence, even personal humiliation. 'What makes me so ashamed is that I've let the Company down,' he grovels, and he adds to Major Raleigh: 'That's what gets me, failing in my duty by you sir. When you've always been so decent to me about everything, and backed me up, and . . . and encouraged me.' All this, of course, is the exact opposite of what Archer actually thinks about the major. But so thoroughly has he abased himself that the court can hardly do more than award him the lightest rap on the knuckles in the form of a reprimand, the same painless punishment as was handed down to Amis. Afterwards Jock Watson cannot quite decide whether Archer was play-acting, as he says he was, or grovelling from the heart. Either way, the process seems to have made Archer older and wiser. 'It was true', Watson reflects, 'that the Army would lick anyone into shape. You could even say that it made a man of you.'

But what kind of man? A groveller? A shameless lickspittle willing to go along with almost anything his superiors tell him? Well, yes. More or less. That seemed to be the sort of young officer the army wanted, at any rate Amis's part of the army. It was tempting to play the game, to go along with the whims of senior officers for the sake of an easy life. Amis explored these themes further in 'My Enemy's Enemy'.

In this story the colonel and the adjutant take a dislike to a technical officer called Dalessio (a surname borrowed from a soldier in Amis's unit) who is good at his job but Italian in origin, which is bad enough, and offhand if not contemptuous about the disciplinary side of army life, which is even worse. The two propose to catch Dalessio out with a barrack-room inspection, which they take it for granted he will fail. Unless, that is, someone tips him off that it is about to happen. Thurston, the signalmaster, is in a position to do so. But since he's known to be sympathetic to Dalessio he is warned off by the adjutant, with the threat that the home leave he is soon due could be cancelled. Thurston keeps quiet but Dalessio is tipped off all the same by another officer, Bentham, a regular soldier from before the war promoted to captain from the ranks. Later, Bentham confronts Thurston. As a regular soldier Bentham has standards even if the others do not. He thinks the army too important for 'this kind of plotting and scheming' and he doesn't like to see 'an able man taken down by a bunch of ignorant jumped-up so-called bloody gentlemen from the Territorial Army'. His main contempt, however, is reserved for Thurston. Wasn't he supposed to be 'the great one for pouring scorn' on the unit's senior officers? 'Yes,' says Bentham, 'you could talk about them till you were black in the face, but when it came to doing something, talking where it would do some good, you kept your mouth shut.' And, he finishes, 'I think you're a bastard.' So Thurston is a crawling collaborator with the very people he claims to despise.

The version of Amis drawn by Frank Coles in a chapter called 'Francis in Sackcloth' in their collaborative novel, 'Who Else is Rank', suggests why it was that Amis/Archer was so prone to fall foul of his military superiors. 'He was easily discountenanced which argued inefficiency, he too readily told the truth and was therefore considered irresponsible, he was honest in his opinions and this made him conceited, he unguardedly made intelligent contributions to the desultory discussions in the mess and was accused of posing, he was polite and considerate to his inferiors and was therefore unable to exercise command.' Amis/Archer was too young – 'he looked little more than a boy' in uniform. He was used to intelligent companions and he was slow 'to adjust himself to the change'. And he 'never learned that the unforgivable thing was to lose face'.

The trouble with Amis/Archer was that he didn't fit into any of the slots thought suitable for young officers. He wasn't what was expected of an Oxford man, which apparently included being uninterested in both sex and politics. He wasn't a gentleman and he did not acknowledge the superior merits of those who were or pretended to be (though the nearest thing to a real gentleman the regiment could boast, Coles thought, would have 'chilled a cavalry mess with horror'). Amis/Archer was not 'superlatively efficient' and he did not display the 'unbounded energy' which would have made him acceptable because then the others could have loaded all the work on to him. Poor Amis/Archer! He wasn't even subservient or unobtrusive. Nor would he ever realise 'that an opinion, however erroneous, if it emanated from a major must be treated with respect and if it emanated from a colonel it had the force of law.' Even after two years in the army he thought erroneous opinions were dangerous and should be combated, which wouldn't have been so bad if had used tact – 'but his superior knowledge was always deployed recklessly and the advantage quickly lost.'

As Frank Coles describes it, he could generally protect Amis/ Archer from being savaged in the mess. Outside, it was different. If Amis/Archer ever put a foot wrong, 'he was for it. Immediately it became a dereliction of duty, a matter for strong censure, his character and efficiency were called into question.' When this happened he was 'always discountenanced and came away thoroughly miserable, humiliated'. Amis/Archer seems to have been no worse than the other officers, no less efficient or hard-working, just more exposed and vulnerable.

Further reasons why this might be so are given in a description of Amis/Archer's general demeanour. He 'played no games; he took less exercise than anyone in the company, though when he could not avoid it he showed agility enough; he philandered in public; he talked freely of his homosexual friends in Oxford; he spoke of intercourse between the sexes much as the rest spoke of football, eating or drinking; he wrote poetry in the Signal Office and used a typewriter for it; . . . he mooched along like [a] student and when he saluted he compelled you to observe how badly he did it; though he studied his appearance the result was never what it should have been.' He did have to his credit undoubted qualities. Above all, he was fun to be with: 'there was no meal that was not duller if [Amis/Archer] was on duty, no incident that was not livelier if [he] had witnessed it too; no topic except shop that you could discuss unless [he] was present.' Amis's mimicry and wit were saving graces but not enough to save him from the bullying of his fellow officers.

When the war in Europe was finally won in May 1945, the mood within the forces changed. Whatever junior officers might have thought of their seniors, or professional soldiers thought of amateur, the overwhelming common interest in winning the fight had kept differences in check. But now the guns were silent conscripted men were eager to go home and resume their old lives, few more so than Amis. Without a common enemy cohesion among the ranks was less easy to achieve, discipline harder to enforce. Having defeated the Germans, soldiers were required not to fraternise with them. There was no one left for the British army to quarrel with but itself. Men ceased to be units in a military machine and became individuals again. Differences of interest, class and politics were free to emerge and divide. And senior officers found they had not many credible sanctions left to brandish at their restless subordinates except a posting to the Far East, where the war against Japan had still to be won.

Women soldiers appeared among the men. It had not been thought proper, so to speak, to bring them into the theatre of war. But now they were fetched from Britain to replace male soldiers due for demob or dispatch to the east. Few places were seen as more suitable for their deployment than a signals office like Amis's, which remained in business providing communications for Second Army's dwindling administrative needs, though less hectically busy than it had been before. Amis fell foul of two of these ATS soldiers in an incident that gave him a lesson in sexual politics to match anything he had learned about men's machinations in the officers' mess and taught him something more about women than he could hope to pick up in a Brussels brothel.

One soldier in the signal office was Private Joy Armitage (a name Amis invented for her later). Her job was to see that messages were directed by the best and quickest route, whether via teleprinter or dispatch rider. Instructions for doing this were clear and simple and there seemed little scope for error. All the same, she managed to send one message by the wrong route – 'some query about surplus three-tonner canopies via a disbanded mobile bath unit in Holland', as Amis caricatured it. He summoned her to his office – 'an area of floor marked off by crates of obsolete line-transmission equipment' – told her how she should have handled it and asked her to make sure she did it that way next time. She said that she would. Amis thought no more about it, being quite certain his reprimand was well deserved and delivered with no more than due severity, if only because by this stage no message seemed so urgent or important that he need get worked up about its going astray. A new colonel had lectured Amis and his other officers sternly on the need for all ranks to treat the newly arrived women with 'perfect correctness', which Amis thought he had done by treating her as he would a male soldier.

That night in the mess, however, he was reproached by Private Armitage's woman officer, who said he had given her 'such a brutal ticking-off that she completely broke down'. At first, Amis insisted he had done nothing of the sort, but then modified his defence to placate the officer, admitting untruthfully that he might 'have spoken a little more sharply than was strictly necessary'. The officer was mollified, though tried to get Amis to apologise, which he evaded doing by pleading that the colonel would forbid any officer to humble himself to a private. Private Armitage and her officer were not the last women – in Amis's life or fiction – who seemed to him willing to get their way by any means, however unfair, including deployment of the fact of their sexual difference as sufficient reason for special treatment.

Amis's Signals section was run down, then finally disbanded. He was sent to a reinforcement holding unit (RHU) at Minden in Westphalia where he found more time to study the batmen's pornography collection than inclination to sightsee in the town. Around August he was sent to another RHU in Bruges. Here, too, he avoided sightseeing, though the attractions of Bruges's bridges and churches were greater than anything in Minden. He tried to write a story and to chase a nurse, unsuccessfully in both cases, though he made a better job of finding drinking companions among old Second Army comrades who had turned up at the same unit.

Amis catches the confused and tetchy atmosphere of these early days of peace in the third of his stories based on army life, 'I Spy Strangers'. Frank Archer is now with the rump of a Signals company, the main part of his old unit having been sent to handle communications at the Potsdam conference. The company is commanded by Major Raleigh, a temporary officer who, in his reluctance to return to civilian life, still harbours hopes of glory, ranging from the instant suppression of a Nazi uprising 'crushed by him in a single prompt and ruthless blow' to the more modest ambition of being made area commandant with a promotion to lieutenant-colonel. He tries to keep his company up to military scratch while he awaits one or another of these improbable triumphs; but he senses 'anarchy mounting . . . discipline and seriousness and purpose melting away'.

The major connects this with politics and the impending general election. He knows of hardly anybody who intends to vote Labour and yet he feels unwelcome change in the air. The story focuses round sessions of a mock parliament he has authorised in which a few enthusiasts of left and right argue their case around and above the heads of the minority who bother to attend. On the left stands Archer, who plays the part of Speaker, and Signalman Hargreaves, the Foreign Secretary. On the right, indeed the far right, stands Sergeant Doll, the smoothly efficient unit clerk and former complaints supervisor at a Leeds department store who is

Opposition spokesman on defence. Hargreaves supports all things Russian while Doll attacks them, even going so far as to propose that the Nazis be enlisted in the all-important cause of resistance to Communism. This is too much for Hargreaves. He storms out of the parliament, which heavily increases Raleigh's dislike for him and also gives Raleigh an extra reason to feel hostile towards Archer, because he is Hargreaves's superior officer.

In another session of the parliament Raleigh is moved to counter the leftish tone of the proceedings by speaking himself. As he waffles on about how the classes in Britain 'have always got on pretty well together', Hargreaves interjects with 'I spy strangers', the formula by which MPs at Westminster can clear the Commons of visitors. Raleigh, as a senior officer, is in the parliament as a visitor, not as a member. Speaker Archer rules that he must leave. This piece of impertinence is absolutely the last straw for Raleigh, who now uses his friendly contact at headquarters to get both Hargreaves and Archer posted east. But his plans are subverted. Sergeant Doll contrives to ease Hargreaves's lot by getting his friend Hammond put on the same transfer list, while Archer is saved from going east altogether by being granted an early release to return to Oxford to finish his degree course. The Signals company is broken up, Raleigh's hopes of promotion and a new command are dashed and, as if all that were not enough, Labour wins the general election. The major's world, for the time being at least, is in pieces.

That story was written years after the war and dramatises the mood of the time rather then describes events that took place. There was, for instance, no mock parliament in any of Amis's units. Another fictional version of the same period, written at the time, has the feeling of a closer version of Amis's political views and his tactless way of presenting them to his fellow officers. It is in the section of 'Who Else is Rank' called 'Parting'. Frank Archer is among a group of officers still hanging about in the mess bar at two in the morning. The colonel harangues them about what a good job their unit has done, then invites Archer to say a few words, calling for silence and introducing him as if he were an MP, the 'member for Enger'. Archer endorses the colonel's praise, and then goes on to say that it's not the officers or even the senior NCOs that have made the unit a success but 'the lance corporals, the signalmen and the drivers'. So far so good. The colonel agrees. But then Archer extrapolates a lesson for civilian life from his praise of the rank and file. It is the 'common men' of the unit who have done the real work, so 'we must see to it after we're demobilised that these common men, from whom we're separated only by a traditional barrier – we're no more than common men ourselves – benefit from the work that has been done, and if the system won't let that happen, well, we shall just have to change the system.'

This is going much too far. Archer is told he is talking a 'load of balls' and that he shouldn't bring politics into the mess. The colonel puts the opposite view – he 'summed up for the government'. He didn't want any talk about changing systems. The present system had produced the 'team work' that had won the war. 'It's tradition, do you see, that makes a nation great.' What he wants to see is the spirit that has made his unit go with a swing carried on – 'the comradeship, the mutual confidence, the old helping hand'. He promises to do what he can to help anyone from the unit who gets into difficulties when they are all civilians again. Archer thinks this is an 'obscene suggestion' and makes his exit proclaiming that 'politics . . . come in everywhere.' Even in an officers' mess, where by tradition politics, along with sex and religion, are not discussed for fear of stirring up arguments and divisions in what is supposed to be a single-minded and coherent body of men.

This little incident carries much of the flavour of 1945. On the right stood those like the colonel who thought that Britain's being on the winning side in the war was self-evident proof that the country was in excellent shape, needing only a bit of goodwill to keep the system well oiled and running. On the left were those like Amis and Archer who thought exactly the opposite, that the war had shown up all the unfairness and inefficiency of the old system which would have to be thoroughly shaken up if there was to be justice for the real heroes, the ordinary unsung privates and junior NCOs.

Although Amis might be at odds with the likes of his colonel, however, the army had diluted his left-wing convictions. He was no longer a Communist. The declarations and slogans of that faith now seemed to him abstract and empty, ideas in the head rather than recipes for getting things done. Having seen something of the huge range of detailed and expert work that had gone into the invasion of Europe, Amis no longer believed that great undertakings could be left to men with ideas alone. For all its inefficiencies and absurdities, its jumped-up officers and put-upon men, the army had got the job done. Would more benevolent-seeming organisations like, say, the

Fabian Society or the General Post Office have done it better? Of course not. The very idea was ridiculous.

The army had taught Amis something else about political reality. Communism and socialism were, after all, about more than ideas. They were about ideals too, requiring some degree of self-sacrifice and altruism for their fulfilment. Amis had seen little of that in the army. From privates in the barrack-room to Colonel the Lord Glenarthur, what he had seen was the relentless play of self-interest and personal gratification, with people jostling for whatever advantages they could squeeze out of the system. The post-war Amis was more sceptical, less starry-eyed, about politics. He made sure that his father delivered his proxy vote for Labour at the 1945 election, but by now he was hopeful of small mercies from politics rather than large. In 'I Spy Strangers' Archer looks forward to an England 'full of girls and drinks and jazz and books and decent houses and decent jobs and being your own boss'. That also sounds very like the England Amis wanted by the end of the war – not put through a root-and-branch revolution, Communist-style, but improved and made jolly comfortable and nice for everybody by Labour. Even in his imagination, however, Amis did not go quite so far as to believe that a Labour government would secure him a constant supply of girls.

Loss of political illusions was not all Amis took away from the army. As he put it in his poem 'A Reunion', he and his fellow conscripts 'were not so much young as new'. He and they were in a ripe state to be marked by whatever happened to them and the army in wartime was an excellent place in which to accumulate experience, one sort often seeming to contradict another.

If selfishness was a conspicuous motive among the characters in Amis's army experiences and the stories he based on them, alongside or beneath it lay something like its opposite – a basic decency, a sense of solidarity, often more likely to be found among the lower ranks than the officers. He paid tribute to this in the same poem:

> So, when one of us had his leave stopped,
> Was awarded a dose of the clap
> Or an extra guard, or was dropped
> Up to his ears in the crap,

Or felt plain bloody browned off,
He never got left on his own:
The others had muscle enough
To see that he soldiered on.

The 'imaginary comrades' Amis saw in the stories of military heroics he had read as a boy did in the real army become something like the 'shadow brothers' he had also seen in those stories. As a means of acquiring companions, of populating his days with the chums that his home life had so conspicuously lacked, the army was far from being a total disappointment. Amis did not, however, make many friendships that lasted beyond his demob.

The army also left Amis with 'a catalogue of fears' which he described in another poem, 'Release', published in his first collection, *Bright November*, in 1947. These included fear 'of the telephone, / Of entering a crowded room by myself, / Of loud voices calling me to interview, / Of typewritten notes, of arriving late at night'. But, he added, 'I have survived them too, and they have caused / Nothing that will ever want to remember them.' The army had done its bit to lengthen the list of anxieties from which Amis has suffered all his life, but, if the army had not supplied them, some other source very probably would.

The army's system of rank and authority – reasonable enough in theory but arbitrary and unfair in practice – also left its mark on Amis. The sense of power having fallen into the wrong hands, of the wrong people being in charge, became a regular theme of Amis's, beginning with the first page of his first novel, *Lucky Jim*. 'No other professor in Great Britain ... set such store by being called Professor,' the young lecturer Jim Dixon thinks of his boss and tormentor Professor Welch – in fact, just like the major in 'Who Else is Rank' whom all junior officers made sure to call 'sir' because there was no other 'field officer in the British army who attached so much importance to not having the Sir left out'. How, Dixon wonders, had Welch become a professor of history at all? 'By published work? No. By extra good teaching? No in italics. Then how? As usual, Dixon shelved this question, telling himself that what mattered was that this man had decisive power over his future.' Again, this was just like Amis's superior officers in the army whose promotion was as much of an uncrackable mystery as Welch's but whose power over Amis was every bit as decisive. Jim's struggle to keep a job he doesn't much want against the capricious, second-rate but godlike authority of Welch has something in it of Amis's struggle to keep on at least tolerable terms with his equally capricious, second-rate but godlike officers in an army he very much didn't want to belong to.

At Thirsk Amis handed in his revolver, picked up his demob clothes and headed gratefully back to Oxford, causing 'a lot of satisfying resentment among Class "A" men' who, though many were older and with longer service, were not judged ready for release, having no interrupted university degree course to which to return. Amis was twenty-three years old and part of Group 19, while some of the others were twenty-five years of age and belonged to Group 46, which in theory gave them a higher priority for discharge. But such were the arbitrary ways of the military and for once the toss of the coin came down with Amis's side up.

Glad though he was to be back in civilian clothes, Amis did not forget nor much resent his wartime stint. The army continued to fascinate him and he was to return to draw on his experience of it much later, not only in his short stories but later still in his novels, *The Anti-Death League* (1966) and *Russian Hide-and-Seek* (1980), and in characters like the two old sweats Company Sergeant-Majors Furze in *The Egyptologists* (1965) and 'Shorty' Shortell in *Ending Up* (1974). His attitudes towards the army were mixed and ambiguous, like most ex-conscripts', a combination of affection, wonder, grievance, respect, overlaid by a powerful desire never to repeat the experience. He continued to admire the military virtues, even though he could hardly claim many of them for himself, and his schoolboy's admiration for romantic heroism survived, though in modified form. 'I know military glory is all nonsense,' he said in 1968, 'it consists of knocking off a lot of other human beings. But it is glorious not to run away, and fortitude is a virtue, and so is not letting one's side down.'

It seems only proper to have Amis exit from this unsettling deviation from the natural path of his life to the march of the Royal Corps of Signals. Or rather to the words invented for it by some anonymous signalman, which Amis still remembers fifty years later. They sum up much of the typical British soldier's feelings about his lot – pride in a job well done mixed with resentment at being buggered about by rule-bound higher-ups who don't understand or care, this all-too-familiar contradiction being resolved in a resigned and weary humour:

Begone dull care,
The links are all working today.
The lines are through
And the S plus Dx* is OK.
But the Branch they don't tell us
We're bloody fine fellows
They send in a stinker to say:
'Your strength returns
Have not been sent here today.'

* The S plus Dx was a state-of-the-art Second World War device for transmitting
speech and teleprinter messages along the same line.

∞ 6 ∞

Peacetime Oxford

The Oxford to which Amis returned from the army in October 1945 was still drab and rationed but Amis did his best to enliven things by following his personal post-war programme of 'not working, getting drunk and pursuing young women'. A setback to these plans soon emerged when Amis discovered that his father had failed to pay his outstanding college bills of close to £100, settling which cost most of the gratuity the army had paid him off with. This roused some resentment, though Amis never did challenge his father over the missing cash.

Amis went back to live in St John's, in a new set of rooms he found handsome, but located in the college's New Quad, which he found dismal. Although rationing and shortages were still the rule for everything from clothes to food, the most intrusive restrictions of Amis's first stint at the university – blackout, fire-watching, military training – had disappeared. So too had many of his friends, who were either still in the forces or had been spared military service on health grounds and had therefore been free to graduate and move on. Philip Larkin was by now a librarian, as he was to remain for the rest of his life, already holding his first job in Wellington, Shropshire.

Distance did not mean that the friendship between Amis and Larkin diminished. Rather, it intensified. When they were both at Oxford they had tended to meet casually, two friends among groups of undergraduate chums. Now when they met it was by careful arrangement, alone or with their girlfriends, at Oxford or Berkhamsted, in Wellington or, from 1946, in Leicester, where Larkin moved to the university library. They got into the habit of writing regularly, once a week or so. These letters were small entertainments, written for each other's amusement as much as to pass on information, often breaking into a private language in which words would be reconstructed by their sounds (as in '*saf*-fic le-*ay*-zong' for sapphic liaison or '*kar*-nal con-*cew*-piss-ents' for carnal concupiscence). Amis's words would be misspelt as though he had bungled his two-fingered typing ('seruously htuohg'), or nonsensical ('And bum-ball and gam-ball and poke-amoke, and knock-bum and pock-bum and shaggy-aggy-schwaeppss'), or in the form of pastiche ('Thank you most kindly for your letter which arriv'd yefternight and ye moft diuerting Booke of Mr. Iohannes Betieman wch dyd come thys Morning'). Larkin replied in similar style.

Their letters were full of plans to meet and assessments of the latest jazz records to come their way. Sex, too, featured strongly. They reported recent developments with the girls in their lives and continued the lesbian stories they had started when they were at Oxford together ('A wisp of yellow hair fell across her brow and one of the straps of her slip dropped from her shoulder, revealing . . .' – but what Marsha revealed Amis left Larkin to guess in his letter of 5 February 1946). Sometimes their letters were even serious.

Their friendship deepened and each became and remained the other's most important male friend for the rest of their lives. Although Larkin was to be as famous a poet as Amis a novelist, literature was not the main basis of their friendship. If that had been all they had in common they might hardly have been friends at all, since Amis disliked most literary talk and in any case disagreed with Larkin on many literary matters. What there was between them instead Amis tried to sum up in two letters in June 1946. In the first, written soon after they had spent a weekend together, Amis said:

I enjoy talking to you more than to anybody else because I never feel I am giving myself away and so can admit to shady, dishonest, crawling, cowardly, unjust, arrogant, snobbish, lecherous, perverted and generally shameful feelings that I don't want anyone else to know about; but most of all because I am always on the verge of violent laughter when talking to you, and because you are *savagely uninterested in all the things I am uninterested in*. At intervals during the past few days phrases like 'very good-natured little wolf', 'a taunting guardian', 'Dogshit, his man' (that makes me crap), 'no beer sir' come into my brain and make my mouth open into a large laugh.

Five days later Amis returned to the theme. 'I have a feeling that what we have to say to each other is more or less inexhaustible – by which I mean that we DON'T feel (a) as one feels with women so often that once one has worked through what has happened to each since the last meeting there will be little left to say; nor (b) as one feels with other friends (even Christopher [Tosswill] of late) that one has been through everything once and all one's conversation is a reworking of former themes. Am I right about that?' Amis most probably was. Conversation has always been important to him, and someone he could talk to on every level of intimacy, someone who was funny, verbally deft, interested and uninterested in the same things and inexhaustible too made the ideal friend for him.

But Amis's appetite for company required friends closer at hand, in Oxford, and he set about recruiting some new ones. Among them was John Wain, who had been spared war service because of his lungs and was already established as a junior fellow of St John's and a lecturer at Reading University. Amis felt some reservations about him because he sported a walking stick and had the general air of a don in his forties, though three years younger than Amis himself. But Wain was a jazz-lover, told stories, was 'worldly-wise'. The two got on well and were soon having occasional lunches at the Eagle and Child – otherwise known as the Bird and Baby – the St John's pub. Amis was to credit Wain with giving him several shoves along his career path: encouraging him to aim for a First, to teach at a university rather than a school, to be a 'proper writer' rather than a 'dabbler'. Wain also gave *Lucky Jim* what was to turn out to be a useful boost on the radio even before it was published. But they were to drift apart over the years through a series of often trivial incidents, culminating in a story, not denied by Wain, to the effect that he believed Amis's first wife would 'break down his bedroom door' to get her hands on him.

Another post-war friend was Kenneth Tynan, who came up to Oxford for the first time in the same term that Amis returned, bringing with him the first hints that pre-war glamour might return – or, as Larkin more gloomily put it, that 'all that was starting up again.' Tynan – later famous as theatre critic, polemicist and first dramaturge of the National Theatre – wore velvet suits and damask shirts, was witty and flamboyant but also generous with his money and his encouragement, buying drinks and trying (unsuccessfully) to help Amis get his first novel published. They later collaborated on a television programme. But his theatrical world was not

Amis's. Indeed, what Amis called in his *Memoirs* 'the general trendy-left-wing, showbiz-Zen-Buddhism, pot-smoking, bullfight-cult, international-America, "happening" sub-culture' which Tynan promoted at Oxford and then in London was very far from the sort of thing Amis ever enjoyed.

It was a point of honour among Amis and his circle not to pursue university women, a case, perhaps, of forging a virtue from necessity. Male undergraduates far outnumbered female at the time and the women who did get into Oxford tended to be hard-working types. Nurses' homes, secretarial colleges, language schools and other such institutions provided girls the university lacked, and Elliston and Cavell's tearoom made a point of contact where undergraduate men went on the chance of picking up women and women went on the chance of being picked up.

If he spotted a woman he liked the look of over morning coffee, a man might send her a message and she might agree to go out with him if she, too, liked what she saw. One day very soon after he returned to Oxford in January 1946, Amis spotted a very young, very pretty blonde across the room. A girl he knew agreed to pass on his interest. She turned out to be Hilary Bardwell.

That morning she received two other messages, one from a man reading chemistry and another from one reading something else, now forgotten, perhaps politics, philosophy and economics. Whatever it was she did not feel she knew enough about it or about chemistry to make for an easy outing with either man, whereas she knew something at least about the English Amis was reading. So she agreed to go out with him, though with some reluctance. She did not think he was particularly good-looking, being romantically keen on the darker, more Latin look at the time. Nor was she impressed by the clothes he wore. His wardrobe included what Amis called his policemen's trousers, together with a hairy ginger suit. She assumed he was still going about in his demob clothes but they were not, though they may have looked like it.

Hilary Bardwell was the youngest of a family of five, two girls and three boys, her father Leonard being one of five children too. He was an unenthusiastic official of the Ministry of Agriculture whose job involved a lot of paperwork to do with the export of animals. It would have been far more to his taste to be an academic or librarian. The Bardwells were originally from Cheshire, but the vagaries of ministry postings had moved the family around the country. Hilary was brought up in Kingston upon Thames, near London, and when her father retired after the war the family home shifted to Harwell, not far from Oxford and later known as the centre of Britain's nuclear programme. She was sent to boarding school at Bedales in Hampshire and was removed from there to Dr Williams's School for Young Ladies in North Wales so as to be safe from wartime bombing. Hilary was expelled from this school for running away and returned to Bedales, which she left again soon afterwards at the age of fifteen and under something of a cloud, having run away from that too. Animals were her great love, then and all her life, and she found a job as a trainee kennel-maid in a set of kennels run by two lesbians at Bracknell.

Art and biology were the only things at which Hilary felt herself to be any good by the end of her abbreviated education. So when the war finished she signed on as an art student at Ruskin College in Oxford. But she did not pursue her studies for long, abandoning her course and becoming a model at the college instead. When Amis met her he was told that she was Ruskin's 'head model', a title that carried some resonance. Surely the head model must be the best, therefore the most beautiful? Perhaps she was but her title was a job description rather than a badge of rank, meaning that she modelled only her head. She was far too shy to strip off her clothes for the benefit of any student life-class. Hilary got by on her earnings from that and the £3 a week her father allowed each of his children.

Amis soon learned that Hilary was not the twenty-two-year-old he'd thought she was but seventeen and hence, as he wrote to Larkin, 'not nearly so depraved as I had hoped'. He also discovered that she liked jazz and collected records, one of which she soon sold to him. It was 'Georgia on my Mind', sung by Hoagy Carmichael, which Amis bought for 7s. 6d. By early March Amis was able to report to Larkin: 'Hilary is coming on nicely. She does really like jazz. Her breasts are concave on top. And she likes me.'

Not that everything went smoothly between them. A week later, Amis told Larkin that, at a party, 'Hilary was stupider and more boring than ever.' But a week after that he was staying on in Oxford beyond the end of term, partly to avoid going home to Berkhamsted and partly to enjoy himself with her, smoking, drinking and 'making (at present) unconsummated love'. He discovered that her nickname was Hilly, and Hilly was what he called her for ever afterwards.

In May, Amis apologised to Larkin for being late with his regular letter, giving preoccupation with Hilly as his excuse: 'We have been arguing for the past week about sleeping in the same bed as each other. First she said no, and I said she would have to say yes, then she said yes, and I said I had forced her into it and what she meant was no, then she said no, and I feel hurt and angry and disappointed and am trying to make her say yes, and there for the moment the matter rests . . . If only one could be ruthless about these things!' On 22 May, however, Amis had better news to report – 'Hilary has yielded.'

In July Amis went to stay with Hilly's family at their home, Abbey Timbers, in Harwell. He liked being with her but not with her family, for reasons he described to Larkin. One brother wore '*sandals* and *saffron* trousers, and NO SOCKS and a *green* shirt, and plays the *recorder* (yes) and likes Tudor music'. Another brother was a conscientious objector. 'And the father does folk dancing (polk dancing? pock dancing? fock dancing?)'. These were just the kind of tastes that Lucky Jim Dixon was to find so bogusly 'artistic' and therefore offensive in Professor Welch and his family.

Towards the end of August, Amis went with Hilly and an Oxford friend, Christopher Tosswill, to France, staying at the Hotel du Cheval de Bronze at Remiremont in the Vosges until the second week of September. The hotel cost each of them 12s. 6d. a day, including three meals. Nothing wrong with that. But the beer, Amis complained to Larkin, 'tastes of sweet cardboard'. In Paris he had not found any jazz records worth buying and in Remiremont there was a shortage of local colour, apart from mountains. Amis did not have enough money to 'get drunk much' and he had failed to bring any books with him. He didn't think he really liked the French, because they expected people to talk to them in their own language. And, as he put it to Larkin: 'I am in continual fear of being thrown 8 of the hotel for ill liss sit seck sew all in dull gents.' The proprietor, Amis thought, 'looks as if he might be strong on ethical and moral standards'.

After a week or so of this, Amis began to think 'that Miss Hilly and myself have been seeing quite enough of each other, and I want to get my life to myself again.' As he put it, 'sir tea er tea has more or less the same effect on a young chap as fruss tray shun. There is the same wanting to be by oneself, and the same not wanting to take the trouble to be nice, and the same feeling that one has let oneself down, and the same feeling that one has been caught in an unpleasant and ineluctable conspiracy.' Amis recognised that this was 'uncharitable' of him because Hilly was 'so nice'. It would probably all clear up 'when we see less of each other: only it will have to be a GOOD DEAL less'. The trouble with Hilly was that she didn't talk. Amis thought he had to be 'the source of everything we share'. It would have been different if Larkin had been with them. Then they would have smoked and drunk and talked and Amis would be laughing a lot of the time and 'enjoying myself ALL OF THE TIME'. There were good things, though, like sleeping and eating and 'idling a lot'. Perhaps he was ill humoured because he had broken out in itchy spots, brought on by eating too much fruit, and the French cigarettes he had been smoking (Caporal Ordinaire) 'that make me feel not well'. But before the end of the holiday he and Hilly made things up and 'we parted quite cheerfully.'

By then it was Tosswill he was getting fed up with. In Hilly's phrase, he talked 'too much like a BADLY-WRITTEN BOOK'. Crossing the Channel back to England in choppy seas, Amis avoided being sick but Tosswill did not. 'I always have to lie down if the water is at all disturbed,' Tosswill said. But why, Amis complained to Larkin, couldn't he just say, 'if it's rough'?

Perhaps Amis tried too hard to keep the conversation going with Hilly. As he was to learn, the effort wasn't really necessary. A year later he told Larkin: 'The great thing about Hilly is that when I don't want to talk to her, she doesn't mind; or at least she doesn't complain, which is the sam[e] thing for our sort.' Women who talked too much and for the wrong reasons were one of Amis's special bugbears. Hilly was sparing with her chatter and so exempt from all such charges. A great many other women were not. Soon after Amis had told Larkin what he thought of Hilly's attitude to silence he wrote again: 'I should like to give you a little lecture on *feminine intuition*.' It had occurred to him that this 'consists of ideas which men, possessing reasoning powers and the ability to see when they would be TALKING PISS, would never bring to utterance, and by the well-known technic of women they pretend that this weakness is their strength; also, they will never admit they are in the wrong and hence utilise another familiar concept, viz. that the truth is what is ferociously maintained and not what is true. And of course it is a kind of universalised substitute for being interesting or amusing . . . This sounds awfully elementary but I felt I had to tell somebody about it.' Amis's tone is facetious, but that does not mean he did not believe what he wrote, then and later. Women who talk too much, who talk 'piss' which is neither interesting nor amusing and who think that what they vigorously assert must be true have cropped up all too often in Amis's life and his fiction too, especially his later novels. But Hilly was never one of those.

Amis's feelings towards her grew warmer. He was still in touch with Elisabeth Simpson, about whom he confessed to Larkin he 'still felt strongly'. But he resisted the powerful temptation to go the whole way with another girl. 'I saw her and removed a lot of her clothing in Xt ch [Christ Church] meadows, but I didn't shag her because I decided it would spoil things with Hilly and that I am not anxious to be at doing.'

On her side, Hilly's feelings grew stronger too, although Amis was not the easiest of young men to get along with. At the end of an evening out together it was she who walked him home instead of he who walked her back to her digs off St Aldate's, such was his alarm at being out alone in the dark. It was around this time that the anxieties which had been accumulating in Amis since childhood became so acute that he began to wonder if he might not be about to go mad. His anxieties expressed themselves in panic attacks in which he became depersonalised – he would 'cease to seem real' to himself. Darkness, being alone or confined – or fears of all three, as of being trapped in an empty lift without lights – were what brought the attacks on. Amis went to see a specialist called Armstrong at the Littlemore mental hospital near Oxford. Armstrong asked him a lot of questions, then gave him a reassuring diagnosis: there could be no guarantee that Amis would never go mad, but there was no sign of it now. What perhaps reassured Amis as much as anything was his learning that he was not unique, that his symptoms were not all that rare and that other people were able to live with them. At any rate, his panic attacks stopped for ten years, though they recurred, and descriptions of attacks similar to his own were to appear regularly in his novels, from *Take a Girl Like You* onwards.

At one point Hilly tried to give Amis up. But he was not easily discarded and maintained his pursuit, though without the traditional inducements to romance like roses, which were beyond his means. He was good at educating her, especially in music, about which she knew something but he knew much more. They went together to the university jazz club; he even went so far as to dance to the music. They stayed at her home in Harwell and his in Berkhamsted during vacations. On the excuse that they were going to visit Amis's old army friend, Frank Coles, they also stayed at the Morton Hotel near Russell Square in London, Hilly paying her share of the bill. 'We went punting, took coach trips to London, had all the fun of being brassily cheated at St Giles's fair,' Amis recalled in his *Memoirs*. 'Hilly impressed me by being rejected as not readily hypnotisable at a so-called hypnotist's stage show, and in a different way by washing her hair and her smalls in the Randolph [Hotel] and, much to my trepidation, in the bath-house at St John's.'

During term-time Amis would often do a bit of work in the morning, then, lectures permitting, meet Hilly in a group for coffee in the Cadena Café in the Broad at eleven. The group included Mervyn Brown, Graham Parkes and Christopher Tosswill, along with other girls from Ruskin. They might sit over coffee until the pubs opened at twelve, have a drink, then go to a 'British restaurant' at Gloucester Green or at the Town Hall in St Aldate's, where shepherd's pie and rice pudding could be had at the subsidised price of ls. 3d. After lunch those who played games did so. Amis did not. There was more work until dinner, eaten in the college hall, unless Amis went with Hilly to the Stowaway, or the Chinese as it was also known, though, as Robin Davies who ate there in *You Can't Do Both* put it, the restaurant 'had never been known to serve any food even distantly associated with that land'. While Amis was still in college he had rooms near Parkes and Brown and since coal was still rationed they got together in whichever of their rooms could manage a fire that night. Things changed the next year when Amis moved out of college into Miss Butler's lodging house at 19 St John's Street where he and three others had rooms – two Welshmen and 'an extravagantly handsome Lothario type' who smuggled girls into his bedsitter and later made a fortune selling children's shoes in Australia.

In time, Amis grew on Hilly as she had grown on him, and in a bit more time she became pregnant. Amis announced this to Larkin in somewhat nervous style in a letter he wrote on 6 December 1947, after four days of post-examination drunkenness: 'Hilly and I are making a man at the moment which is worrying me rather.' Amis worried more about this over Christmas at Berkhamsted. At this stage in his life, with little money and no job, becoming a husband and father was not what he'd had in mind. However, as he argued to Larkin, he was 'determined to keep' Hilly. 'Since I enjoy living with

Hilly better than I enjoy living anywhere else, it's difficult to believe that I shan't enjoy living with her all the time.' She was 'overjoyed' at the prospect of having a baby. If they were to be together she would want a baby soon, so why not now? The obvious answer was marriage. Amis went to Abbey Timbers and put the question to Hilly. She said yes.

In the abbreviated version of events Amis gave Larkin on 12 January 1948, he told his parents what had happened 'and they now accept the position.' But they were not as content as that suggests. Nor were the Bardwells. After William Amis's lengthy campaign to keep his son wholly innocent of sex, and his rage and disappointment on discovering Kingsley's wartime affair with Elisabeth Simpson, he could scarcely have been expected to welcome this new development, confirming as it did all his worst fears about his son. He even went so far as to issue threats of excommunication from the family, until his wife Peggy told him to stop being a fool. Then he agreed to come to the wedding, though only to save face all round, making it clear to his son that he would continue to hold his immoral conduct against him.

The Bardwell parents, Leonard and Margery, seemed likely to be even more intractable. They, too, were sufficiently shocked by their daughter's condition to threaten a boycott of the wedding. Peggy came to the rescue once again, telephoning and telling them they would regret it for the rest of their lives if they did not turn up. 'I don't want to find myself some time in the future wishing I'd been there when it's too late,' Robin Davies's mother – also Peggy – tells the parents of Nancy Bennett in *You Can't Do Both* when she is trying to coax them to come to the wedding of their daughter, whom Robin too has made pregnant. As did the fictional Bennetts, the real Bardwells gave in, even urging that the marriage go ahead as soon as possible.

The wedding took place on 21 January 1948 in the Oxford registrar's office, then lodged above the Lamb and Flag public house next door to St John's College. 'The marriage ceremony was funny,' Amis reported to Larkin: 'Nick [Russel] came along, as well as both lots of parents, and laughed at the back while a Hitchcock fat character with bursting eyes and lips like wizened Paris sausages (he resembled a broker's man) said the words over us. We only had two sentences each to repeat after him, and then there was nothing to do but signing and paying money.' There was a family tea at the Randolph Hotel with a special cake on top of the teatime menu. In the evening Russel gave them dinner at the George, and then Hilly climbed on the bus for Harwell while Amis returned alone to his digs. There was no money for a honeymoon.

By the end of the month, though, they were installed in a flat at 14 Norham Road in North Oxford, found for them by Hilly. It was rather small and very expensive, Amis told Larkin, consisting of 'one all-purpose room and a kitchen opening from it' and with a shared bathroom. Hilly brought pots and pans, bed linen and a carpet-sweeper, along with a handy sum of between £200 and £300. Amis reckoned that between them they would have an income of about £550 that year. They acquired a ginger kitten called Winkie which hunted down moths relentlessly and sometimes followed them on walks like a dog. After a game of halma, at which Hilly would regularly win, Amis went into town for a lecture or an exam while Hilly stayed behind, happy in her new role as housewife and mother-to-be.

They soon began to look for somewhere larger to live, a one-room flat being too small for a baby. A flat at Iffley fell through, but a cottage at the village of Eynsham was found instead. Marriner's Cottage had three bedrooms, cost the same as the flat at Norham Road and was only twenty minutes by bus from Oxford. It stood at one end of a short row of cottages, at the other end of which, conveniently to hand, stood a pub. On the cottage's right, set back from the road behind a wall, was a fairly grand mansion. Behind the cottage was a walled garden with 'rambling roses, hollyhocks, a walnut tree and the best gooseberries [Amis had] ever tasted. Some evenings, encouraged by a saucer of milk, a hedgehog came visiting.' They moved there on 23 June – 'and very nice it is', Amis told Larkin. Now that they lived in the country, Hilly acquired a dog called Mandy.

The Amises had two names ready for their child, Philip and Sally. Amis had first thought of another name if the child was a boy – Humphrey. But Hilly talked him out of that by asking him to imagine how embarrassing it would be to have to call out 'Humphrey!' on a crowded bus. So they settled on Philip, after Larkin, instead. Hilly gave birth to Philip Nicol on 15 August 1948 at the Radcliffe Maternity Home, and on the same day Amis wrote to Larkin with the news – the first person he told outside the family. Hilly's labour had taken 'rather a long time and must have been very nasty. However, she is completely all right now, though a little weak.'

115

Amis described his new son as having 'very fair hair and a conical head (it will not stay conical, they said), and a face like that of an ageing railway porter who is beginning to realise his untidiness has meant that he'll never get that ticket-collector's job he's been after for twenty years.' Hilly, Amis said, was 'very happy and glad, as I am, to have something to name after you'. A few days later, Amis was able to reassure Larkin that 'the shape and arrangement' of his son's face 'seems more usual than formerly'.

In *You Can't Do Both*, Robin Davies reflects on the possibility of marriage to the pregnant young Nancy. He discovers himself to be 'almost an ideal non-husband' for her, 'being selfish, self-indulgent, lazy, arrogant and above all inextinguishably promiscuous by nature'. Earlier, when he is starting off with Nancy, he had told himself that 'he deeply wanted not to be the sort of man who, when just getting into his stride with a satisfactory love-affair, nevertheless seriously contemplates getting started on another.' But it also strikes him that 'that was the sort of man he at least as deeply wanted to be.' Even on the steps of the register office a passing girl arouses in him 'a fleeting sense of excitement and mischief'. These are signs of danger ahead for Robin and the same dangers lay ahead for Amis. But the early months of his marriage were happy enough in the traditional way of young couples making their first home together.

While Amis was courting, marrying and becoming a father, he was also pursuing his activities at the university. Although the City of London School magazine's 'Oxford Letter' of April 1946 claimed that 'Generalissimo Stalin and Fats Waller vied for pride of place in Mr Amis's affections', it was right about Fats Waller but wrong about Stalin. Amis kept up the enthusiasm for jazz he had acquired in his first four terms at Oxford, but he had given up Communism for good and resisted all efforts made to woo him back into the party. He voted Labour at the 1945 general election, considered himself on the left wing of the party and believed he would be a Labour man for the rest of his life; but politics were no longer his main serious enthusiasm. The army had tempered his idealism, and so far as he was concerned the new Labour government could be left to get on with bringing in whatever socialism was needed. Amis had serious work to do, degrees to acquire: first his BA, later a B.Litt.

His tutors – dons in general – did not make a very favourable impression on Amis. He was later to divide them into two types.

The first were old, 'port-and-walnuts men who were courteous and had once written a book, but who said little about literature except that Chaucer was a perpetual fountain of good sense and Milton had organ music'. The second type were young, 'brusquer figures who were, or had been, often absent in places like the [wartime] Ministry of Information and who knew a tremendous lot which they did not ordinarily divulge'. Both types were impressive in their way – 'there was no knowledge but they knew it, or anyway could get at it by stretching out a hand.' Yet they also tended to be upper class, and this reinforced Amis's impression of 'British culture as the property of some sort of exclusive club', which was something he very much deplored.

Gavin Bone, his first tutor, Amis described as 'a very nice, very tolerant man'. But his special subject had been Old English poetry, hardly calculated to endear him; and, besides, he had seemed to combine both types of donnish character, young and old. Bone had died of cancer the year Amis left for the army. His tutor when he came back to Oxford was J.B. Leishman, a man of 'great sweetness and extraordinariness' and 'a shining exception' to the general run of dons. Amis wanted to know him better but found no way of doing so. The only thing he knew that Leishman thought about him was that he was punctual. Leishman, too, died young, in an accident while walking in the Swiss Alps, and Amis was subsequently farmed out to other tutors.

Amis went to lectures too. Like most undergraduates with an eye on exam results, he tended to divide lecturers into the kind that might be useful because they set and marked papers and the kind that did neither, attending the former, however painful their lectures might be, sometimes at the cost of the latter's, however much more enjoyable. As Amis further categorised them, there were hard lecturers and there were soft. Hard men gave you information, while soft ones offered civilised discourse. The hardest Amis came across was the 'repulsive' J.R.R. Tolkien, later to be famous as author of *Lord of the Rings*, but to be listened to by undergraduates in spite of his unattractiveness as a lecturer only for his expertise in Old English, since something of what he expounded in the lecture room was likely to show up in exam papers. The softest of the soft was Lord David Cecil. In between came C.S. Lewis, the 'only reputable hard-soft merchant' and 'the best lecturer I ever heard'.

Amis dutifully 'bashed' his way through his reading list –

Shakespeare, Spenser, Milton, Wordsworth and the rest, though leaving out John Donne, whom he found 'intolerably convoluted' and anyway reckoned he could do without, since there was bound to be a question he could answer in the relevant paper about a more congenial poet, like Marvell or Herbert – correctly, as it turned out. He prepared his exam strategies with some care, becoming, he thought, a good examinee and later a good examiner. One important rule he took care to observe was the simple one of answering the question asked, as opposed to just putting down everything you knew about, say, Keats, if confronted by a question about him – a practice he found woefully common when he came to mark exam papers himself.

When the time arrived to walk down the High to the Examination Schools, at the end of the Michaelmas (autumn) term of 1947, Amis went fully armed with the gist of what he needed to know for each exam jotted down in a notebook, to be studied on the way and discarded at the door: well-chosen quotations, for instance, or the three points about Pope that were considered important – his social eye, the merits of his couplets and his proto-Romanticism. In his Middle English paper Amis resorted to the old trick of finishing with an uncompleted sentence, adding below that he had run out of time, whereas what he had run out of was anything more to say about the 'dream convention' in the literature of the period. He tried to lay an ingenious trap in one paper for the oral exam – the viva – that often followed final written papers at Oxford. Amis wrote to the effect that one good thing about Dryden was the memorability of his couplets, hoping that the examiner would pick up on that, saying something like, 'Well, if they're so memorable, perhaps you can remember some.' Whereupon Amis would let loose a flood of verse, for which he always had a prodigious memory, stopping only when the examiner interrupted him – 'Thank you, Mr Amis, you've made your point.' But he needn't have bothered. His written papers were good enough to get him the desired First without need of a viva. Ironically, his best paper was in Old English, the subject he liked least but for which he got his only A, in other papers doing no better than AB.

He was less lucky with his next degree, a B.Litt., which he failed. The B.Litt. involved attending lectures and taking more exams but it also required the production of a 40,000-word thesis, and here Amis fell foul of Lord David Cecil, or so he was always to believe.

Lord David, a rare full-blown aristocrat among dons and an eccentric but much admired performer in the lecture hall, was designated as Amis's supervisor and as such was supposed to be in contact with Amis twice a term at least. But he made no contact at all and was nowhere to be found when Amis went looking for him at New College. Amis therefore proposed transferring his loyalties to another don he had got to know, F.W. Bateson. Between them they concocted a subject which reflected both their interests – Amis's in Pre-Raphaelite poetry, Bateson's in how the public's taste for poetry was formed – to be called 'English Non-Dramatic Poetry, 1850–1900, and the Victorian Reading Public'. All Amis had now to do was locate Lord David and get his approval for the switch of supervisor, and when Amis ran into him in the street by chance Lord David admitted he had forgotten all about him and agreed to sign the necessary paper without notable fuss.

Amis was to finish his thesis after he left Oxford, completing the text in May 1950 and returning to the university in November that year to be examined on it orally. He was somewhat disconcerted to find that Lord David himself was chairman of the examining board, which was composed of only one other don besides him. Lord David clashed with Amis over his first question, which was about whether poets always wrote in the first place for an individual or a few friends rather than for a wider public. Lord David queried the word 'always'. Amis suggested that he had dealt adequately with this point in the very next sentence. Lord David, however, took offence, perhaps at the implication that he had not read the thesis properly, and Amis sensed at once that he was done for.

A few days later he wrote to Larkin: 'They thought I hadn't borne out my main contention. O Lor!' A week before Christmas he wrote to Larkin again: 'They've turned my thesis down, the direct descendants of sexual perverts. No explanation yet, if ever: I may have to sweat semen re-doing it.' But he never did. Bateson advised him not to bother. As Amis reported to Larkin, though Bateson 'was very indignant about it and thought I had been unfairly treated, it wouldn't be much use rewriting my thesis.' Amis never did get an explanation for his rejection and ever afterwards, when he considered the point at all, believed himself the victim of some inscrutable 'academic crossfire'. Not that it mattered much. Failure to get a B.Litt. had no perceptible effect on the remainder of his academic career.

*   *   *

After the move to Eynsham in June 1948, Amis took the bus into Oxford most days, leaving Hilly on her own and increasingly lonely. People came out from Oxford to visit – people like Kenneth Tynan, James Michie, John Postgate (physicist and son of Raymond, the *Good Food Guide* pioneer) and J.B. Leishman, who cycled out with a map strapped to his thigh. Others, like Philip Larkin and his girlfriend Ruth Bowman, came to stay. But with little money and no car, one child already to look after and another on the way, Hilly was on her own for long stretches. She made some friends among other mothers in the village, including an Irishwoman from the local council estate who would show her on walks what things you could pluck and eat from the hedgerows. Still, that was not enough company to overcome Hilly's feelings of isolation.

However, there was far worse pain soon to follow, for Hilly discovered that her husband was already being unfaithful to her even before the birth of their second child. Proof came in the unequivocal form of entries in a diary which she happened to come across, Amis having made no effort to hide it. He seems to have thought that marking the diary 'Private' was all that was needed to keep it secure. But what secrets could a young husband have from his young wife and what young wife could resist the temptation to look inside her husband's 'Private' diary to find out? At any rate, Hilly did not resist. And the evidence of infidelity she found inside horrified her, all the more so since she found that she knew some of the girls involved.

So, early in their marriage, Amis established a pattern of infidelity which was to last as long as the marriage itself. In this, as in other things, he was determined to indulge his appetite to the full. Any girl who caught his eye and was herself attracted to him was fair game. Often the girls were from his and Hilly's circle of acquaintances, in Oxford and later in Swansea. Hilly knew and sometimes liked them as friends of her own. But what could she do about her husband's infidelity with them, profoundly painful though it was to her? Motherhood and lack of money narrowed her options. Besides, she still loved him, still wanted to keep him. She thought, too, that he still loved her and that, somehow, where other women were concerned he couldn't stop himself. It was part of the nature of the man she loved and if it could not be changed it would have to be lived with. This, at least, was how she reassured herself that it was worth settling for acceptance.

*   *   *

Amis left no great mark on Oxford. He was not an athlete, so won no sporting blues. He joined the Oxford Union but quit after one term without opening his mouth, on the ground that the quality of the debates was not worth the high price of membership, £1 10s. per term, or three weeks of his pocket-money. Acting did not interest him and he did not join any of the stage clubs. He did join the St John's Essay Society but attended no meetings. Where he did leave a vivid trace was in his college's junior common room suggestions book.

Here college members were supposed to make comments about the facilities provided, along with proposals for improvement. But the book turned into something else – a kind of noticeboard or Chinese wall newspaper on which undergraduates inscribed their opinions about whatever came into their heads. Philip Larkin, for example, gave this verdict on one of the undergraduates in a pair of briskly cutting couplets, known as a London rhyme:

Mr A.J. Wilton
Has not the skill of a Milton;
But he seems to have mastered
The art of a Bastard.

No member of St John's was more assiduous in his use of the suggestions-book pages than Amis. In verse and prose he peppered them with wit and obscenity, denounced some of the icons of Eng. Lit. ('Geoffrey "Smoking Turd" Chaucer'), mocked the dons, debunked the whole business of literary scholarship – even, now and then, made a suggestion of some relevance to college affairs.

In the Trinity (summer) term of 1947 he took aim at fashionable methods of interpreting literature in relation to myth and began by setting out the full text of a 'mythos', thus:

There was a young choir-girl of Chichester
Who made all the saints in their niches stir:
Her beauty of form
When the weather was warm
Made the Bishop of Chichester's breeches stir.

Beneath this limerick (author unknown) Amis offers an 'apparatus criticus' complete with variant readings, in the manner of a textual scholar being scrupulous with his audience (of this he was later heartily ashamed): 'choir-boy' for 'choir-girl' is one such variant. Then he proceeds to interpret the real meaning of the verse, like some relentlessly snuffling academic symbol-hound. The choir-girl represents the moon-goddess, while '"breeches stir" is a beautiful image of the stirring of all earth's creatures beneath the rising sun.' The scholars Kock and Krapp are invoked to bolster the Amis interpretation with their bogus authority. According to Kock, 'it is not an exaggeration to say that the hero of all myths is the Solar Hero or god of sun, day, light etc.' Tongue firmly in cheek, Amis concludes: 'for a fiercer treatment the reader is referred to the struggle of Hercules and Cacus (Gk *kakos*, the bad one, the evil god of night) or, according to Krapp, the Arthur-Mordred story.'

The previous term Amis had thumbed his nose in verse at some of English poetry's most honoured names. No doubt his intention was first of all to shock and amuse undergraduates brought up to accept without question the rank order of the nation's poets as laid down by the literary-academic establishment. But in doing so he reveals, too, his own rebellious feelings about that established order, his dislike of being told what to like.

Speak, Echo, Partner of the Earth and Skie.

> *Echo*: I.

In Heav'n how fare our Poets? Do they live?

> *Echo*: Live.

Live they in Blisse, or die in torment frightfull?

> *Echo*: Rightfull.

Where pains him CHAUCER, prince of entertainers?

> *Echo*: Anus.

The deepe-brow'd MILTON, where is he interr'd?

> *Echo*: Turd.

DRYDEN's baye Garland, what has it become?

> *Echo*: Come.

Say what the Realm of honie-tongu'd POPE is.

> *Echo*: Piss.

What ails WORDSWORTH in Nature's mystick Lap?

> *Echo*: Clap.

Hyperion – where does KEATS finish it?

> *Echo*: Shit.

The fierie SHELLEY – what does he endure in?

Anus, Turd, Come & Piss; there is no end?

Clap, Shit and Vrine, but shall these persever?

Earlier in that term Amis had struck out at a very different target in a harsh little diatribe against the practice of undergraduates queueing up behind the college fellows to go into hall for dinner: 'The idea of our waiting for the dons to go into Hall is supposed, I imagine, to invest them with further dignity. Well, all I can say is that any procession of those dons of ours, who as you know include Will "The Butcher" Moore, Bill "Toss-Rag" Costin and Cyril "Dead Face" Norwood, has got *much less* dignity than a procession of syphilitic, cancerous, necrophilic shit-bespattered lavatory attendants and I *object* to waiting while the buggers waddle their oozy way into Hall.'

An entry in 1946 links up the enthusiasm for jazz Amis had acquired since he came to Oxford to his longer-standing enthusiasm for poetry. He describes a literary combo – 'Big Ben Jonson and his Globe Theatre Tribesmen (actually Jonson fronting Jack Donne's Metaphysical Stompers)' – and reviews its performance in the critical jargon of the jazz pundit: 'Ben blows a nasty, mean horn . . . Dick Crashaw (trombone), spotlighted on the over, takes three smeary choruses and plays impeccable tailgate on all 4 sides. Star of "Volpone Drag" is Raffish Jack, maybe a bit fanciful but listen to that crashing left hand . . . Jonson junked sweet orch to appear, and he and Donne, old-line feuders, seemed to fit ideas admirably.'

Amis also left his mark in the St John's Kitchen Book, where college members were invited to say what they thought about the food. He demands that 'a men, you' should be published early in the day so that he can know what will be on offer before he makes up his mind whether to turn up for dinner in hall. He complains about the price of college meals, insists that brown porridge as well as white should be available at breakfast and denounces the college for not giving its members their full rations of sugar and cheese – the war was over, but wartime austerity still prevailed. He worked out that each undergraduate is entitled to a piece of cheese measuring 2 x 1 x 1 inches twice a week and he wants it served up in two weekly lumps instead of being dissipated in macaroni and cheese dishes. For once an Amis suggestion earns a small blizzard of supporting signatures.

Here and there through the pages of the junior common room suggestions book he reveals that he votes Labour and demands that the common room subscribe to the Communist *Daily Worker* (because it is 'now the only non-Tory newspaper'), gives tips on walking in the snow ('take short steps and go slow, don't you hurry for no bugger, not you'), covers two foolscap sheets making nineteen suggestions about how to get in everybody's way while walking along Oxford pavements, devotes the same amount of space to advice about getting off buses, speculates on whether 'two small but fully represented parts' on a Benares brass monkey owned by his landlady could have disappeared because of the cold weather, and invites undergraduates to apply for jobs as 'samplers in a camp of young ladies awaiting passage to South America'.

Sometimes the Amis invective rose to a pitch that even his usually tolerant fellow undergraduates found unacceptable. One volume of the suggestions book for 1942 had been removed so that his victims should not be hurt by what he had written in it (the volume has since disappeared). But however abusive or abrasive, even childish, his ideas, he always expressed them with great care and precision. Amis's entries were not spontaneous jottings but well-prepared utterances: neatly written, more legible than most, with scarcely a blot or a crossing out. He took care to vary his tone from page to page. One day he would be the mock academic, another just plain rude. He played with words ('I can't find the news tate's manna ndnay shun n.e. wear') and accents ('don't you hurry for no bugger, not you; you'll go arseing else'). As he had entertained his school chums by pulling faces, now he learned to amuse and outrage his fellow St John's men by making faces on a page. And as his friends from CLS thought they recognised Lucky Jim in the boy they had known at school, so St John's men believed they had read him first in their college suggestions book.

Leonard Bardwell (Daddy B).

Amis's parents
(second from
left in front and
back rows).

Lieutenant Amis,
(second from left,
middle row)
somewhere in
Europe. On his
right, his co-
author, E. Frank
Coles.

Hilly Bardwell, at the time she met Amis.

Above: Hilly, pregnant with Martin, at Marriner's Cottage with Mandy the dog.

Below: Amis and Hilly with Christopher Tosswill and Philip Larkin at Cambridge.

Amis and Hilly relaxing in Swansea in the 1950s.

Above left: Hilly and Amis with Mrs Barley (who became Mrs Oates in *I Like It Here*) in Estoril, Portugal in 1955.

Above right: Amis making his Evelyn Waugh face.

Below: The filming of *Lucky Jim*. Amis instructs Ian Carmichael, who played the lead, on the art of making faces.

On the wrong side of the bar for once, Amis helps Terry Thomas
at the Edinburgh Festival of 1957, where *Lucky Jim* was being
shown.

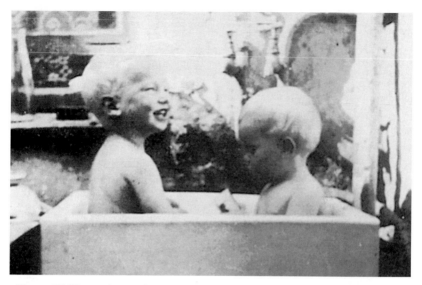

Above: Philip and Martin, aged two and one respectively, bathing in an all-purpose sink at Mumbles.

Below: Amis with Philip aged 12, Martin aged 11 and Sally aged 8 at West Wratting, near Cambridge.

Above left: In France in 1957, suffering the afflictions of abroad.

Above right: Amis and Hilly with their Princeton neighbour, John McAndrew, on vacation in America.

Below: Amis picking fruit at Soller, Majorca in 1961.

Swansea

Amis began his first serious attempt to write a novel on his own in 1947 and announced its completion to Larkin in September the next year. He called it *The Legacy*. Its inspiration was partly Philip Larkin's, or rather a book Larkin gave Amis with an enthusiastic recommendation to read it. The book was Julian Hall's *The Senior Commoner*, which Amis in his *Memoirs* called 'a wonderful marsh-light of a novel' but whose influence, he admitted, was to help render *The Legacy* 'unpublishable'. *The Senior Commoner* is set mostly at Ayrton College, which not only sounds like Eton but, with its proliferation of titles (Provost, Head Master, Captain of the School), its elaborate hierarchies (the Captain of the School is also the Senior Scholar while the Senior Commoner is the Second Senior Scholar), its own variety of football and its 'tradition of careless mastery' resembles Eton in some detail too.

The novel covers four college terms. Things happen but there is no story, unless it be the record of the feelings towards the college, intermittently revealed, of Harold Weir, the Senior Commoner of the title. The spectre of homosexuality looms now and then, only to be shoved back out of sight when it threatens to loom too large. There is not much more than one barely discernible joke, a dry one-liner, which has a schoolmaster punish a boy in these terms: 'I think you might write an essay in words of one syllable on the disadvantages of motoring.' The style is flat, typified by laconic descriptions, like this: 'He was slightly built, with sleek black hair and an olive complexion. There was ripeness in his face but no indication of character.'

The Legacy, too, is not only flat in tone but short on description, giving, for instance, little or no account of how its characters actually look. Although a failure as a whole, the novel contains hints of the fiction with which Amis would succeed. It tells the story of a nineteen-year-old called Kingsley Amis who stands to inherit a £30,000 legacy from his father Lionel providing he marries a girl approved by his brother Sidney and joins the family firm. Kingsley wants to write poetry and marry Jane, of whom Sidney does not approve. For a while Kingsley holds out, but in the end settles for the acceptable Stephanie and his legacy.

Here is the sort of situation that was to become familiar in Amis's novels from *Lucky Jim* onwards, in which a man is faced with a choice between two women while conflicting considerations about them, such as desirability and obligation, animate the plot. A number of Amisian themes also make their first appearance: fear of loneliness and old age, the need for love and affection, conflict between the better and worse sides of one's nature, and of course conflict between the sexes. (Giving his own name to his hero, though, is the kind of modernist trick the older Amis would surely have deplored.)

Kenneth Tynan helpfully arranged a lunch in London with Mark Longman of the publishing firm, to which Amis brought three specimen chapters of his novel. But Longman turned them down. Amis reported this to Larkin in March 1949: 'Longmans sent the Legacy back, of course, saying it was "altogether too slight". That's true in a way, I suppose, but there were plenty of other things to its discredit I would have said before that . . . But I am not discouraged much yet, I hope, and will try again.' He did, but with the same result. Amis tried Gollancz and early in May quoted 'Golly's note' to Larkin: 'I don't think THE LEGACY quite comes off.'

Amis remained undeterred and kept his typescript circulating among publishers for two more years, but to no avail. Anxious to know where he might be going wrong, Amis extracted via his Oxford friend James Michie a two-page answer from Doreen Marston, a reader for Collins and the mother of Adrian Marston, another Oxford friend. On New Year's Day 1951 she typed out a 'completely frank report on my reactions'. She told Amis bluntly that she did not believe he would get the novel published and that it would be a waste of time and money to try again. His two major faults, she said, were 'redundancy and lack of conflict'. By redundancy she meant repetition. 'If I tried to count the number of times your characters repeated themselves, the number of times they light a

126

cigarette, pour out tea, pass plates of food etc: this letter would be an essay in statistics.' By lack of conflict she meant that the characters were too predictable. 'There is no suspense; there is, in its place, a good deal of boredom.' Amis's minor faults, Mrs Marston thought, were 'weak feminine characterisation', 'ungrammatical writing' (but perhaps, she added uncertainly, this was 'a trick') and 'total lack of humour'. There were better points, though they didn't make up for the worse. Amis was good at male characterisation and atmosphere. 'But, even so, all this suffers from repetitive detail.' She thought Amis would have learned a lot from writing *The Legacy* and added: 'I am quite sure that you should try again.' Mrs Marston encouraged him to send her his next effort if he made one, and she was to be deeply disappointed that he did not do so when his next effort turned out to be *Lucky Jim*.

Amis ruminated gloomily to Larkin on her judgements – he was in a particularly glum mood writing this letter, which also contained the news that his B.Litt. thesis had been rejected and, even worse, that the War Office had been in touch just in case it wanted to call him back into the army to fight the Korean War. Mrs Marston, Amis thought, was 'QUITE RIGHT in about half of what she says, but the rest of the time she's missing the point isn't she? . . . I mean detail's the point isn't it? Now if she can't *see* that (I'M NOT ASKING HER TO LIKE IT) what chance have I got old boy? . . . An original writer who isn't very much good, that's what I am: I'll never be Joyce or Warwick Deeping, so where do I stand? In the brown stuff, it seems.'

A decade later, when *Lucky Jim* had made Amis famous, he and Mark Longman were at a crowded dinner in the London Planetarium in Marylebone Road. Longman mouthed his regrets across the hubbub but Amis mouthed back, 'You were quite right.' By then he knew very well, and was not afraid to admit, that publication of *The Legacy* would have led him down a blind alley from which it might have taken him ten years or more to escape and that the drawer in which the manuscript lay for the next thirty years was the best place for it. Amis came to think of himself as lucky, too, in not having any important literary connections. Suppose he had been, say, Stephen Spender's nephew. Then an obliging publisher would surely have been found to take on *The Legacy* and Amis's career would have had a false, perhaps irrecoverable, start. Having to find a publisher willing to take on his book in the hope that it would sell and make money was an indispensable step in Amis's career and, in retrospect, vindication of his belief in the market as the best mediator between novelist and public.

In poetry, too, Amis's beginnings were hesitant. In the 1940s he published in little magazines like the Oxford-based *Mandrake* and the annual anthology, *Oxford Poetry*, whose 1949 edition he co-edited with James Michie. (This co-editorship was achieved by a piece of effrontery that Amis liked to think was not characteristic of him. He suggested to Michie that they apply for the job on the perhaps disputable grounds that they were both poets, and they marched together into Basil Blackwell's, the Oxford bookseller and publisher of the anthology, to make their case. It worked, Amis remembered, 'chiefly because neither of the men we spoke to could think of a reason for turning us down'.) In 1947 Amis's first poetry collection was published, under the title *Bright November*, by the notorious Reginald Ashley Caton, who was rewarded for his trouble with a thinly disguised appearance in *The Legacy* and subsequent insertion into minor and unflattering roles in Amis's first five published novels as L.S. (for Lazy Sod) Caton, before being killed off in *The Anti-Death League*.

Caton's main interest was in property. At his death in 1974 he owned ninety-one houses in his home town of Brighton, 'not a bathroom among them', as he liked to boast. But he also had a publishing business with a base in London, which specialised in pornographic titles like *Boys in their Ruin*, *A Brute of a Boy* and *Chastisement across the Ages*. The business was known first as Fortune & Merriman, later as the Fortune Press. Quite why Caton added a sideline in literature to his pornographic list nobody knew. Perhaps it was to create a cover of respectability, perhaps it was a tax-dodge. There might even be a chance that some unknown writer would take off and earn him a lot of money. Whatever the motive, it seems not to have been love of literature. One poet, Alun Lewis, accused him of publishing poems without even bothering to read them. All the same, Caton is credited with having done more than any other small publisher to promote poetry during the difficult war years, and he had some distinguished names to his credit, Dylan Thomas, Roy Fuller, Cecil Day-Lewis, Julian Symons and Philip Larkin among them. It was Larkin who advised Amis to take his first collection of poems to Caton, as Larkin had done with his own first collection and his first novel, *Jill*.

Caton was a difficult man to do business with. He demanded

financial guarantees, requiring Amis to take fifty copies of his own book at cut price. Amis's mother helped take care of the cost by selling some copies on to her friends in Berkhamsted. But Caton was elusive, too, devious and prone to procrastinate. He liked to travel, disliked the telephone and was in any case much preoccupied with his Brighton properties. He was the model for the publisher to whom the fictional Kingsley Amis took his poems in *The Legacy*. There is no sign at his door, only a notice in red ink. Piles of dust-covered books offer 'little of any literary interest'. The telephone is disconnected. The publisher dodges questions about the book of poems, citing difficulties with getting paper and such costs as binding, glue and promotion, and demanding £20 up front. In the end the exasperated poet cancels the deal and berates the publisher over the telephone. 'Perhaps you'll begin to get wise to yourself and to your miserable, muddling, dishonest, rotten little concern, what, what, what, what, what, what, what, what?'

Amis complained to Larkin for months about 'that Sod Caton' and described a visit he made to Caton's London office in September 1947, by which time *Bright November* had at last been printed. Caton disagreed when Amis said the weather was too hot, complained about the bad language in Larkin's *Jill*, lectured Amis on the importance of book options and denounced the person who rented him his office: '("Irishwoman of course, never trust 'em; don't like 'em; as bad as the Welsh; Alun Lewis double-crossed me; can't stand the Welsh or the Irish what?")'. When the phone rang, Caton made Amis answer it '("tell 'em Mr Caton's gone out for a few minutes").' 'The girl was foreign,' Amis told Larkin, 'and beside herself with passion at having vainly tried to get him for a week.' Caton was unabashed, making out that it was another instance of what he had to put up with from people rather than what they had to put up with from him. He surprised Amis by giving him some extra free copies to send out for review – though leaving it to Amis to pay the postage and taking care to remind Amis that he still owed the £10 for his fifty copies. Still, there was satisfaction in being between hard covers at last, though in later years Amis came to think not much of the volume's thirty-three poems, including only six of them when he published his collected poetry in 1979. Copies of *Bright November* are now hard to find, a fact which causes Amis no grief.

Amis's early writing was eclectic and derivative, like most young writers', depending heavily on friends for publication and on models

129

for inspiration. Models included *The Senior Commoner* for his first attempt at a novel and Auden for his early poems. Among helpful Oxford friends were Philip Larkin, who steered Amis towards Caton, and John Wain, who arranged the publication of his second volume of poetry, *A Frame of Mind* (1953), which was the second in a series published by the School of Fine Art at Reading University where Wain taught (the first volume in the series was a collection of his own poems). James Michie had co-operated with Amis in his first try at anthologising. Less happily, another Oxford acquaintance put him in touch with a professor from the Argentinian University of Tucuman, Jack Rush, who in 1948 commissioned Amis to write a book on Graham Greene for 1,500 Argentinian dollars. Amis was a bit dashed when his father told him that this generous-sounding sum added up to a mere £75 7s. 6d., but he duly delivered his manuscript. The book, however, was not published, the manuscript was not returned and the promised dollars failed to materialise.

Yet another Oxford friend, Bruce Montgomery, who wrote thrillers under the name of Edmund Crispin, also composed music, later becoming successful at providing scores for films like *Doctor in the House* and some of the *Carry On* series. Amis wrote librettos for two operas for which Montgomery was to write the music, *To Move the Passions* and *Amberley Hall*. The first of these was commissioned by the Arts Council for the 1951 Festival of Britain, but it suffered a process of steady shrinkage until it shrivelled and disappeared. Montgomery and Amis 'fudged out' a plot with five characters at the end of 1950. The five characters became four, then three, the cast being progressively reduced to save Arts Council money.

In March 1951 Montgomery announced another change of tactic: the piece was to be a ballad opera which meant, as Amis told Larkin, 'spoken dialogue . . . so [Amis's words] will be audible instead of drowned by a lot of filthy music.' This suited Amis well enough. But by June the project had collapsed. This seems to have been at least partly due to Montgomery's being too clever by half. Turning the piece into a ballad opera altered the balance between words and music, requiring less from him and more from Amis. The Arts Council seemed ready to go along with six or eight lines of dialogue between songs but not the 150–200 lines that composer and librettist actually produced. The trouble was, Amis told Larkin, that Montgomery 'WAS WRITING FLITHY FLIM SCORES AND STINK-KING STORIES FOR THE POPULAR PRESS'; and also, as Montgomery later confessed, that he was diverting some of the music he had written for the opera into a film score. The second piece, a one-act chamber opera, was completed but never performed. The two had better luck with their Coronation ode, 'The Century's Crown', which was performed once by the Glasgow Choral Union and the Scottish National Orchestra in June 1953.

Quite inadvertently, Philip Larkin had as early as 1946 provided Amis with the key which was to open the door to his first published novel, *Lucky Jim*. By that time Larkin had moved on from his first librarian's job at Wellington to another at Leicester University. Amis went there for a weekend visit and on the Saturday morning Larkin left him in the senior common room for half an hour while he went off on some unavoidable university business. Looking about him at the circumambulating professors and lecturers, it struck Amis that there was a good subject here, as yet unreached by any novelist. But it was to be several years before he was able to put solid fictional flesh on this insight.

Such contacts and connections as these were later to give colour to the notion of a movement of young writers, later still elevated into 'the Movement'. But in the late 1940s and early 1950s Amis was not trying to be part of any such thing. Friends, especially Larkin, were important for their support and encouragement. But he was not attempting to become a member in any chorus of voices. He was struggling to find a voice of his own. Sometimes he would have several projects simultaneously on the go: poem, novel, short story, libretto. At others he would have nothing but frustration to tell Larkin about in his letters. It was not until the mid-1950s that his writer's identity began to take firm shape from among the false starts and failures and the occasional successes.

By 1949 Amis had to look seriously for a job. There was no choice, since he had a wife to support and one child, Philip, with another on the way, while his government grant was on the point of running out. He scanned *The Times Education Supplement* for vacancies, sent in his applications to universities in Scotland, Liverpool, Birmingham, Durham, Manchester – even for a British Council job in Prague – and went to a succession of interviews, at each one of which he encountered the same small circle of hopefuls, minus the one who had been successful at the last attempt. He was interviewed a second time for the Prague job but still didn't get it.

By the beginning of October, things were looking fairly bleak. The Amis family had increased by one, Martin, born on 25 August ('Blond as P[hilip],' Amis reported to Larkin two days after the birth, though 'less horrifying in appearance'). But the family was now without a home of its own. The Amises left Eynsham, moved in with Hilly's parents at Harwell, and began looking for a flat in Oxford. Just before they left Marriner's Cottage a telegram arrived from Swansea. It came on the first Monday of the month and invited Amis for an interview on Friday of the same week. Amis was unwell, recovering from a dose of what he called 'trench mouth', and asked if he could be spared until the following Tuesday. But Swansea wired back that it intended to make a decision on the Saturday. There was, after all, not much time left before term began. So Amis made the trip.

He told Larkin about these developments in a letter from Harwell dated 5 October. His next letter was dated 17 October and the address from which he wrote it was 134 St Helen's Road, Swansea. He had got the job, surely the very last one to need filling that academic season, seeing off in the process four rivals, three of whom he was 'glad to defeat: Evans, a big, very short man with tonsured fair hair, a convoluted tobacco-pipe and a face like an old Rugby football converted, by a tyro hand, into the semblance of a red pumpkin; a nameless cadaver with castellated teeth who talked in baying cockney of his qualifications; and a Manchester man, his face sored but not shaven by shaving, who talked with his mouth almost shut.'

After spending most of the 1940s at Oxford, Amis planned a last day which would be 'a gently melancholy communion with all I know and love so well', as he put it to Larkin. Instead, 'I said goodbye to all the people I didn't want to say g-b to and hardly any I did.' There was, too, a feverish round of chores to be got through, which included 'finishing up in Bodley' and final packing and inventory-taking at Eynsham, before, with Hilly and Martin, he finally caught the bus for Berkhamsted, where Philip was already installed with Amis's parents. 'I was glad to see Pinetop [Philip] . . . he is walking and looks very pretty.'

Amis had mocked and derided many things about Oxford while he was there, but he would very likely have done the same to any institution of learning in which he might have spent those years. He never warmed later to writers he learned to dislike then. To have to go through Chaucer all over again, he wrote in his *Memoirs*, 'would kill me'. But he changed his mind about what were and were not proper objects of English studies. The linguistics which had so bored him when he had to get to grips with *Beowulf* and Chaucer were, after all, valid subjects of study, whereas the post-Chaucerian literature he and his contemporaries far preferred were not. They should be approached 'in the spirit of self-cultivation and entertainment' rather than laboured over as fodder for exams.

He saw more, too, in some of the literary dons he had thought so little of when they taught him. Men like Neville Coghill, Maurice Bowra, C.S. Lewis and even Tolkien were 'big enough to be worth laughing at'. Could the same be said about their successors? Amis thought not. He even came to think more kindly of the glamorous side of Oxford life, the Oxford 'that I hardly knew', which he admitted to 'mildly coveting' while at the same time disapproving of it. In his *Memoirs* he wrote that 'to sneer at glittering comes rather too easy to those like myself who could never have afforded to go in for it on any kind of regular basis.' And he concluded, wryly, 'Poor moralist, and what art thou?'

In his poem 'Their Oxford', he describes his university days as having once seemed 'so various' but having now become 'all one', 'A block of time, which like its likenesses / Looks better now the next such has begun'. Saying this, he acknowledges the familiar trick of seeing one's own youth as better than the generation's that followed. But he finishes the poem, 'Looks, and in this case maybe really is.' So perhaps it was not an illusion. Perhaps Oxford really was a better place in the 1940s, however much he may have criticised it at the time.

The University College of Swansea which Amis now joined had been founded in 1920 as the newest of the (then) four constituent colleges of the University of Wales. Student numbers had increased ten times since that first year but were still small, barely topping a thousand in 1949, more than half of them ex-servicemen. The college was housed at Singleton Abbey, a 250-acre estate, once the home of Lord Swansea and, with its view across Swansea bay, the most attractive university setting Amis had or has ever seen. As a campus, though, it was anything but grand. Much of the teaching was still done in huts put up for soldiers in the First World War.

Swansea itself was drab, like most of the rest of the country at the time, not yet able to put behind it the depredations of the war. Its main square was crossed by a Bailey bridge above a bomb crater.

Rebuilding was not complete until the 1950s. Amis celebrated, or at least marked, the result in two lines of his sequence of poems, 'The Evans Country':

The journal of some bunch of architects
Named this the worst town centre they could find.

On a starting salary of only £300 a year (equivalent to £5,000 now – only a third of the salary a lecturer in Amis's position then would expect today), plus £50 in child allowances, the outlook was drab for Amis too. His pay was set to rise in annual increments of £50 but money was a constant problem until the later 1950s when his college salary became a shrinking proportion of his earnings. To help make ends meet, Hilly washed up at the local Tivoli cinema and Amis corrected exam papers (making a useful £55 from the Higher School Certificate in his first summer) and lectured Workers' Educational Association classes. Shortage of cash was made most irritatingly manifest in shortage of cigarettes. Like Lucky Jim, Amis was never 'able to smoke as much as he wanted to'. Eking out the tobacco supply became a mild obsession. Rationing and other strategies had to be resorted to, including cigarette-rolling machines and herbal mixtures. One of his similarly obsessed colleagues, David Sims, was driven to rounding up the fag-ends from the common-room wastebin.

Prosperity peaked when the monthly pay cheque came in and dwindled thereafter. Getting by involved constant juggling. Sometimes the cash simply ran out. On 20 June 1950 Amis had to write to Larkin for help:

Now here's something that won't make you smile: I am very short of money – in fact H[illy] and I have 2/6 between us to last until I get paid on 30th June ... Can you send me a cheque for £5 by return? I will let you have a cheque dated 30 June back. If you can't manage £5, £4 would be very nice; so would £3 and so would £2 and so would £1, in descending order of niceness. Honestly old boy I shdn't ask unless I was broak; I haven't smoked since 9 p.m. last night. Not that that matters to you. It's a question of eating, d'yous spyous d'you see? and Hilly wants to go on a little holiday for which abt. £3 is necessary. I shd. be v. grateful for the money, only send it a.s.a.p. if you will.

Larkin obliged with the desired £5 and Amis expressed himself 'supernally grateful'.

Poverty also meant, to begin with, the most modest accommodation. In the rush of his appointment Amis had to leave the family behind, though he wrote to Hilly every day until she was able to join him in Swansea. In the meantime, he had to take digs at a house in St Helen's Crescent overlooking the civic centre which he shared with Willy Smyth, a fellow lecturer (in classics), a Dubliner and a considerable eccentric. It was Smyth's face, minus its spectacles, that Amis saw first every morning as Smyth made his way to the bathroom, awkwardly placed on the wrong side of Amis's bedroom – 'no great gift of the gods' to be roused by this sight, as Amis put it in his *Memoirs*.

Hilly came down for a few days in November and a flat was found at 82 Vivian Road in the suburb of Sketty into which the whole family was able to move in time for Christmas 1949. In January they moved on to another house in Sketty which they shared with David Sims and by May they were living in the first of two flats on the Mumbles Road which ran west from the city along the curve of the bay. The flats, at numbers 644 and 388, were both fairly similar and about as primitive as each other. In its then condition Number 388 would perhaps not now be thought fit for a family to live in at all. At Vivian Road the twin pram had to be negotiated up steep steps featuring right-angle bends to the front door. Martin slept in a drawer, a tea-chest next to the fire with a hot-water bottle in it did for an airing-cupboard, and the electric stove gave off shocks. Like the flat in Amis's short story 'Moral Fibre', his own living arrangements included a 'wash-hand basin that did, but only just did, as a sink', while the sitting-room 'doubled as dining-room and lunching-room' when people called.

The trigger for that story was a bizarre incident which incidentally underlined the depth of Amis's hard-upness. Through a social-worker friend, Margaret Aeron-Thomas, the Amis family acquired for a time the services of a home-help, Betty, who was held to be in need of reform. At the age of fifteen Betty had gone on the streets and at sixteen married a Norwegian sailor called Johansen (Arnulfsen in the story). He left her with a baby, since when she had had three more illegitimate children and reverted to whoring, supplemented by theft. In the story Betty offers herself – free – to John Lewis, who politely declines. The real Betty made the real Amis a similar offer, as he described to Larkin. 'She was quite good-looking, but very dirty, and smelt a little, so when she offered me an evening out, with all drinks paid for and a shag at the end gratis, I declined, and persisted in my refusal even when she offered to pay our coal-bill in addition, a matter of some £2–15, out of her earnings.' Amis thought this was 'one of the nicest compliments I've ever had paid me'. But he might also have felt a twinge of dismay that his respectable job put him at such financial disadvantage to a whore, however amiable.

Unlike a schoolteacher, a university lecturer was not considered to require any further training beyond his degree before he faced his students. Amis was therefore pitched straight in to teach what up to now he had only studied, his initiation being all the more precipitate as term had already begun before his appointment was finally settled.

Teaching methods, derived from Oxford and Cambridge tradition, involved lectures and, for the honours class, weekly essays and one-to-one tutorials. This set Amis a fairly punishing routine for his first year or two at Swansea. He had to give half a dozen 50-minute lectures a week and he knew he could not just get up and ad lib them. Nor did he, unlike his senior colleagues, have a backlog of lecture material to draw on. Others might be able to speak from notes but each one of Amis's lectures had to be carefully prepared from scratch and written out in full for delivery.

Amis was well thought of for his professionalism by his students and by others on the college staff, though not always for his opinions. The word most often used to describe him is conscientious. His lectures were thoroughly prepared, usually lively, often amusing. He would read out passages at dictation speed if he thought there was something special in them the students should get hold of. Amis seemed iconoclastic, too, shocking some of the students with the irreverent opinions he had picked up at Oxford about writers generally thought to have a secure place in the literary canon – and even shocking a little his own mild-mannered head of department, Professor W.D. Thomas, when word of Amis's views percolated through to him. Early in 1951, while Thomas was preparing a report on him for the college council, he told Amis that he was 'rather concerned about the way in which I seemed to do rather more denigration in my lectures than he thought was helpful'. Having seen some student answers to questions on Browning, Thomas thought Amis had been 'less than fair to him'. As Amis told Larkin, he personally believed that any lecture time not spent 'firing dollops of starched shit at Browning was being more than fair to the bastard'. But he told Thomas, more soberly, that he would watch this denigratory tendency in himself.

Amis continued to hold his opinions on the likes of '*filthy Jane Austen, stupid bloody snorting Browning, shrieking Dickens, et. al.*' and he was happy to talk freely about these in the tutorial and the pub, where he mixed with his pupils on terms as level as relative intellectual power allowed. This being Swansea, Dylan Thomas was the natural hero, particularly admired by students from the local grammar school. But Amis thought little of his poetry, didn't mind saying so and succeeded in swiftly converting at least some Thomas loyalists to his view.

There was no 'Swansea school' or special literary orthodoxy in the English department. Professor Thomas had run the department since being imported to do so in 1921 from the University of Saskatchewan. The college history describes him, with all possible tact, as 'distinguished-looking, courteous and very much a gentleman', also 'a brilliant teacher', but perhaps 'over-critical' of himself, as a result of which his writings, though of course 'full of grace and polish', were regrettably 'few and far between'. A more down-to-earth judgement, given by one of his colleagues, was that he was a person 'everybody liked but nobody thought much of'.

Amis's own first impressions were favourable and remained so. Within a couple of weeks of his arrival in Swansea he described Thomas to Larkin as 'a smiling, smoking, light-suited man of about 58, who doesn't drive me hard. When you tell him anything, he says regularly "Yäs . . . yäs . . .", like one of those Grgr [Graham Greene] priests who have heard all the sins that men can commit a long time ago . . . listening to a confession.' When he died in 1954 Amis told Larkin that 'he was a lazy fucker, but a very decent chap, which seems remarkable in a professor.'

Thomas told Amis that the most he could hope to do for most of his students was to help them understand what it was that, say, Byron was on about, though for just a few students it might be possible to do more. This seemed a bit tepid to Amis, who had arrived with what he thought of as more advanced ideas about the teaching of English based on the then new New Criticism which advocated, broadly speaking, that attention should be paid to the text and not to the author's life or his presumed intentions. In time, though, he saw the sense of Thomas's modest ambition and developed a powerful dislike of the imperious kind of critic or academic who stands between the reader and literature and deceives himself that what he has to say is more important than the poetry and fiction he pontificates about.

Amis got on well with Professor Thomas and the other members of the English department. Isabel Westcott was described by the college history as 'for forty-one years ... the devoted linchpin of the department', by Amis to Larkin as 'a smiling, apple-cheeked Anglican lady who does a lot of good works' and in Amis's *Memoirs* as 'a middle-aged spinster of sweetness and natural propriety'. James 'Jo' Bartley was an Ulsterman, but born and reared in Dublin (and therefore, as Amis told Larkin, 'both pro- and anti- every Irish question'); he was also an enthusiast for smoking, drinking and women. David Sims ('small, balding, laughing, ex-Jesus') became Amis's closest friend at Swansea, both of them actually liking literature – 'a rarity then as now', according to Amis – and sharing interests in drink, cigarettes and having a good time.

With a fourth lecturer, Sam Dawson, Amis was to be less comfortable. He thought of Dawson as an altogether too slavish devotee of the school of F.R. Leavis, the Cambridge critic, down to having on his shelves all the writers Leavis approved of – not only Austen, Dickens and Lawrence but even Elizabeth Myers, whose *A Well Full of Leaves* had earned its place on the shelves through being endorsed by Leavis in a footnote. Besides, Dawson was judged guilty of the quintessentially Amisian sin of being reluctant to stand his round. When he did once offer to do so his colleagues immediately ordered doubles of the most expensive drink they liked, Amis opting for Green Chartreuse. One of those colleagues remembers Amis suggesting that a day a month should be set aside for Dawson-hunting, though allowing that Dawson should not be seriously damaged if caught.

Professor Thomas was succeeded by Professor James Kinsley, which put Amis into 'a shuddering rage', as he described his feelings to Larkin in January 1954. Two candidates had been interviewed for the job, the other being John Holloway. Holloway 'impressed most of those concerned as pleasant and intelligent', while Kinsley seemed 'ugly and pompous'. So Larkin 'would have no difficulty in guessing which one got the job'. Kinsley was 'hated by everyone' at his previous post in Aberystwyth. He smoked a pipe and was 'a Scottish Methodist lay-preacher'. On top of which he was an expert on Dryden! Amis was convinced that his appointment had nothing to do with his qualities and everything to do with his being a Scot and a Balliol man – like the college principal, John Fulton. 'I feel like launching a poison-pen campaign, going on strike, hanging out of the window a short summary, well tricked out with obscenities, of my views on the matter,' Amis helplessly fumed.

Kinsley came to reciprocate Amis's antipathy. Amis learned that he had enquired of the professor of pure mathematics, Rowland 'Tug' Wilson, how you went about getting someone sacked, having Amis clearly in mind. Wilson was by now the senior professor at the college and knew the ropes. He told Kinsley that sacking people was very difficult. It had only happened once to his knowledge. That was back in the 1920s – and the sacked person was mad.

Frustrated in this tactic, Kinsley tried another. He proposed that Amis's pay should be stopped at the 'efficiency bar', which meant that instead of proceeding in annual £50 steps to £1,100 Amis's salary would be left marking time when it reached £900. Kinsley's grounds for this punishment were that Amis had not published enough work. Amis had by this time published several pieces in *Essays in Criticism*, edited by his old supervisor F.W. Bateson, which Kinsley seems to have missed, and he had also published his first novels and poems. But these did not appear to count as serious business for an academic. Professor Kinsley's logic may have been no more than to deduce from the fact that Amis was writing fiction the conclusion that he could not have enough time left over to give his best to the English department. Amis insists that he never missed a lecture or a tutorial to make time for writing, but concedes that a novelist-lecturer was then such a campus rarity that Kinsley may have had cause for suspicion. At any rate, nothing came of Kinsley's campaign against him, and Amis's salary continued slowly to rise.

A potentially more serious threat to his security came when he was summoned by the principal, John (later Lord) Fulton, to explain why he had said the disparaging things he had to the journalist Jill Craigie (wife of the future Labour leader Michael Foot), who wrote about him in a newspaper article. Amis claims he believed he was just having a sociable drink with Craigie and Foot and did not realise his criticisms of the college would be quoted, a claim Craigie denies, insisting that she has always been scrupulous about clearing her quotes with interviewees. Fulton, at any rate, seems to have felt that he could not ignore Amis's quoted remarks but that he couldn't do much about them either. To have fired the by now celebrated writer would have made Swansea 'the laughing-stock of the world', as one fellow lecturer put it to Amis. Whatever Fulton might have wished to do, a rap on the knuckles was probably as far as he could dare to go and Amis survived. Even so, Amis was alarmed enough by the dangers to alert the local representative of the Association of University Teachers, so that his union would be ready to come to his rescue if needed.

∞ 8 ∞

Lucky Jim

By the early 1950s Amis had not made much progress as a writer, or not as much as he would have liked. There had been at least as much frustration as success. He had published one volume of poetry, *Bright November*, co-edited a poetry anthology, worked hard on his abortive thesis and his equally abortive Graham Greene book for Tucuman University and his opera libretti for Bruce Montgomery. And he had written a novel, *The Legacy*, which was still doing the dismal round of London publishers. Amis had brought to Swansea several draft chapters of another novel, but much more time and work would be needed before that turned into the triumphant typescript of *Lucky Jim*. In his first year as a lecturer Amis had written only a single poem and, in his own opinion, not a very good one.

In August 1950 Amis spent a few days typing out all his poems and found they numbered exactly forty, 'which doesn't seem many for a nine-years career', as he told Larkin, though he had only included in his collection poems written since 1945. Amis also happened to have been turning out his drawers and to have found and re-read some of Larkin's letters. He thought Larkin's letters were much funnier than his own. But reading them and reflecting on his poems made Amis think of more than jokes. It made him look back over his writing career so far and the part Larkin had played in shaping it and Amis himself:

Do you remember, old trencherman, reading my sheaf of poems in John's (that room I had in the North Quad) sometime in late '45 or early '46, and sang [saying] you liked them on the whole, and thought they were quite good? And I put IWDAFY ['I Would Do Anything for You'] on the gramophone, and you said 'I knew you were going to put that on.' It's hard to say this without arousing both our laughters, but it was then I decided I was *gong on* [going on] writing po [poetry] – I knew you'd never thought much of the po I wrote in 1941–2, and I'm inclined to think I shdn't have tried too much more when the war and EAS [Elisabeth Simpson] packed in, if you hadn't said what you did. *Now*, I know it hasn't *got me anywhere*, and I'm still *not as good as both of us would like me to be*, but at least I *went on*. And so for that encouragement I am *deeply grateful*.

Amis felt grateful to Larkin for more than his poetry. *The Legacy* 'only got written in response to your suggestions' and without *The Legacy* Amis thought he would have written no more prose.

Also, dear man, I have to thank you for *stopping me from being a shit* and *encouraging me to be funny* in the right way and *getting me interested in modern po* (all this in various intangible ways) . . . Today, you are my 'inner audience', my watcher in Spanish, the reader over my shoulder, my often-mentioned Jack, and a good deal more. I feel it's a shame that our friendship only began after we'd stopped seeing much of each other. (Again you were largely responsible for beginning it. (do you remember '6477599 Fuc. P.A. Larkin, Excrement Boy, 1st B.U.M., Mond's Lines, Shatterick Ramp, Forks'? – I have just laughed like necrophily, writing that).) But I'm very glad we've got it now. I felt I had to get this said some time, and hope it hasn't embarrassed you.

This notion of an 'inner audience' was very much on Amis's mind, for it was at the heart of the Oxford thesis he had completed not long before. A writer, Amis thought, could benefit greatly from having a private audience of one particular reader or a small group of them, and he believed this had been characteristic of some of the better Victorian poets. This reader or readers would also be the first to see and respond to the writer's work and it would be after their scrutiny that the work passed on to a larger audience. Larkin and Amis were to perform this service for each other, passing their poems back and forward – criticising, proposing, supporting. And Larkin was to play a particularly important part in helping Amis get *Lucky Jim* into publishable shape.

In the same letter that Amis talked of his 'inner audience', he promised to show Larkin 'D & C' when they met. 'D & C' stood for 'Dixon and Christine', Amis's working title for what was to become *Lucky Jim*. By February of that same year, 1950, Amis had already written 250 pages of the novel and in November of the following year he announced to Larkin: 'I've finished Dixon now.' In between, Amis had had a stroke of literary luck, more prosaic than a sudden flood of inspiration but every bit as useful. He had acquired a room of his own in which to write. Hilly had received a legacy of £2,400 in January 1951, enough to buy a house in the Uplands district of Swansea and still have £200 left over. The Amises moved up in the world, from a flat on Mumbles Road to 24 The Grove (sometimes The Grave in letters to Larkin). They also acquired a car, a washing machine and a refrigerator. Amis celebrated by purchasing a modest trophy for himself, an elaborate metal machine for rolling cigarettes, though more to symbolise his new well-being than to actually use.

But the room was his most important acquisition. It was not a very grand room – it looked out on the blank wall of the house next door – but it was big enough for a table, a chair and a typewriter, and it possessed a door that could be closed against domestic distractions. Over the years, he grew to become, Martin Amis remembers, 'a kind of affectionate presence' around the house. His room was 'very much off-limits to the kids', while he 'managed to abolish all responsibility for the domestic side of life', making sure that 'other people looked after all that for him'.

Larkin's contributions to *Lucky Jim* were more than playing 'inner audience' to its author. It was through him that Amis had first picked up the book's theme, when he went to visit Larkin at Leicester in 1946 and spent half an hour in the university senior common room there. And if Amis's earliest ideas had worked out the novel would have been a story about Larkin himself and Larkin's girlfriend, Monica Jones.

Back in that common room, Amis had seen something he had never seen before: 'Professors and lecturers sitting, standing, talking, laughing, reading, drifting in and out, drinking coffee'. And he thought: '"Christ, somebody ought to do something about this." Not that it was awful – well, only a bit, it was strange and sort of *developed*, a whole mode of existence no one had got on to, like the SS in 1940, say.' He had stumbled on a whole new world of provincial university life which had never been explored or described by an English novelist. Here was a ready-made and virgin scene, ripe for a chronicler.

Since Amis had been a lecturer at Swansea for more than four years when *Lucky Jim* was published, people assumed that his novel was based on his own experience there. But this was only so to a slight degree. He had begun the book before he ever went to Wales or became a lecturer. The town where Jim Dixon teaches was at the early stage called Hamberton, an imaginary place located in northern England with a name Amis had invented for use in 'Who Else is Rank'. And imaginary is what the setting of the novel essentially remained. There is only one point of physical resemblance between Jim's college and any real place: a cemetery, which lies just across the road beyond the college railings. There was such a cemetery at Leicester University but none at Swansea or any other college of which Amis was aware. As he saw it, he had put sufficient touches in the novel to make the background 'compatible' with a provincial town in Britain, rather as the background to Sherlock Holmes's stories is 'compatible' with Victorian London, which is very different from a lifelike portrait of a real place in either case. There are Welsh touches in *Lucky Jim* but they were inserted, as much as anything, to provide in-jokes that would amuse Amis's friends. A character called Athro Haines is mentioned though he never appears. His name sounds like a Welshman's and stands roughly for 'professor of history' in the Welsh language.

What began to take shape in Amis's mind after his Leicester visit he summed up in note form long after the novel was written: 'University shags. Provincial. Probably keen on culture. Crappy culture. Fellow who doesn't fit in. Seems anti-culture. Non-U. Non-Oxbridge. Beer. Girls. Can't say what he really thinks. Boss trouble. Given chores. Disaster. Boring boss (a) so boring girl (b). Nice girl comes but someone else's property. Whose? etc.' If the plot did not come together as neatly or as easily as this condensed account suggests, it was near enough the way the pieces eventually fell into place.

Amis at first intended his hero, Jim Dixon, to be like his friend Larkin but in the course of pondering the character Larkin dwindled and disappeared. His traits simply did not fit the needs of the story. By the time it was finished the only faint whiff of Larkin left was the name Dixon, taken from Dixon Drive, the Leicester street where Larkin had lived. (Contrary to speculation among Amis's friends, the name Jim was not borrowed from any real person but chosen for its popular ring and because it could be extended to the formality of James when formality was required.)

The version of 'Dixon and Christine' that Amis completed towards the end of 1951 began a brief circuit of potential publishers, beginning with Michael Joseph, but met the same melancholy fate as *The Legacy*. A publisher friend of John Wain's turned it down for 'not having alive or exciting enough characters', as Amis told Larkin at the end of June 1952. Larkin was by then working at the Queen's University in Belfast and Amis himself applied for a vacancy there. A few days after he wrote that letter he went to Belfast for an interview. He did not get the job. But he took 'Dixon and Christine' with him for Larkin to read and criticise. 'I'm jolly glad you're taking such a decent interest in the thing,' Amis wrote towards the end of July: '*don't let it slip*. We should be able to fudge up something good between us.' Amis thought Larkin's most acute criticism so far had been that 'D[ixon] should sod up the romantic business actively'. He made some suggestions of his own and repeated his plea to Larkin: 'Don't let it slip.'

Amis had by now decided to regard what he had sent to the publishers as a first draft and to start all over again. He and Larkin continued to exchange ideas over the summer, part of which the Amises spent on a caravan holiday by the Thames near Pangbourne in Berkshire. 'Well here we are nicely settled in as you might say if you were a bloody fool,' Amis wrote to Larkin on 11 August, 'living in our flithy caravan. Really it's not so bad, not nearly so bad, as the words "caravan holiday with the flithy kids" would indicate. It is dry inside, it's warm enough, you don't have to shit in a field – especially *that*. The Calor gas smells like the farts of someone with cancer of the large gut.' The only new idea for the novel that had struck Amis at his caravan was a change of title. Now he thought it should be 'The Man of Feeling', because 'D[ixon] has more to feel about than he had before'.

Amis and Larkin continued to exchange ideas and synopses. 'Send me another line pronto,' Amis wrote from Swansea on 21 August. 'I'd like to get stuck into draft 2 as soon as I can.' A week later he wrote again: 'Shoot me a further revision when you have time; I yearn to re-start.'

One small but ticklish problem of identity had to be resolved between them. Although Larkin himself had disappeared from the story, his girlfriend Monica Jones had very much not, aspects of her character still seeming more appropriate to the plot than any aspects of Larkin's. The story called for Jim Dixon to be caught between two girls, one highly desirable but seemingly out of reach,

Christine Callaghan – the other hardly desirable at all but to whom Dixon nevertheless feels obligations, at this stage called Veronica. This undesirable girl continued to resemble Monica Jones, even when her boyfriend Jim no longer resembled Larkin. As seen by Amis, at any rate one version of Jones's relationship with Larkin was of a 'suffocating intimacy'. She had, Amis recalls, 'an awful chumminess' about her. There was to be a good deal of that in Jim Dixon's unwanted girlfriend, who is given to saying things to him in pubs like, 'Can't we talk about ourselves? We've got so much to say to each other, haven't we?', which makes Jim want to 'give an inarticulate shout and run out of the bar'. There was something of Monica Jones in the girlfriend's appearance too. She wore 'a sort of arty get-up of multi-coloured shirt, skirt with fringed hem and pocket, low-heeled shoes, and wooden beads', which sounds not so very different from that of Monica Jones in her Oxford undergraduate days, as described more sympathetically in Larkin's biography – wearing 'brightly coloured, hand-made, unconventional clothes'.

There were surely other sides to Monica Jones or it is unlikely that Larkin would have remained her regular (not the same as faithful) companion for the rest of his life. But he could see the truth in Amis's version, once even asking him, 'you weren't actually there taking notes [of our conversation] were you?' Perhaps surprisingly, Larkin did not resent Amis's appropriating his girlfriend's less charming characteristics for his novel. He did, however, draw the line at Amis borrowing her name too. Though known as Monica, her full name was Margaret Monica Beale Jones. In September 1952 Amis wrote to Larkin: 'I think the best thing about Veronica's name would be to change it to Margaret Jones.' This would create another in-joke between them: Amis would tell his Swansea friend Margaret Aaron-Thomas that he was 'cutting at' a girl in Leicester, while Larkin could tell Monica (who lived in Leicester) that Amis was really getting at a girl called Margaret in Swansea. 'How would that do? It's a common enough sort of name, God knows.' But Larkin rejected so obvious an identification. 'I'm sorry you feel strongly about Margaret Jones,' Amis wrote in October. 'Would you pass Margaret Street?' In the end they settled for Margaret Peel, which sounds like the Beale in Monica's name but was in fact the name of a man in Leicester Larkin had known and disliked.

Amis borrowed other characteristics from real people, though never in such quantity or detail as to add up to anything recognisable as a portrait from the life. Much of it was about little more than names. Michie is the name of a student Jim Dixon steers particularly clear of. James Michie, Amis's Oxford friend, was convinced that this Michie was meant to resemble his own brother. But Amis intended no such likeness, although he did make his fictional Michie a man who had commanded a tank troop at Anzio, unlike the real Michie, who was a conscientious objector. Evan Johns, an office worker at the college who shared Jim's digs, was named after Evan John, an acquaintance of Larkin's in Belfast, where Amis had met him briefly. Alfred Beesley, another inhabitant of Dixon's digs, resembled another Oxford acquaintance, Alan Beesley, but in name only.

Professor Welch, head of the history department that employs and torments Jim Dixon, has more complex origins, as does the professor's family. Bardwell is the family name of Amis's wife Hilly, so her father and mother became known as Daddy B and Mummy B, while Amis's were by similar logic Daddy A and Mummy A. Daddy B was 'a townsman by birth and a countryman by adoption', according to himself. Among his eccentricities, he was keen on folk-dancing and had learned three languages which had nothing in common except the fact that they were spoken by very few people – Welsh, Swedish and Romansh (found only in the Swiss canton of Grisons). Hence, by a roundabout route and with the change of only one letter, the name of Welch (which was also the way Amis, following army style, often spelt Welsh in his letters to Larkin).

Amis worked up powerfully derisive feelings towards his father-in-law. As early as 1949 he thought he must put him in a book, if only to stop himself murdering the old boy. After spending six days at his Harwell home in January, Amis wrote to Larkin:

I have jotted down a few notes for my next book about Daddy B; I don't see how I can avoid doing him in fiction if I am to refrain from stabbing him under the fifth rib in fact. I have been thinking of a kind of me-and-the-Bardwells theme for it all, ending with me poking one of Hilly's brothers' wife [sic] as a revenge on them all. I heard a hell of a lot more about Sweden, some anecdotes twice, and a hell of a lot more about folk dancing and folk cultures, and a hell of a lot more about the language of the Romanschi, and a hell of a lot more about all his friends in London. The best time was when I was lying in a partially filled bath with him in the room underneath accompanying on the piano, his foot regularly tapping, folk tunes which he was playing on the gramophone, there being a difference in pitch between the two sources of sound of approximately one-3rd of a full tone. As one vapid, uniformly predictable tune ended and another began I found that the hot tap was now dispensing cold water, and, getting out of the bath, began drying myself.

In July Amis returned to his derisive theme. 'My hatred of Daddy B has reached a new high,' he told Larkin. He and Hilly had been with friends to a carnival at Eynsham and they returned to Marriner's Cottage for tea when

the old ape-man turned up – he had 'come to see the country dancing' ... WITHOUT THE APE we should have been a merry party, but THE APE WENT ON SITTING ABOUT AND GRINNING and *spoiling everything by his presence* and *not realising it* and *doddering about and missing his bus* so that he stayed till *seven* BUGGERING THINGS UP FOR SIX PEOPLE *JUST BY BEING THERE.* I hate him; I *hate* him; I HATE the old APE's BASTARD. I have almost stopped being amused by hating him, too ... I shall swing for the old cockchafer unless I put him in a book, *recognisably*, so that he will feel *hurt* and *bewildered* at being so *hated*.

Amis needed to make his Professor Welch keen on something similarly twee and awful to Daddy B's folk-dancing but thought for reasons of tact it should be at least a little different. So he made Welch a fan of madrigals instead (though, just to keep the connection alive for those like Larkin who could recognise it, Professor Welch does threaten to send Jim to a folk-dancing conference). 'Poor old Daddy B,' Larkin said when he read these passages in Amis's novel.

The arty flavour of the Welch household resembles that of the Bardwells' too. Amis described the Whitsun weekend he spent there in 1949 to Larkin:

We had a nice Whitson, I hope you did, wasn't the wheather marvelous, we went over to the Mrss' people over at Harwell, proper scream the old boy os osod is with his fokl-danceing, and old Bill [Bardwell, Hilly's brother, a composer] with his conchertoes and sympathies, little Marion Partin[g]ton [Hilly's niece] was their two. I had all the usual kicks, an account of a blind fiddler, over eighty he is, at Adderbury, a meritless short story shown to me by mummy b–, a talentless and flavourless sonatina played me, gratuitously, by Wm. . . . Bill B. has *given himself the shits* by his own *filthy French cooking* – that'll teach him, with his I'm-more-at-home-in-France-than-in-England *balls* – YES, you *stab rad* [bastard], you ARE, you *star dab*.

Only one incident in *Lucky Jim* was taken from an event that actually occurred. When the Duke of Edinburgh visited the college at Swansea, a reception was held at which drinks were handed round. The registrar, Edwin Drew, took care to see that this unusual service was cut off after only a round or two, certainly before Amis had had his fill. The same thing happens at the reception before Jim Dixon gives his disastrous lecture on Merrie England. Amis also borrowed some words from Dylan Thomas, whom he met one evening in 1951 in a pub. After a glass or two of light ale, Thomas announced, 'I've just come back from Persia, where I've been pouring water on troubled oil.' (There had recently been a crisis over Iranian threats to nationalise British oil interests.) Amis made 'his stock retort to the prepared epigram. I said: "I must go and write that down."' Eventually Amis did exactly that, putting some of the same words into the mouth of the odious Bertrand, son of Professor Welch, who says that the government's foreign policy might have been a good deal worse, 'with the exception of their spectacular inability to pour water on troubled oil'.

When Jim Dixon had to be provided with a subject for an academic paper he is writing, in the hope that one L.S. Caton will find room for it in a shadowy new journal he is said to be publishing, Amis asked his friend Esmond Cleary, an economics lecturer at Swansea, if he could think of a suitably dim and dreary topic. Cleary came up with 'The Economic Influence of the Developments in Shipbuilding Techniques, 1450 to 1485' – 'a perfect title,' Jim thought, for the way it 'crystallized the article's niggling mindlessness, its funereal parade of yawn-enforcing facts, the pseudo-light it threw upon non-problems'.

These, then, were some of the ways in which Amis accumulated some of the details of his novel: happening on a scene that sparked his imagination (the Leicester senior common room); borrowing aspects of character and appearance from real people (Daddy B, Monica Jones); incorporating small incidents and places from life (the cutting off of drinks at a reception, the cemetery near the college); helping himself to names (Dixon, Michie); making in-jokes for the amusement of friends (Professor Welch and his folk-dancing). But none of these were incorporated into the novel for the sake of any kind of autobiographical realism. Amis's intention was to write a good, readable, funny story, and all his material had to be bent to that purpose. So, for example, Larkin simply disappeared from the novel because his is not the character Jim Dixon needs if he is to play the part assigned him by the plot. The cemetery across the road from Jim's college is inserted not for any reasons of geographical authenticity but to be the basis for a joke about 'the Honours class over the road'. And if Professor Welch was based on Daddy B the likeness was not so close that Daddy B ever recognised himself in the professor – although perhaps transforming Daddy B into Welch had exorcising effects on Amis. At any rate, he never did stick a knife into his father-in-law.

But there is another matter of realism or otherwise: how far does Lucky Jim represent Amis himself? 'All my heroes', Amis has written, 'start from me and in a sense stay with me.' As always in an Amis novel, though, the signals are confusing, often deliberately so. The most obvious resemblance between Jim and Amis is that both lecture at a provincial university college. But Amis had thought of this job for his hero long before he became a lecturer himself. And he gave Jim other characteristics that were not his own. Jim lectures in history, Amis in English; Jim comes from Lancashire, Amis from London; Jim is single, Amis married; Jim hates Mozart, whom Amis adores; Jim appears drunk at a public lecture, which Amis is far too law-abiding – and, as he himself put it, 'cowardly' – ever to do.

Old school and Oxford friends of Amis had another reason for thinking they recognised him in Jim. Jim's talent for pulling faces matches Amis's; indeed the two share much the same repertoire, including a Chinese mandarin face, a crazy peasant face, Martian-invader, lemon-sucking, Eskimo, Edith Sitwell, Evelyn Waugh, lascar, mandrill – even a Sex Life in Ancient Rome face. But Amis and Jim pulled the same faces for different reasons. Amis's faces were for entertaining his friends. Dixon's are done mostly behind people's backs and act as 'the covert protests and tension-reducers of a man in enemy territory without effective allies'.

There were, then, enough similarities between Amis and Jim for friends to think they recognised the author in his hero, as did others who knew no more about Amis than the bare outline of his biography. It was an easy step from there to investing Amis with some of Jim's other, less pleasant, qualities, especially his aggressively philistine attitudes. Dixon's policy is 'to read as little as possible of any given book'. He tells Margaret, 'you know as well as I do that I can't sing, I can't act, I can hardly read, and thank God I can't read music.' 'Haven't you noticed', Dixon asks his friend Beesley, 'how we all specialise in what we hate most?', meaning in his own case medieval history. Most notoriously, Dixon talks of 'filthy Mozart'. It was talk like this that upset Somerset Maugham and led him to include Dixon in a new class, 'the white-collar proletariat', which he labelled 'scum'. Maugham took care not to attach the label to Amis, but by a familiar process of osmosis others did, turning him by association into someone held to believe in the filthiness of Mozart too.

This is the opposite of the truth. Amis loves music, considers it first among the arts and believes Mozart pre-eminent among composers. In July 1951, Amis wrote to Larkin that he had 'just bought 3 more Mozart piano concertos and some other longhaired stuff which would all be a bit beyond you I fear'. He called Mozart 'divine' and urged Larkin to 'go into a gramophone shop and play to yourself the first side of the D minor pno concerto, or the slow movement of the C maj one (K.467) and see if you can go on thinking what you do YOUFFFFOOL.'

It is not even clear that Jim Dixon himself dislikes Mozart. What he certainly does dislike is hearing a snatch of Mozart's music sung badly by Professor Welch first thing in the morning, when he is a guest of Welch's, has destroyed his host's bedclothes by falling asleep while smoking and has the most devastating hangover. Immediately after thinking of 'filthy Mozart' Dixon 'very slowly, like a forest giant under the axe . . . heeled over sideways and came to rest with his hot face on the pillow.' At a moment of supreme crisis like this almost anybody might think Mozart filthy.

Where Amis is present in the novel it is often in ways least likely to be recognised by his readers. He is there in the Dixon who struggles to keep inside his smoking ration and wonders how he is going to make £3 last the nine days until pay-day. He is in the Dixon who worries about 'the awful business of getting on with women'. He is very much in the Dixon who regards himself as 'a boredom-detector . . . a finely-tuned instrument' who could be sent ahead like a canary down a mine so that 'the boredom co-efficient of any gathering could be read off' him. And in so far as Dixon's dislike of anything that could be called culture is really dislike of pretension and exclusiveness – of culture as the possession of horrible snobbish people like Professor Welch and his hateful sons – then Amis is with him there too.

In spite of his own firm belief that what he had written was pure fiction, Amis did have some qualms about the possibility that people might see things in his novel which weren't there – might even see themselves. When *Lucky Jim* was being prepared for publication by Gollancz, he wrote to his publisher, Hilary Rubinstein, Victor Gollancz's nephew: 'I don't know whether your firm makes a practice of putting in the "all-characters-are-fictitious" formula at the front of their books. If so, it might make my life smoother if something were appended to it in my case, along the lines of ". . . in particular, the University College depicted, together with its members, must be clearly understood as having no connexion whatever with any such institution in real life."' Amis wanted this 'just for the sake of common room comfort'. It wasn't that he thought he had produced recognisable portraits of his colleagues but that they might mistakenly think they recognised themselves.

The trouble was, as he put it in another letter to Rubinstein soon afterwards, 'some of the men may hate me here for ridiculing aspects of college life in general.' But – this was the important distinction – 'I'm confident there's nothing *libellous*', that is, nothing derogatory of any specific individual. To make his point that Gollancz need have no fear of being bombarded with writs, Amis set out the few passages where the novel connected with reality:

The college described in the book has a couple of points in its physical set-up in common with Leicester (to lead people away from Swansea), e.g. the cemetery opposite, but nothing else in common with either college. Everything is quite unlike Swansea, town as well as college. As regards the characters mentioned: Professor Welch is based slightly on my father-in-law (between ourselves) but on nobody in college here, and [Evan] Johns, in point of name and appearance *only*, resembles a lecturer in music I met once for about 5 minutes in Belfast. Bertrand Welch, Margaret Peel and Carol Goldsmith are as fictitious as any fictional characters ever are, I should imagine. No *incidents* are based on anything that happened here, or anywhere else as far as I know. Apart from the cemetery mentioned not a single place corresponds to reality: there is no such town and city . . . as far as I know, [Professor Welch's] house . . . is all made up, so is the ballroom, the digs etc. The stuff on p. 173 or so about exam-passing [which implied the results were rigged to look good] could be taken as referring to Swansea, I suppose, but equally to dozens of other provincial colleges, and I hardly think the Swansea Senate would condemn themselves by corporately prosecuting us. The ex-registrar at this place used to cut off the drink at sherry-parties . . . but I imagine the same applies to that. That's really all I can think of.

Amis was less than entirely candid about Margaret Peel, omitting to mention her resemblance to the real Monica Jones, presumably because there was no risk of her suing. But otherwise the list appears complete – Amis would hardly have put his first novel at risk of destruction through libel cases by lying to his publisher about its contents. The rest of *Lucky Jim* is fiction: Amis made it all up. This has never stopped people believing that there must be some 'real' character lurking behind every fictional one, if not Amis, then somebody else. At least once a year for the nearly forty years that Amis's colleague Sam Dawson continued to lecture at Swansea after *Lucky Jim* was published, it would occur to one of his students that he might be the original of Jim Dixon simply because both he and Jim come from the north of England. Such is the general inability to believe that novelists can and do actually invent their characters, something that has irritated Amis all his fiction-writing life.

On 3 March 1953 Amis wrote to Larkin that he had finally settled on a title. 'I've called it *Lucky Jim* now, to emphasise the luck theme – epigraph Oh, lucky Jim, How I envy him [these two lines from an old song were to appear at the beginning of the book] . . . I'm afraid you are very much the ideal reader of the thing and chaps like you don't grow on trees, course not.' On 30 March Amis announced that the book was done. 'Well, my old bandolero, I have just this moment finished my last task on that Dickson thing, so that it is all ready to send to a lot of men who don't know a good thing when they see one, and by way of celebration am letting you hear the good news at once . . . It comes to 349 pages of quarto, which is about 87,000 words, which is a lot.' Amis had just been through his typescript for the last time, changing things like 'arod' to 'road' and 'his Indian beggar face' to 'his Evelyn Waugh face'. The only thing he felt with any certainty about the 'finished draft is that it is the finished draft; no, or almost no, bugger is going to make me do anything to it ever again.'

It was entirely appropriate that Amis should let Larkin know he had finished *Lucky Jim* as soon as he had done so. Larkin had played the part of 'inner audience' to the full. He had made a great many proposals, some of which Amis accepted, others not. It was Larkin who suggested that Margaret should have another admirer besides Jim Dixon. This became Catchpole, who was able to clear up Dixon's suspicions about Margaret's supposed suicide attempt, which she had made in order to win the sympathy of both Catchpole and Jim, the discovery that she had not been serious about it giving Jim the justification he needed for abandoning her for Christine. Larkin thought Professor Welch should go in for pottery and weaving, but Amis stuck to madrigal-singing and recorder-playing, rather to Larkin's annoyance. And Larkin had helped Amis get rid of some surplus characters and put greater emphasis on others. He thought Amis was too soft on Bertrand, Professor Welch's artist son, and persuaded him to make Bertrand even more hateful. Less tangibly, Larkin's continuing interest and support had helped Amis to keep on working at his story when he might otherwise have run out of conviction and the steam to carry on. Larkin had earned his recognition as *Lucky Jim*'s dedicatee. But in helping Amis get launched as a novelist he had done something else: he had helped to secure Amis's independence as a writer. After this first novel, Amis never again needed so active an 'inner audience' to guide him. *Lucky Jim* was the only novel he was to show in draft form to a third person and ask for advice.

Lucky Jim's publication went as well as any first novelist could hope. This time round, Amis had no difficulty finding a publisher willing to take his book on. Hilary Rubinstein, whom he had known slightly at Oxford, had gone on to work for the publishing firm of Victor Gollancz, his uncle. Rubinstein spotted a biographical note Amis had written about himself for a PEN poetry anthology in which he said he was in the middle of writing a novel. In November 1952 Rubinstein wrote to him saying, 'If you are not committed elsewhere, would you care to send it to us when it is finished?' Amis replied that he would be very glad to do so. 'It would never make an author's or a publisher's fortune,' he said modestly, 'but I think it is quite funny; that, at any rate, is its aim.' He promised it for February or March the following year but in the event posted it off on 15 April – two weeks after he had finished the typescript and the day before his thirty-first birthday – with a note which said, 'As you'll see, serio-comedy is the formula really, though if it gets by at all I imagine it'll get by chiefly on the score of the comic angle. Still, that's all up to you.'

The following day, before Rubinstein had had a chance to read the typescript, Amis sent him another letter with the news that a long passage from his novel was going to be read out on the BBC Third Programme. 'Since I still believe (in the face of much contrary evidence) that acceptance by the BBC is a recommendation rather than the reverse, I pass on this information in the hope that someone at your end shares my belief.'

The radio broadcast was a stroke of luck. John Wain had been given a series called *First Reading* with instructions to 'liven things up by selecting young contributors'. A previous series, run by John Lehmann, the former editor of *Penguin New Writing*, had been judged dull for its dependence on an older generation of writers, which may perhaps have been unfair, there not yet being much sign of a new generation around. Wain opened his series on 26 April with the extract from *Lucky Jim*, described as a 'novel in progress', which was taken from the incident in which Jim Dixon sets fire to his bed while a guest in Professor Welch's home. So the reading public was introduced to Amis's novel even before its acceptance for publication was confirmed.

The day after the broadcast, Rubinstein gave Amis his verdict: 'I think it is a really brilliant book and got more fun out of it than anything else I have read for a long time.' He had put off saying so, he wrote, until his own opinion had been confirmed by the firm's chief reader, J.R. Evans, and confirmation had come in only that morning. 'I am happy to say that he shares my great enthusiasm for the book. Now everybody in the firm is itching to read it.' Amis wrote back by return that he was 'overjoyed' by 'all the nice things' Rubinstein had to say. 'Somehow the thought of anyone actually *liking* it came as a great surprise.'

Rubinstein was careful to insert a slender get-out clause in this first response, saying that he was '99.999% certain' Gollancz would be making him an offer. In fact, there was some resistance inside the firm, from its chief, Victor Gollancz himself. He never took to *Lucky Jim*, thinking it 'vulgar and anti-cultural', and never really liked the other Amis books he published. His daughter Livia and Rubinstein persuaded him he was wrong about *Lucky Jim* and when the novel sold well Gollancz became an enthusiastic supporter of Amis's work, publishing everything he wanted in order to keep him, even poetry that did not look likely to be much of a seller.

One minor point remained to be settled: the title. Rubinstein reported unhappiness with 'Lucky Jim' for having a 'faintly novelettish sound to it'. Amis made no objection to a change and suggested his original title, 'Dixon and Christine'. But that seemed not to have the right ring to it either, and, nobody having a better idea, *Lucky Jim* it stayed, and a contract to publish it was agreed, including payment of £100.

At the end of May 1953 Amis told Larkin that he felt 'a bit let down' now that *Lucky Jim*'s fate was settled. 'In a way it wants to make me want to *stop writing*, like a man who's satisfied with one seduction because he's proved he can do it and wanted reassurance, not pleasure.' But he told Larkin not to take him too seriously. 'No doubt I shall be fudging something up by the end of the summer, about a Welsh provincial university or something.' When he got a proof of *Lucky Jim* a fortnight later he told Larkin that he had read it through 'with mounting admiration'. For the first time he felt he had written a book. But he also remembered and saw the point of a remark Larkin made long before 'about feeling "embarrassed and only slightly interested" at the idea of your first novel coming out'.

Amis got on quickly with the business of correcting his proofs – 'an efficient device for making you hate what you have written', as he described it to Larkin – and got momentary cold feet about the problems of recognition. 'I am terrified now of the probable local effect of the thing . . . I keep finding more and more bits that might be taken as referring to people here.'

Rubinstein sent a proof copy to C.P. Snow, the novelist who also reviewed for the *Sunday Times*, along with an effusive letter in which he aimed to whet Snow's interest by saying that Amis's novel had something in common with the work of Snow's friend, William Cooper, whom Amis did indeed admire. Both Cooper and Snow read the book and liked it enough to want to meet its author. Amis came up to London for a lunch at the beginning of July. Snow was not available, but he met Cooper, who gave Amis 'some well-intentioned and useful advice', as Amis told Larkin. 'I know that's bound to sound frenziedly ironical, but it isn't meant to be.' The encounter stirred Amis's enthusiasm to make writing another novel his 'first priority' and he told Rubinstein he hoped to get it done by the beginning of the next year.

Snow came through handsomely, providing a two-sentence 'puff' for the jacket which Rubinstein thought would greatly help the book's chances of being reviewed. The puff said, 'LUCKY JIM is humorous, self-mocking, hopeful and endearing. For promise and achievement combined, it is the best first novel I have read in the last two years.' Further puffs were forthcoming, from Cooper, Lehmann and J.D. Scott, the literary editor of the *Spectator*.

Although Rubinstein had wanted to get the book out by Christmas 1953, Victor Gollancz decided otherwise, preferring to rely on a favoured tactic of his own. January would be better, he thought, especially for a new writer, since it was a quiet month in the publishing trade and *Lucky Jim* would stand a better chance of getting noticed when there was less competition for reviewers' column inches. The delay was annoying for Amis because under his contract his £100 advance was not due until publication date. In July he sent Rubinstein 'that embarrassing thing, a begging letter'. He had been on a holiday trip and 'the avarice of Oxford hotel-keepers and publicans has reduced us temporarily to a state approaching beggary.' Could Rubinstein let him have £25 'advance on an advance'? Rubinstein could and promptly did.

In November Amis wrote again along the same lines, this time wondering how soon he might get his hands on some of the money from the American publisher, Doubleday, to which Gollancz had by then sold rights in *Lucky Jim* for $2,500. The previous month Amis had written to Larkin: 'Why don't doubledays pay me the money they owe me, filthy cheating dollar imperialists?' A few days later, on 17 October, Amis calculated that after drawing his last £10 from the bank and handing over £7 to Hilly he would have £3 15s. 6d. left, or 'an average of 5/5½d *per diem*'. To Larkin he went on, 'pretend youre happy when youre blew becos theres fuckall'. His letter to Rubinstein was more circumspect. 'I don't know whether I'm more avaricious than most of your authors, but recently I've found my mind turning more and more often to the question of money – the Doubleday kind in particular . . . I'm sorry to sound *grasping*, but there it is.' Rubinstein told him that his money was not due until July, which was 'rather dismal' news to Amis, who was worried now about the impending arrival of a third child in the family. He asked Rubinstein if he could let him have 'some of my dollar-equivalent sent on to me as soon as convenient after the amount comes to you'. About half the total '(a good bit over £300?) would just do us nicely, equip the baby with nappies and buy me a couple of drinks and a shirt as well.' Rubinstein agreed to get Amis his money as soon as possible.

Meanwhile Amis had another stroke of luck. John Wain wanted to publish his poems in a new series being produced by his university, Reading. Amis had to get a selection together as quickly as possible and on 27 July he wrote to Larkin for help. 'Since I'm less confident of my ability to discriminate among my own work than some chaps, I wonder if you'd be good enough to glance through these.' There were to be twenty-four pages of text in the book with a maximum of thirty-four lines to a page. Larkin made his choice promptly, and

Amis wrote to thank him on 3 August. Of the eighteen poems which were finally selected for the book, Larkin had approved fourteen and queried two, leaving only two he had evidently not thought much of. One of those went in because Wain liked it and the other because Amis did – 'though I've never met anyone else who does'. Amis asked Larkin for help in choosing the title for his poetry book too. He agreed with Larkin that a title should be plain but found it hard to think of one that was not 'pretentiously plain'. Among those he suggested were 'A Temperate Zone', 'Thoughts on Paper', 'A Train of Thought' and 'One-Man Charades'. In the end he settled for none of these but found a new one. The book was published under the title *A Frame of Mind*: 'a bit dowdy certainly, but plain without being very pretentiously so . . . and descriptive of the contents to some extent: these poems are a way of looking at things & the result of getting into a certain fairly definable mental attitude.'

While Amis fretted at the delay in publishing his first novel, he got started on his second, though he found progress on this held up by rehearsals for a college production of *Hamlet* in which he had agreed to play Osric as well as choose the music and put the records on (since Ophelia dies before Osric appears, Amis was able to hand over this task to the female student who played her, Frances Climie, before he went on stage). The production in December got good write-ups, Amis told Larkin, though they did not mention him – 'expect they were all shocked by my straight-as-a-die playing of a pansy: I did my homo laugh at least 4 times.' But there was comfort in reports that a lot of the girls were 'saying I was marvellous and an absolute scream'.

In spite of its later start, *A Frame of Mind* beat *Lucky Jim* to publication. 'It's nice and warm in here now,' Amis wrote to Larkin eight days into 1954. 'The cat and dog lie couched at my feet. My wife reads H.E. Bates and eats a red apple. Near her is copy no. 1 of *A Frame of Mind*.' Sally, the Amises' third child, arrived before *Lucky Jim* too, on 17 January. The birth took place at home. When 'a pink fuzzy head' began to appear, Amis 'went quickly downstairs and read a page of *Lucky Jim* with great attention. After 2–3 minutes there was the sound of a baby crying.' Hilly, Amis reported to Larkin, was 'very fit, apple-cheeked as ever, and very pleased with herself for producing a girl. So am I.' Sally, Amis thought, 'looks no worse than might be expected. Rather better, really.' Larkin responded by writing a poem for the new child, 'Born Yesterday – for Sally Amis', in which he wished her something nobody else would, that she might be 'dull – If that is what a skilled, / Vigilant, flexible, / Unemphasised, enthralled / Catching of happiness is called'. The poem arrived very soon after Sally, and Amis replied on the 23rd: 'Sodding good and touching was the poem, moving me a great deal as poem and as friendship assertion. I think it's about the nicest thing anybody could do for any new-born child.'

Two days later, *Lucky Jim* was at last published in an edition of 750 copies, price 10s. 6d., and Amis celebrated by taking the swiftly recovered Hilly into Swansea for dinner and opening a bottle of champagne. Rubinstein had sent him the standard six free author's copies and Amis wrote back to thank him. 'I am absolutely delighted with the edition from every point of view. A most handsome job.' Amis also announced that he had reached page 40 of his second novel, for which he already had a title, *That Uncertain Feeling*. This might seem to Rubinstein a step back rather than forward, since Amis had earlier told him he had written more than a hundred pages; but it was retrogression in appearance only. 'This is the final draft and as before is progressing swiftly. I thought I could do it in one draft, but it seems I am a two-draft man and must just accept the fact.'

Which Amis did. He remained a two-draft man for the rest of his life, putting each of his novels through the same process of first and final versions. He was unwilling to give Rubinstein a firm date for his 'fair copy' of *That Uncertain Feeling* but aimed for March. 'There have been one or two distractions here recently, of which the arrival of a new Amis – female – at 2 a.m. to-day is perhaps the most important. Incipient fatherdom is a great inspiration in the long run no doubt, but not in the short run.'

'The fame that *Jim* has brought me is sending me giddy,' Amis wrote to Rubinstein on the first day of February. 'I am to judge the Oral English Competition in the Eisteddfod at Llywn-y-Bryn Girls' School, Walter Road, Swansea, and there is to be a piece about me in the students' paper. They wanted to head it "The Kingsley Report", but I soon killed that.'

In the same letter Amis acknowledged the receipt of two cheques from Gollancz in the previous fortnight and the 'great feeling of *bonhomie* [that] swept through me as I took them out of the envelopes'. Two weeks later he had yet more encouraging news to respond to. *Lucky Jim* was already into its fourth printing. 'My oath, I never expected this kind of thing. Just shows the great British public aren't such fools after all.' Yet these rapid reprintings were not quite such impressive evidence of British taste as they sounded. It was Victor Gollancz's cautious habit to print relatively few books at a time, particularly books by unknown writers. Then if the book didn't sell he hadn't wasted too much money. But if it did sell he could slap 'second', 'third' and 'fourth impression' on every new order of a thousand or two copies he made, thus giving the appearance that the book was going with even greater zip than it was. In fact, there were about 7,500 copies in print by mid-February. A week later *Lucky Jim* went into its fifth impression. The momentum was continuing to build.

An almost uniformly good press had helped it on its way. The *Observer*'s Sean O'Faolain was first off the mark on 24 January, saying Amis had made 'the gayest of bricks with the most common straw'. In the next week's *Spectator*, John Metcalf said it was a very funny, very human novel. Anthony Powell in *Punch* declared the novel had 'energy and form, and a real power of presenting the academic world it describes'. Walter Allen in the *New Statesman* called Amis a 'novelist of formidable and uncomfortable talent'. The anonymous reviewer in *The Times* called Dixon a 'genuinely comic yet credible character'. And in the *Daily Telegraph* John Betjeman compared the novel to 'a Harold Lloyd film or a Buster Keaton film in prose'. (Amis had sent a copy of the book to Betjeman, who told him in a letter that he had sat up until 3.30 in the morning reading it and that he had 'certainly never read such a consistently enjoyable book'.)

Amis got letters too. 'Fan mail not too bad on the whole,' he wrote to Larkin on 19 February. 'One from Geoffrey Gorer. One from Tynan. One from that FUCKING MEAN PONCE Roger Sharrock, asking me to send him a copy. One from a painter. One from a boy at school I wanted to bugger; too late now. None from cultural nymphs; unless you count MummyB.' Daddy B ran true to form. He wrote to Amis after reading the only two unfavourable reviews so far, in the *Sunday Times* and the *News Chronicle*, and began with a lengthy paragraph about the correct pronounciation of the Welsh name Myfanwy, which had been given the infant Sally as her middle name. Amis did get a letter from a 'nymph' a few days later, in fact from Elisabeth Simpson. She addressed him 'Dear Bill' and signed herself 'Sincerely, Betty Simpson', and she said: 'May I congratulate you on the publication of your first novel – I think perhaps I may.'

Amis recognised the handwriting and told Larkin that it 'still had power to make me pant and tremble slightly . . . Funny, all that. Sometimes these cogitations still amaze. O lyric love, half angel and ½ bum.'

By the middle of March, Amis reckoned he'd spent 14 shillings on stamps answering fan letters over the previous twenty-five days, a sum which would now buy postage for three letters but in the mid-1950s would have sent sixty-seven letters on their way, or eighty-four postcards. The fans included 'one ex-army acquaintance and a handful of Oxford craps, the rest painters, American agents, script editors drama (sound), features producers (Wales), actors wanting to play Dixon on the films . . . sexual maniacs (male)'. Amis felt encouraged and discouraged too. 'I feel in a sense that "they can't stop me now",' he told Larkin, 'except when I take up my new novel and feel how easy it will be for me to stop myself.'

More success was on the way. On top of book sales came proposals from the BBC to broadcast a radio adaptation by 'the wonderfully-named E.J. King Bull', while the Boulting Brothers optioned the film rights for £200. In fact there had been interest in film rights even before *Lucky Jim* was published, much to the amazement of Amis, who 'thought I'd seen to it that no one could visualise a single page of *Jim* on screen'. For an exhilarating spell it even looked as though Alfred Hitchcock would make the film, but he failed to take up his option. Foreign publishers began to take an interest in translation rights. 'Looking forward to seeing you at the premiere of *Lucky Jim on Ice*', Amis ebulliently wrote to Rubinstein in April. The early news from America was less good. Doubleday had made a money-back offer – if readers didn't find *Lucky Jim* uproariously funny their $3.50 would be returned. Copies came back in shoals. But Amis was too elated to let himself be depressed. 'Rather a gloomy letter from Doubledays,' he scribbled at the end of a letter to Rubinstein – 'good reviews, poor sales. What do I care, though, with 7/8 of $2500 in the bag.'

Amis thoroughly enjoyed his success – now he could buy a television set and have drink *in the house*, the Larkin test of affluence – but let himself get rather carried away by it all, anticipating his earnings by spending them before the cheques came in. On 4 April he wrote to Rubinstein saying that 'if you could let me have a cut of the option-money fairly soon it'd come in very handy. We are being more or less compelled to buy a *car*, of all things, next week, and this

161

will leave my bank account in rather a debilitated condition.' Four days later he wrote again asking Rubinstein to 'hurry up that £180 like a good fellow, or I shall be reduced to smoking toilet-paper. I bought the car yesterday, or rather paid the deposit. Mad, mad.'

In October he wrote again: 'hope is not dead in my bosom – it's my bank account that makes me sweat a little, and this is where I begin to feel embarrassed. Can any money be winkled out of your concern in the near future? I ask this in a hushed tone and inside a "Personal" envelope, so that you can handle this just as you see fit (handling of course includes instant destruction of the letter).' Amis admitted that on the prospects of a film deal he had 'plunged into spending rather, and am now facing some unpleasant bills plus . . . a shocking income tax demand. Now naturally I blame no one but myself for this bit of pre-natal chicken-counting, but if you could manage to get together some cash for me (the more the better) . . . you would save your old pal a heap of worry.' Rubinstein quickly obliged with a cheque which, Amis said, would 'help tremendously towards keeping the Amis family afloat until I get that £10,000 advance from you for my next novel'. But Amis's second novel was not progressing as fast as expected. The typescript he had hoped to deliver in March was still a work in progress in October.

By the end of its first year *Lucky Jim* was in its sixteenth Gollancz-style impression and had sold some 12,000 copies. Sales continued to swell in Britain, America and elsewhere. By 1972 it had sold one and a quarter million in its American paperback form alone. The book was eventually translated into twenty languages under such French and Italian titles as *Jim la Chance* and *Jim Il Fortunato* – but also into more uncommon tongues like Czech, Hebrew, Korean and Serbo-Croat. It was filmed by the Boulting Brothers and Jim was later turned into the central character in a television sitcom. In 1992 it was republished as a Penguin Twentieth-Century Classic. It was to remain Amis's best-known novel and became a landmark in post-war English fiction. Though now *Lucky Jim* looks like nothing more – nor less – than the propitious start to Amis's long career, in the 1950s it looked as though it was going to be the start of something bigger or wider. It was seen as the signpost to a new trend in English writing. Or even two trends.

∞ 9 ∞

Movements

In early 1954 a squally correspondence broke out in the pages of the *Spectator*. It began with a hostile review in the 8 January issue of Edith Sitwell's new book of poems, *Gardeners and Astronomers*, by the magazine's poetry editor, Anthony Hartley. Feeling his way through admittedly tricky poetic terrain, Hartley came down on the side of 'our young academic poets, the University Wits (Kingsley Amis and Donald Davie, for example)', and against what he called 'the Neo-Symbolists', to whose camp he assigned Sitwell.

Two weeks later she wrote (from Hollywood) to defend herself and to denounce, no doubt with Amis and Davie in mind, 'little Mr Tomkins . . . this week's new great poet'. In the next issue, who should pop up in the letters column alongside John Wain and Elizabeth Jennings to defend Hartley but 'Little Mr Tomkins' himself – '(Name and address supplied)'. Taking issue on a point of Sitwellian imagery ('Personally I think the sap of a tree is more like Double Diamond than peridots and beryls'), Little Mr Tomkins signed off, 'It just shows how we great poets differ.'

Sitwell came back fighting with another letter the next week, and on 15 February J.D. Scott, the *Spectator*'s literary editor, rang Hilary Rubinstein to tell him that Sitwell had sent yet another letter, saying that however much she might disagree with Hartley she agreed with him on one thing: 'I have read Kingsley Amis's most remarkable, most distinguished first novel "Lucky Jim" with enthusiastic admiration.' She had already written to Amis himself, care of his publisher, and to Victor Gollancz directly, expressing her enthusiasm for the novel. 'There is no doubt Queen Edith is crazy about your book,' Rubinstein wrote to Amis two days after Scott's telephone call; but, he added, Scott had told him 'the joyful irony of this, which made me sick with laughter'.

The letter from Sitwell containing her unsolicited testimonial for *Lucky Jim* duly appeared in the next *Spectator*. The following week the irony that had made Rubinstein sick laughing was revealed – in a letter from Amis. 'I am sincerely grateful to her,' he wrote. 'At the same time I feel I should point out that I myself am "Little Mr Tomkins".' It was Amis who had abused his influential fan under pseudonymous cover, though he had not known of Sitwell's enthusiasm for his novel when he first adopted it. When he knew what she thought he became uneasy about not owning up. He told Larkin that he felt he would 'get branded mean sod, bloody fool' if he did not express some regret.

Sitwell forgave Amis and even proposed inviting him to lunch, at which she promised he would find her oblivious to what had happened between them. The lunch duly took place in the summer at the Sesame, Imperial and Pioneer Club. Amis described it to Larkin a few weeks later: 'Had a corking fine lunch with my pal the Dame. Actually it was rather on the dull side, thank God; rather like lunching with a kindly maiden aunt who wants to show you she's interested in all that writing you're doing.' Others present included the wheelchair-bound John Hayward (T.S. Eliot's long-term companion – Amis found it 'curious, almost macabre', to hear the Dame ask Hayward to give 'Tom' her love), John and James Pope-Hennessy and Alan Pryce-Jones, the editor of *The Times Literary Supplement*, who was 'very oily' to the Dame. Things passed off peaceably enough, though there was a 'horrid moment' when Sitwell, looking at Amis, declared: '"Of course, there is just one thing which is absolutely unforgivable." (Pause.) "That is to attack someone anonymously."' Amis felt 'like an Indian brave whose horse runs away with him at the moment he realises the white men he is charging have repeating rifles'. But he need not have worried: the Dame was referring to a 'monstrous profile' of herself in the *New Statesman*. 'I fair ped myself in relief,' Amis told Larkin.

This little literary teacup storm is amusing though not in itself important. But it says something about the casual way in which Amis got caught up in literary circles and became identified with the 'movements' of the 1950s – the Movement itself and, later, the Angry Young Men. Here is not the place to disentangle the intricacies of these movements, or even to discuss whether they had sufficient substance to deserve any such title. But both had one thing in common at least. They were not the result of young writers sitting down to concoct a literary manifesto, then following it passionately through. Things happened more the other way about. Movements were declared to exist, then writers thought suitable were recruited to join them. They were the products of publicity (with a good deal of chance thrown in) rather than of any concerted literary programme.

In the early 1950s literary circles seemed imbued with an air of disappointed expectancy – like people waiting at a station platform where the train did not arrive at the time expected or, when it did, failed to disgorge any glamorous passengers. This was not how it had been in the first half of the century when literary movements had turned up regularly and in forms readily categorisable by decade or common purpose. There were, among others, Edwardians and Georgians, the Bloomsbury Group and the left-wing writers of the 1930s. Admittedly there had been a bit of a gap in the 1940s with nothing much new to fill it up besides Dylan Thomas. As a source of literary inspiration, the Second World War itself had turned out more disappointing than the First, its more promising young poets having either been killed early – Sidney Keyes, Keith Douglas, Alun Lewis – or not been called up to fight – Philip Larkin, John Wain, John Heath-Stubbs – or dodged conscription, like Dylan Thomas. But that decade was past and the question now was: where were the new writers of the 1950s and what was their message?

It seemed there was none of either. In the farewell editorial Cyril Connolly wrote for the last edition of his magazine *Horizon* at the end of 1949, he moaned with weary fastidiousness: 'it is closing time in the gardens of the West and from now on an artist will be judged only by the resonance of his solitude or the quality of his despair.' Bidding farewell to *Horizon* in the *New Statesman* soon afterwards, T.C. Worsley noted: 'Five years after the war there is still no sign of any kind of literary revival; no movements are discernible: no trends.'

Here was a vacuum waiting to be filled and much casting around went into filling it. John Wain's appointment to the BBC's *First Reading* programme was one attempt to find someone who might be able to do so. Various claims were made that a movement had been detected, though its membership and identity tended to fluctuate according to who was making the claim. Amis himself was aware that some sort of movement was in the air. But what was it all about and who was in it? 'There's no doubt, you know, we are getting to be a movement,' he told Larkin in March 1954, 'even if the only people in it we like apart from ourselves are each other.' Personally, Amis didn't 'give a pinch of shit for old Al [Alvarez]'s stuff, nor [Donald] Davie's, nor old John Barry Wain's really much.' That more or less left the movement to be made up of himself and Larkin, and Larkin wasn't pulling his weight in it because he was failing to produce any fiction. 'For GOD'S SAKE, MAN, turn out something in the fiction line . . . YOU, you flat faced pullet.' But Larkin never did write another novel after *Jill* and *A Girl in Winter.*

John Wain described the policy for his radio programme as one of 'consolidation', a reaction against modernism, a return to traditional forms. But this was of its nature vague and hardly much of a rallying cry for new writers. Other banners were tentatively hung out: Wain himself tried a 'new Elizabethan era', Hartley the 'University Wits' and 'the Metaphysicals', Donald Davie the 'new Augustans'. But none of these caught on. Then, in an editorial in the *Spectator* on 1 October 1954, J.D. Scott adroitly capitalised the hitherto lower-case letter 'm' and announced 'the Movement'.

This was, Scott later acknowledged, a more or less calculated publicity stunt, an attempt to liven up the literary pages as part of a campaign to boost the magazine's declining sales. Nevertheless, this label stuck as others had not, and Amis was, inevitably, stuck with it. Wain was also enlisted as part of the Movement. So was Iris Murdoch, whose first novel *Under the Net* appeared the same year, to a highly favourable review from Amis. Other members seemed to be Donald Davie, John Holloway, Elizabeth Jennings, Philip Larkin, Robert Conquest and Thom Gunn. Membership became largely a matter of getting poems accepted in collections like D.J. Enright's *Poets of the 1950s* or Conquest's *New Lines.* Sympathetic writers who did not succeed in getting into those volumes, like Philip Oakes and Anthony Thwaite, were classed as allies or associates.

It was by another trick of chance – one even more obviously the result of the vagaries of publicity – that the Angry Young Men appeared from nowhere. This followed the production of John Osborne's play *Look Back in Anger* at the Royal Court in May 1956. Although well received by some critics, especially Kenneth Tynan, the play did not do well at the box office – until, that is, the theatre's press officer, George Fearon, was struck by a flash of apparently off-the-cuff genius. Asked by a journalist what he thought of John Osborne, Fearon replied that he supposed Osborne was 'a very angry young man'. Osborne himself repeated the phrase on television soon afterwards and it began to take off, as did Osborne's play.

Amis was the only writer to find himself a fully paid-up member of both groups. Unlike other Movement members, he attracted publicity beyond small-circulation magazines or the review columns of the quality press. His first experience of this sort was with the journalist Dan Farson, who wrote to ask if he might interview him for the London *Evening Standard*. They fixed to meet in Cardiff – seen by Farson as an adequate compromise between London and Swansea, with the extra advantage, as Amis advised him, of a decent restaurant to lunch at – and he turned up with his friend Francis Bacon, the painter, in tow. After lunch, while Farson went for a stroll around the docks, Bacon showed Amis a magazine of homosexual soft-porn and asked if he was interested. Amis said no, wondering whether perhaps Bacon had come all this way to meet him on the offchance that he might share Bacon's sexual bent. Farson's article eventually appeared on 4 May 1956.

Two months later, Farson took things a significant stage further. He had become friendly with Colin Wilson, author of the hastily acclaimed work of quasi-philosophy, *The Outsider*, and he arranged for John Osborne to be present at a celebration party of Wilson's, along with another new playwright, Michael Hastings, and Mary Ure, the actress who had the part of Alison in Osborne's play. She got drunk, as did Hastings. Nevertheless, in two articles for the *Daily Mail* Farson contrived to construct from this unpromising start a new movement, consisting of the three partygoers, Wilson, Osborne and Hastings, plus Amis. Two weeks later the *Daily Express* joined in, with an article by John Barber which lumped the same four together for the first time as Angry Young Men.

So were launched the two groups into which Amis was corralled willy-nilly. To the press, popular and quality, the groups had the advantage of vagueness in both identity and membership, enabling them to be endlessly examined and discussed, and approved or deplored according to taste. New names could be added to the list of Angries as they arrived, as were John Braine with his first novel, *Room at the Top*, and Arnold Wesker with his first play,

Chicken Soup with Barley. Others could be quietly dropped from the running when deemed to have failed, as was Colin Wilson when his second book flopped. New angles on the youth theme could be struck by finding even younger writers to publicise, as happened when, following the lead of the French writer Françoise Sagan, whose novel *Bonjour Tristesse* was published when she was nineteen in 1954, the English went one better by discovering Jane Gaskell, aged sixteen and already author of *Strange Evil*. Yet more tenuous connections could be made – with the emergence of home-grown rock and roll stars like Tommy Steele or with the American Beat writers like Jack Kerouac and William Burroughs. Protest, foggy and formless though it might be, was in the air, and it came as no surprise that Angries like Osborne and Braine – though not Amis – should take part in the demonstrations of the Campaign for Nuclear Disarmament, launched in 1958.

For Amis, there was more literary substance to membership of the Movement than in being an Angry Young Man. The Movement at least contained friends who could help each other get published and broadcast, or give genuinely useful literary advice, as Larkin did with *Lucky Jim* and as Amis reciprocated by criticising Larkin's poems. The Angries were not his friends, though John Braine became one later, and there was to be more than a little of him in the Yorkshire-born Pope of Amis's 1976 novel, *The Alteration*. Amis met Osborne more often at the Garrick Club in the nineties than he ever did in the fifties, though he had little more to say to him then than earlier. To Colin Wilson's *Outsider* Amis gave an unfavourable review in the *Spectator*. Wilson wrote him a two-page letter by return and called on him in Swansea. But Amis did his best to avoid him thereafter, finding his interest in serial murders scarily sinister. (Wilson, however, sometimes proved hard to elude. Amis once gave as a reason for not meeting him a party he had to attend at Brown's Hotel in London, only to find Wilson turning up there uninvited.) Amis also avoided Wesker, though for different reasons. He disliked Wesker's plays very much indeed and found Wesker's self-assured indifference to his dislike hard to take.

Celebrity Amis thought of as no more than a useful means of promoting book sales and, considering how well known he became, remained surprisingly cautious about publicity through the fifties. He declined opportunities to broadcast, for instance, except when he could do so on his own terms. Responding to a 1956 invitation from the BBC to take part in a discussion on the Bloomsbury Group, Amis took care to set out his stall: 'a) I am against Bloomsbury, though not (I hope) violently or crusadingly so. b) I know most about, and most dislike, Virginia Woolf of the Bloomsbury set . . . I promise to be reasonable about her, though hostile.' More often, though, he turned down invitations altogether. In January 1957 he refused a chance to discuss a Fabian pamphlet he had written on the air. 'I'm no good at discussions,' he wrote; 'and on this sort of topic the other chaps would just make mincemeat of me.' In July he refused a chance to appear on another discussion programme, *Give and Take*. 'I am afraid the programme you describe', he wrote to Huw Wheldon, its producer, 'is not for me . . . I can never think of the really telling retort until afterwards.' In November he accepted an invitation to talk face-to-face with Anthony Powell but turned down one to appear on *Woman's Hour.* 'live microphones terrify me and I have sworn off them for the next quarter of a century.' This was an exaggeration. He was to become at ease in front of the microphone and later the camera, even in the sixties interviewing a series of pop stars for television.

Having a job, a home and a family in Swansea helped Amis avoid the worse pitfalls of celebrity. Life, though, became fuller. He began to keep a diary; and one friend, Clive Gammon, remembers Amis's slight air of embarrassment when he rang to fix a date for a drink and Amis asked him to hold on while he looked in his 'book'. Amis liked going up to London and did so more often than before. But he kept his sense of proportion. 'As a means of shortening your life-span, but good, I heartily recommend [London],' he wrote in the *Daily Express* at the end of 1957. 'Nowadays, after about 72 hours of as much richer food, stronger drink, etc., as I can stand, I beat it for the Welsh Marches with all the enthusiasm of an ancient Celt pursued by Saxons.'

Reflecting on the business of celebrity in his *Memoirs*, Amis wondered if he had done the right thing when he started the ball rolling with Farson. Doing it once meant doing it again – 'and things like appearing on television, giving quotes over the telephone about how you will be voting at the next election or spending Christmas, helping to advertise a brand of beer, all that.' The alternative was to say nothing for twenty years, 'after which you may condescend to be photographed with your face mostly hidden, preferably in a foreign magazine, and expect to be taken seriously by some sorts of people.' Without knowing it he was casting a die when he met Farson, but, he concluded, 'not a very important one'.

Perhaps, though, it mattered more than he thought. By becoming a publicly known figure Amis acquired an identity and an image which were easily confused with his work and blurred the intended distance between himself and his novels. The temptation to see real people in fiction grew all the stronger when readers believed they knew what the author was really like. On the face of it, much of Jim could be found in Amis: provincial lecturer, fond of cigarettes, booze, girls, a puller of funny faces, irreverent, a Lefty – and a bit of an anti-intellectual, if not an outright philistine, to boot. Thus in the public mind Kingsley Amis merged into Lucky Jim – they became one another. When in 1967 Amis published an article which had been given the headline 'Why Lucky Jim Turned Right', nobody had to be told that it was about himself and his own change of politics, not Lucky Jim's.

With articles like this his image began to change and eventually to change radically into more or less the exact opposite of Jim: Amis as clubman, curmudgeon, anti-trendy, Tory, misogynist, blimp – with only a fondness for drink connecting him to the old image. One way or another Amis seemed always to project a stereotype of himself, even to relish promoting the more extreme versions of his own caricature. He often got irritated at the way he was identified with characters in his fiction so that people thought he was, say, as much of a misogynist as Jake Richardson of his 1978 novel *Jake's Thing*. Why can't people realise that novelists actually make up their characters, he will querulously ask. But such confusions, always possible, are perhaps inevitably intensified by the fitful and uncertain illuminations of publicity.

Amis's horizons expanded steadily through the 1950s. He went to London, more often after *Lucky Jim* was published, and made new friends there. The first was Robert Conquest, the Anglo-American poet, later celebrated as the historian of Communist crimes, who endeared himself to Amis by reciting his own long sequel to 'Eskimo Nell' at a party in Chelsea in 1952, held to celebrate a new PEN anthology of poems in which both Amis and Conquest appeared. Anthony Powell wrote to Amis after he had his first book review published in the *Spectator*. It was about James Thurber and in it Amis remarked on Powell as a serious writer who was also funny.

170

Amis told Larkin that Powell found his style 'very sympathetic'. This out-of-the-blue approach from a writer he greatly admired bucked Amis up considerably and he found himself imagining for Larkin's benefit a conversation fifteen years hence, by which time Powell would be dead and Amis would be able to report him as having said, "'. . . Kingsley, you're the one we all look to to keep the English novel going when we go.'" Amis met the journalist George Gale at a *Spectator* party in the middle of the decade. He began to correspond with John Betjeman when *Lucky Jim* was published and got to know him early in the 1960s.

Increasingly often his new friends would come to visit Amis in Swansea and he, like Robin Davies in *You Can't Do Both* (who also becomes a lecturer at a provincial university), would visit their homes or make short trips which 'would include one night or possibly two in London and an assortment of meals, drinks, recordings with publishers, agents, BBC men, not forgetting friends'. Amis described such a trip to Larkin in 1954. First he went to Oxford, to a party at All Souls and dinner with friends afterwards. Next day he went with Hilly and Bruce Montgomery to London. There he met 'the Spectator lot who all seemed very decent'. Then it was on to a 'tremendous lunch' at Montgomery's club, after which they met Tony Hartley and John Scott of the *Spectator* in a pub. From there, 'Hartley, Hilly and I went and saw Ken Tynan and his wife [Elaine Dundy], but I was so pissed I could hardly speak.' Black coffee was produced and Amis proceeded to another pub where he found himself 'advancing towards Harry Hoff [William Cooper], Bruce and C.P. Snow sitting in a line. Hoff greeted me by saying, "Here's Amis, but too drunk, I see, to say much."' Amis, however, managed to get quite a lot said, though he could not remember much of it afterwards.

The next day he and Hilly drank with P.A. Newby, 'a small, rather nasty man with apostolic eyes', and picked up Powell at 1 p.m. – 'the nicest man we met'. Powell variously expressed 'admiration' for Proust, 'affection' for Graham Greene (though not for his books), 'sorrow' for Evelyn Waugh and liking for David Cecil, 'whose fag [Powell] was at Eton' (though no liking for Cecil's books either). Later they returned to Oxford with Montgomery for a 'brains trust' whose members included C.S. Lewis, G.S. Fraser, James Kirkup and Amis – 'I was sullen, jerky and facetious by turns.'

It was out of these trips that Amis's weekly lunches at Bertorellis' in due course emerged. These became mildly notorious as the 'Fascist' lunches in the 1960s and 1970s because of the right-wingers who often went to them, but they started out as a convenient arrangement for Amis: if he turned up in London on a certain day he could be reasonably sure of finding sympathetic people to lunch with.

These trips also provided useful cover for assignations with women, for both Robin Davies and Amis. Not that Amis had to go to London for that. He had affairs with women in Swansea too. To one, whom he called Oligadora, he wrote a rather bitter short poem of that name, in which he describes her reaction to a gift: 'This is too little? / 'Am I in default, / Or are you grasping? / Which of us shows more / Inadequacy?' Contemplating the attractions of some of his female students, Robin Davies thinks that the only reason he doesn't go after them is fear of the sack, to avoid which he follows the advice 'never [to] lay a finger on them till they graduate' given him by 'a veteran Ulsterman with teaching experience in India' (a man who sounds very like Amis's Swansea colleague Jo Bartley, described in his *Memoirs* as 'an Ulsterman who, with a Second to live down, had taught in India', though Bartley did not give Amis the same advice). But Robin concedes to himself that his resolution on the point might 'wilt one day'. Amis's certainly wilted, women students not being excluded from his conquests.

Amis made little effort to avoid temptation. Quite the opposite: he seized every chance of sexual adventure he could. 'Will alert you to my next visit,' he wrote to Conquest from Swansea in November 1956. 'May want to borrow your flat at times like the early evening for an hour or two, to entertain a young lady. You bachelors have got to stand by us married men, you know.'

A year later he involved Conquest in an elaborate alibi:

Visited London briefly last week and meant to see you, indeed rang you Thurs and Fri evening. However, I used you as an alibi on Friday afternoon while I was in fact with – guess what – a young lady. But, if it ever comes up . . . please note that on Friday, 13th September, you and I

(a) met at lunch-time
(b) 'wandered about' in the afternoon
(c) parted about 6 p.m.
(d) were accompanied by 'a vague cousin or brother-in-law' of yours – a nice circumstantial touch, I thought
(e) were going to meet again in the evening, but when I phoned by arrangement c. 8 p.m. you said you couldn't make it.

Okay? you know I'd do the same for you any old time, eh?

At the beginning of 1958 he again involved Conquest in a cover-up, this time over an affair with a Swansea woman. 'As you can see,' he wrote, 'there's a little chore I'd like you to do for me, if you would be so good as to put the enclosed in the post and keep your trap shut with reference thereto thereafter. I'm sure you see the point: a London postmark is insufficiently specific to catch the husbandly eye, but a Swansea one points straight to yours truly. Do the same for you some day.'

Sometimes these little stratagems of deception did not quite work out as planned, with results that can now seem hilarious, though surely did not seem so to Amis at the time. Writing from the address of the English department at Swansea in December 1960, Amis asked Conquest for use of a bed, even as he urged him to be careful.

Now listen . . . Can you fix me up with a bed for tomorrow week (19th Dec.)? From say 11.30 a.m. onwards . . . Reply, if you will, to the above address, not Glanmor Road, because a fault in my security system has led Hilly to connect a Conquest letter with an impending Amis screw. And you at your end can help by not telling Desmond Williams this time (or anyone else at any time, for the matter of that) what I am up to . . . So please, chum, no leaks to anyone on this. & christ, as you'll readily understand, what bugged me was having to serve my sentence without having had a chance to commit the fucking crime, so to speak.

Infuriatingly, Hilly had caught him out in an act he had not got around to committing.

Amis was not unhappy or unsatisfied with Hilly. But he could not or did not turn down any promising chance of going to bed with another woman that presented itself. His opportunism in these matters was prodigious throughout his married life. Once, much later, while staying with other couples for the weekend, he found himself having to give a warmly ingratiating smile to all the women at the breakfast table, being boozily unable to remember which one of them it was he had made love to the night before. Another time he was invited to dinner by a married woman with whom he was having an affair. The woman's husband was there, as was Hilly, and a third couple made up the dinner party. Amis contrived to make a date with the wife in this third couple, behind the backs of his own wife, his hostess-lover and two husbands. The subsequent fling started out in farcical style when he had to see off a window-cleaner who was about to get to work on the windows of the maisonette Amis had borrowed for the occasion. And it continued bizarrely when the woman revealed herself to be an enthusiastic masochist and her husband a sadist. This came as a surprise to Amis because both seemed utterly conventional in outward appearance and behaviour, indeed in intimate behaviour too, since the wife's revelation came immediately after a perfectly straightforward and thoroughly enjoyable sexual act. But that other side of sex did not appeal to him. He once tried an experiment with his second wife involving a hairbrush but when neither got any kind of thrill out of it they quickly desisted and resumed conventional relations. Amis may have been a Don Juan in his day but he was never a de Sade at any time.

The Don Juan character is sometimes thought to be motivated by dislike, even hatred, of women, and there have been times when Amis has speculated whether this was his motive for chasing them too. But in his earlier years at least he seems genuinely to have liked women as well as wanting to go to bed with them. Women friends from his time in Swansea remember him as being unusually interested in them as people – as human beings – instead of as the sexual prey which was the role into which prevailing male attitudes preferred to cast them.

Amis certainly thought liking women was an important thing for a man to do. In a 1963 BBC *Woman's Hour* broadcast under the title 'Letter to Someone's Son', he chose to make this his theme. 'Yes, I know it sounds absurd,' he declares in his script, 'every normal bloke likes women, doesn't he?' But his own answer to that was 'a very emphatic no'. He had some difficulty defining what he meant by liking women, beyond it being 'obviously to do with responding to all feminine qualities, not just the specifically sexual ones'. Nevertheless he had no difficulty defining how men who liked women behaved towards them: 'They treat them with genuine and continuing affectionate interest.'

There were more men who disliked women, he thought, than women who disliked men. And there were other kinds of difference which told against men. When a woman misused her sexual qualities she did so to get what she wanted – as when a woman driver fluttered her eyelashes at an irate policeman. Yet a man often used his 'as a way of being nasty to a woman', by turning on the charm, say, making a date and then failing to turn up. 'For all sorts of reasons (economic, social, biological, psychological) men have it in their power to damage women far more severely than women can damage men. A man who realises he has this power and never uses it is a man who likes women.' Ordinary kindness, he thought, was a much more significant part of man's attractions than more showy displays. 'A girl dislikes unkindness more than lack of skill in the Twist, prefers affection to an expensive dinner and would rather get a letter from you than a pair of sharp earrings.'

Amis found another kind of difference between men and women which he described in a poem, 'A Bookshop Idyll'. Imagining himself leafing through an anthology of new poems, the poet thinks them immediately divisible by their authors' sex. Men's poems have titles like 'Landscape near Parma' or 'The Double Vortex', whereas the women's are called things like 'I Remember You' or 'Love is my Creed'.

> Man's love is of man's life a thing apart;
>   Girls aren't like that.
>
> We men have got love well weighed up; our stuff
>   Can get by without it.
> Women don't seem to think that's good enough;
>   They write about it.

Never mind 'the awful way their poems lay them open' –

> Women are really much nicer than men:
>   No wonder we like them.

In their poems men like to show off how clever they are, while women try to get to grips with love. But perhaps men are not confident enough to write about love and so find it more convenient to 'forget those times / We sat up half the night / Chock-full of love, crammed with bright thoughts, names, rhymes, / And couldn't write.' Neither men nor women understand love, but women at least realise its importance while men think it's enough to be clever.

Amis liked women, so why not enjoy as many of them as possible? There is a sort of logic to that, which Robin Davies takes to its furthest extreme. 'It had even occurred to him that sticking firmly to one girl could be unethically used to obtain exemption from the sometimes gruelling task of promiscuity.' So you can, if you really work hard at it, make sleeping around seem like the tougher, therefore ethically sounder, option. In one part of his mind Amis seems genuinely to have thought of sex-on-the-side as a child might think of a slice of cake: as a treat, and in an adult no more heinous a crime than having, say, an extra couple of large ones in the pub beyond the usual ration. This sense of sex as an enjoyable perk to be taken as a side-dish to the main meal of a happy marriage is caught in a letter Amis sent to Robert Conquest in November 1957: 'Sex recedes a bit, though maritally I have nothing whatever to complain about, and I had a very funny (ha-ha) extra-marital one the other day which I'll tell you about – you'll be the ideal audience for it. Oh, why isn't there *more* of that kind of thing?' Why, Amis might have been asking, couldn't a man have his cake and eat it too?

But in another part of his mind, which Amis did his best to keep at arm's length from the first part, he knew perfectly well that there was more to sex than a bit of fun. In a poem, 'Nothing to Fear', Amis describes his feelings on turning up at a flat for an assignation (in his own parallel experience a flat loaned him by Conquest):

All fixed; early arrival at the flat
Lent by a friend, whose note says *Lucky sod.*

(When Conquest read this poem he wired up his flat with a tape recorder so that the next time Amis borrowed it for sexual purposes the words 'Lucky sod' sounded alarmingly the moment he switched on the light. Sometimes Amis found Conquest's practical joking a little hard to bear.)

Drinks on the tray; the cover-story pat
And quite uncheckable; her husband off
Somewhere with all the kids till six o'clock
(Which ought to be quite long enough);
And all worth while: face really beautiful,
Good legs and hips, and as for breasts – my God.

So far so good. Everything is in order and a good time is about to be had. Then the poet has a momentary qualm, which he quickly brushes aside.

What about guilt, compunction and such stuff?
I've had my fill of all that cock:
It'll wear off, as usual.

But something else suddenly takes the place of guilt: 'slight trembling, / Dry mouth, quick pulse-rate, sweaty hands'. It isn't that the woman he is waiting for is the first, nor that he is impatient for her to arrive. Nor is it 'fear of failure, thank you, Jack'. He tells himself he's impervious to the dangers of beauty. Its 'touch will burn, but I'm asbestos, see?' So what has brought on this feeling of terror?

sitting here, a bag of glands
Tuned up to concert pitch, I seem to sense
A different style of caller at my back,
As cold as ice, but just as keen on me.

Waiting for his girl to arrive has led him to imagine death calling in her place. The poet tries to write this off as 'a dead coincidence'. But is it? Or is it a supremely acute sense of guilt surfacing from some unfathomable depth, from deposits laid down heaven-knows-when and made up of God-knows-what ingredients – perhaps something to do with his parents' prudishness, or his feelings about Hilly? Who can say what forces have combined to forge this new anxiety, this completely unlooked-for association between death and adultery?

Death as the associate of sex-on-the-side is an extreme expression in Amis's work of his anxieties about his own behaviour but it is by no means the only one. Such anxieties, indeed, show themselves in his work from beginning to end. Relations between men and women are his most persistent theme, in particular the way men treat women, more often than not badly. Adulterous husbands are prominent in Amis's novels from his second, *That Uncertain Feeling*, to his latest, *You Can't Do Both*. And it is possible to see Amis examining his own concerns about his own behaviour in many of them.

That Uncertain Feeling, published in 1955 and set in the not-un-Swansea-like Welsh town of Aberdarcy, is the story of a librarian, John Lewis, who has an affair with Elizabeth Gruffydd-Williams, the wife of a local councillor who has the clout to get Lewis the promotion he wants. Having been faithful to his own wife Jean throughout their five years of marriage, John Lewis keeps his distance for a while but eventually succumbs after a beach party on the evening of the day he has been interviewed for the promotion by a committee headed by Elizabeth's husband. He is worried about his chances but, after they've made love, Elizabeth tells him he needn't be. Her husband has fixed it up for him to get the job, whatever the committee thinks of his interview. Whereupon Lewis summons up the scruples to turn it down, on the grounds that it is 'a dirty deal'. Elizabeth is contemptuous: 'I suppose adultery isn't a dirty deal according to your way of thinking,' she says. 'You and your conscience are quite happy about that, eh?'

When Lewis reveals to his wife that he has both turned down the job and slept with Elizabeth, about whom Jean has had suspicions for some time, she is contemptuous in her own way too. Don't she and the children need the extra money that goes with the promotion? But she is hurt far more than she is worried about cash. 'Whatever you told me [about having Elizabeth] couldn't be as bad as what I felt when I realised what you were thinking about that Elizabeth and preferring her to me. That was the worst time.' She goes on to pour scorn over his whole attitude to sex. 'Yes, I know a bit of chasing round after other women now and then doesn't matter, according to you. As long as there isn't too much of it. Well, according to me a bit does matter, a bit's too much . . . Anything at all of that sort matters. According to me.' She even admits, falsely, that she has had a fling of her own with Gareth Probert, the local stage-Welsh poet-playwright.

Her anger and despair frighten Lewis. He rushes out of the house and gets involved in an absurd brawl. When he calms down he asks himself what he is going to do. 'Since I seem to have piloted myself into the position of being immoral and moral at the same time, the thing was to keep trying not to be immoral, and then to keep trying might turn into a habit.' But he knows this will not be easy. 'I was always, at least until I reached the climacteric, going to get pulled two ways, and keeping the pull from going the wrong way, or trying to, would have to take the place, for me, of stability and consistency. Not giving up was the important thing.'

The plot of this novel is entirely fictitious. No wife of an influential local councillor able to do Amis a favour came his way in Swansea. But the feelings of John Lewis and his wife Jean and the arguments between them reflect those of Amis and Hilly, she bitter at his infidelity, he all too well aware of her bitterness. There were rows in the Amis household too, real ding-dong affairs, which on occasion the three Amis children would sit on the stairs to hear with tears in their eyes, not quite knowing what it was all about but hoping it would soon be over.

John Lewis, however, is careful not to commit himself never to straying again after his adultery has been discovered. The most he will commit himself to doing is to try harder. So it was with Amis. He, too, thought he must try to behave better while at the same time recognising that he could give no guarantees. As a forecast of Amis's own behaviour until he reached his own climacteric, John Lewis's permanent tug-of-war between his worse and better selves was to turn out all too close to the truth, with Amis's better self losing the struggle more often than not.

When he is wondering whether to marry Nancy, Robin Davies imagines their future, with Nancy passing round cucumber sandwiches to 'other fecund young mums' while he 'craned out of the boxroom window with a telescope pointed at a girls' school playground with a stand up to his neck'. This obstinate randiness has a ring of Amis in Swansea, even in detail. Writing to Conquest in April 1956, soon after moving to a new house at 53 Glanmor Road, Amis said: 'Got a nice big study, but the girls from the school up the road keep coming past and taking my mind off my work. Never mind: one's got to keep in touch with one's μανία [the Greek word *mania*, meaning 'passion'] – not the same as mania, see.'

But Hilly had found her own way of striking back. In Swansea she began to have flings of her own. One of these seemed likely to pose a serious threat to the Amises' marriage when what had started as an affair turned into something more. In October 1956 divorce itself emerged as a possibility. A decision would be reached, Amis told Larkin, 'in the course of the next few months, so that the amorous pair can decide whether they are the victims of an "infatuation", or of "something more important"'. Amis thought his marriage had 'about one chance in four of surviving till next summer'. He was not pleased with this, especially as it might very well include separation from his children, 'about whom I feel strongly'. 'During the interlude while a decision is being awaited on divorce, there will be none of the old marital rites engaged in here, naturally. This has all been rather diminutory of the gaiety of nations, inhibitory of general merriment, &c. As you can imagine. Having one's wife fucked is one thing; having her taken away from you, plus your children, is another, I find.' Amis had not told anybody else and he swore

Larkin to secrecy. He reckoned that if it came to divorce he would resign from Swansea and perhaps move to London. 'It'll be odd to be a bachelor again.'

Matters swayed to and fro for several weeks but by 20 November Amis felt able to tell Larkin that Hilly's affair had 'quietened down a lot since I last wrote. My marriage now has the odds heavily on its side.' He had been planning to visit Larkin in Hull a couple of days later but sent him a telegram to cancel, explaining in a letter dispatched on the same day that it was not only Hilly who had been giving him sleepless nights: 'my young lady whom I was rather attached to . . . has just decided to give me up', and all this 'sort of stuff gives me great physical fatigue apart from anything else'. He simply could not face the twelve-and-a-half-hour journey to Hull. By 6 December, though, Amis was declaring himself to be 'less of a misery now'. Hilly had agreed to give up her lover for six months, marital relations were to be resumed and there was to be 'increased cordiality' between husband and wife – 'and that is sodding good-oh, believe me, sport.' A bonus to this new harmony was also thrown in. 'As a consequence (though I can quite see how you can't quite see that this can be so) I have got my girl-friend back too.' Amis, with both wife and girlfriend quiescent, pronounced his love-life to be 'quite near an even keel at the moment'.

Amis's idea of an even keel might not have been everybody's. Still, the marriage kept afloat. In August 1958 Amis felt able to tell Larkin that he and Hilly were on 'superlative terms'. By the end of the following year, though, the relationship had plunged back into stormy waters, as he explained to Larkin:

Tackling [talking] of adultery, my activities in this direction, which have ramified some in the last few months, received a severe check when Hilly found some letters I had received and initiated the most strenuous and painful row we have ever had. I am to give all that up, it appears, with an exemption clause covering my Yank girl [acquired during a year spent teaching at Princeton] when she turns up over here in the summer. Trouble is it's so hard to give all that up, habit of years and all that, and such bloody good fun too. Especially just after finding the most splendid busty redhead in an ideal location. But being walked out on by H. (and kids), the sure-fire consequence of any further discovery, is a rather unwelcome prospect too. You can't have it both ways, you see.

But both ways was how Amis did try to have it until, not very long afterwards, he discovered the truth of his own judgement: that having it both ways really was impossible.

Amis joined the army a Communist and left it a committed socialist supporter of the Labour Party. Thereafter, he moved to the right, though in fits and starts. But he was shaping up to be a free-marketeer in matters of art long before he became one in politics. From an early age, Amis believed that the writer should satisfy his audience or go to the wall, just like the businessman.

His attitudes were shaped partly by his own tastes and partly by his origins. Amis genuinely liked the popular work – films, science fiction, jazz – which he had started to enjoy at school and continued to enjoy for the rest of his life, though the forms he took pleasure in were to vary over the years. Science fiction, for instance, fell from favour while television rose into it. Coming, too, from lower-middle-class Norbury, Amis was consistently on his guard against anything that looked like the takeover of culture by a self-styled superior class. The traditional upper class was the obvious enemy here, but it was not the only one. Any group that tried to seize hold of art for itself and defend it against the hordes of plain people was just as offensive to him. The upper class marked itself off from the herd by its money, speech, clubs, schools, and so forth. A literary upper class might achieve the same effect by obscurity, complexity and other bars to common understanding which fenced off their literary property from ordinary readers. This, to Amis, was simply snobbery by other means.

Something of his feelings shows through in a letter Amis wrote in May 1947, in which he told Larkin he was 'no end pleased to hear' that his friend had 'got something together for Lilliput', a popular men's magazine of the period. Amis was beginning to think that Larkin had 'got a bit high-and-mighty' when the *Girl in Winter* [Larkin's second novel] went over so well with the highbrow critics on its publication three months earlier. Larkin, Amis feared, might be going to turn into 'one of those people like Aldous Huxley or D.E. Lawrence [*sic*; a deliberate ambiguity?] who go off into an ivory tower and consider the ordinary reading public beneath their notice.' Amis pronounced himself 'very bucked' to think that Larkin was not going to give himself 'airs like that' and that he'd seen that 'writing for the man in the street is the only decent ambition a man of letters can have these days.' This was not only a question of 'the artistic side of things'. Larkin would never secure his 'bread-and-butter by obscure, sexy, clever stuff'. Amis recommended that he 'get a reviewer's job on a decent paper', and then he might end up as the Frank Swinnerton or Hugh Walpole of the 1960s. 'Think it over,' he urged his friend.

If Amis set his sights a bit low here – he surely aimed higher for Larkin and himself than Walpole or Swinnerton – he was pointing at what was for him the right target. A writer in his view must aim to be read, to avoid snobbery and exclusiveness and, not least, to make a living. Amis changed his mind about how a writer should set about achieving these things over the years. He himself was to spend thirteen years as a university lecturer before he became an independent freelance writer, while Larkin remained a librarian-poet for the rest of his life. But Amis never changed his mind that those were the things a writer should aim to achieve. And he never became too grand to refuse work at the *Lilliput* level, provided he could write what he wanted and get a decent return for his efforts.

Amis's literary tastes were strictly in line with these ideas. He was choosy but he chose from the mainstream and rejected the experimental, the wilfully obscure and the snobbishly exclusive. In December 1952 he rearranged his bookcase in order to assemble what he described to Larkin as his 'vital books' on a single shelf. The list reveals the man. It consisted of

volumes of poetry by W.H. Auden, John Betjeman, Lawrence Durrell, piss John Donne, Andrew Marvell, Alfred, Baron Tennyson, William Wordsworth, W.B. Yeats and piss Robert Graves; works of fiction by John Dickson Carr, Cyril Connolly, Graham Greene, Julian Hall, Christopher Isherwood, James Joyce [Amis liked the early fiction but not *Ulysses* or *Finnegans Wake*], Henry de Montherlant, Flann O'Brien, Anthony Powell; and works of general interest by Cyril Connolly, William Empson (add piss William Empson to the list of poetry), Graham Greene, Arthur Hutchings (*A companion to Mozart's pianoforte concertos*), piss Robert Graves, Q.D. Leavis, W. Somerset Maugham and Stephen Potter (NOT FARCKING GAMESMANSHIT ETC but *The muse in chains*).

Larkin's *The North Ship* would be added 'if I ever get hold of another copy', as would Larkin's novels *Jill* and *A Girl in Winter* 'when I get them back from loan'. Alexander Pope's poetry should be on the shelf too, as should George Herbert's. Writers like T.S. Eliot, Ezra Pound, Dylan Thomas, D.H. Lawrence and all of Bloomsbury were missing as well, but by Amis's deliberate decision.

The choice he had agreed with his supervisor of a subject for his Oxford thesis showed something more of his concern for the proper relationship between artist and audience. His subject was 'English Non-Dramatic Poetry, 1850–1900, and the Victorian Reading Public', and in 156 pages his thesis explored the ways in which poets in that period reached their readers. In 1952, he abbreviated part of the thesis to a dozen pages for publication in *Essays in Criticism*, under the title 'Communication and the Victorian Poet'. (*Essays in Criticism* was then edited by F.W. Bateson, who had supervised Amis's thesis.)

In both essay and thesis, Amis divides his poets into two broad groups. The first is of those anxious to communicate with an audience and usually keen as a first step to get the critical approval of a circle of friends; the second is of poets more or less indifferent to their readers. Dante Gabriel Rossetti is the hero of the first group. Rossetti wrote, 'Above all ideal personalities with which the poet must learn to identify himself, there is one supremely real which is the most imperative of all; namely that of his reader.' And Rossetti was assiduous in trying out his work on his friends.

Gerard Manley Hopkins wanted an audience but failed to find the right inner circle for a sounding-board. George Meredith, in the opposite group, aimed 'only at satisfying [his] own taste', and this may help to explain why his verse is 'unnecessarily obscure' and failed to find many readers. William Morris suffered from being too keen on writing for its own sake – 'Morris sacrificed the arduous and irritating task of communication to the unmixed pleasure of putting words down.' Swinburne, like Morris, was a fluent writer who didn't even pause to let a second person read his poems before sending them straight to the printer. His favourite book of his own poems, *Songs before Sunrise*, was also the only one Swinburne put 'through a crucible of revision' by showing it first to friends, including the Rossettis.

Some poets do not fit the hypothesis, as Amis acknowledged. Christina Rossetti, for example, 'consulted nobody, and solicited no advice', but her poetry is admitted to be none the worse for that. And then there is Tennyson. Nobody could accuse him of not seeking an audience – quite the contrary. But didn't his poetry suffer as a result? In his published essay, Amis issues Tennyson with a muted rebuke, tucked away in a footnote. 'The suggestion can also be made that concern to communicate with a very large audience is bad for a poet's work. The later career of Tennyson is the example here.' In his unpublished thesis Amis went further: 'It seems logical that a poet should write for his readers, but dangerous if he writes for readers who are also his intellectual inferiors.'

These pieces now seem to be just as revealing of Amis himself as of his Victorian poets. It is as if he had picked his way through their lives and works to find evidence that would support his view of how writers should behave. Signs of the literary values Amis was to pursue all his life can be seen peeping through the rather stiff prose and tentative scheme of ideas. In the essay, Amis 'precariously' suggests that 'a poet who is concerned to communicate with an audience is more likely than one who is not to produce work which will survive the passing of its original readers.' And the essay concludes that, if this connection could be established, then 'perhaps the modern practitioners of a chap-fallen Romanticism may give up exhibiting themselves before their readers and at last set about telling them something.'

Here is Amis taking an early crack at modernism and its obscurities. Other pieces of his in *Essays in Criticism* point in the same literary direction. In one, published in 1952, Amis joined in a controversy started by the *Observer* over the breakdown of communication in modern literature. Ivor Brown blamed the writers, Philip Toynbee blamed 'The Laziness of Readers', but admitted cases of wilful obscurity and gave Edith Sitwell's 'Emily-coloured hands' as an example (what she actually wrote in her poem was 'Emily-coloured primulas'). Amis set about dissecting the possible colours that might be associated with the name Emily, then the wide range of colours real primulas could be. But this makes him none the wiser and he concludes that Sitwell 'isn't much interested in communication'. Her readers were simply 'overhearing her personal fantasies'. And so 'for her purpose any "easily-come-by freak" is as good as the product of thought and imagination.' In another piece he accuses the Ulster poet W.R. Rodgers of 'a partial abandonment of the usages of articulate language'. Amis examines Rodgers's use of words like 'niggling' and 'core' and concludes that they 'have no discernible sense-connection with the context'. He thinks Rodgers's poetry is 'deficient in both meaning and sound' and finds that it has more affinities with 'some kind of word-puzzle' than with poetry. 'The attentive reader, who is incapable . . . of surrendering his experience of the language, will react with irritated bafflement.' The final effect, Amis declares, 'is the abandonment of all decent seriousness'.

Something like the outline of a literary programme can be made out from all these, roughly as follows. The writer must avoid obscurity. He should be amusing – not the same as comic. He must not show off or be pretentious. But neither should he write down to his audience or sink beneath his own intellectual level. He should be modest enough to try out his work on friends. Otherwise nothing should come between the writer and his readers – no Arts Council, for instance (though Amis's concern about that came later). For if writers can be said to have any common purpose it is above all to communicate with an audience. Throughout his own writing career, Amis always worked with an imaginary audience in his mind, constantly looking over his shoulder and asking them: does this make sense, is it boring, does it work? These views were shaped entirely by thinking about literature and had nothing to do with politics. But it is not difficult to see how in time they might infiltrate Amis's political views: if an untrammelled market is the best means by which to produce and distribute good poetry, why not other things too?

Politics

Following what became jokily known among Robert Conquest's friends as his First Law, Amis became recognisably reactionary about education before he was reactionary about anything else. According to Conquest's law, people start becoming right-wing or conservative about things they know most about. As a teacher, what Amis knew most about was education. At Swansea he didn't like what he saw, and much of what he didn't like was summed up in the person and views of his college principal, Sir John (later Lord) Fulton. He thought of Fulton as 'an Oxford-Balliol-Lindsay-sociological-philistine man' and blamed him as one of the architects of university expansion which, in Amis's view, far from bringing improvement, very nearly ruined the whole university system.

In 1960 Fulton made a speech in which he proposed that a new university needed a faculty of 350 to 400 professors and lecturers. The ratio of students to staff should be 8:1. Ergo, the university must have 3,000 students. Soon everybody seemed to be saying the same, and Fulton went on to become the first vice-chancellor of one of the new crop of universities, Sussex. Four years after making his speech, Fulton felt able to say that he had been wrong only about one thing. He had thought it would take ten years to create a new university along the lines he prescribed; now he thought it could be done in six. These were just the sort of arguments that Amis detested, since they reduced everything to numbers and left out everything about what a university should actually be doing or what it was really for. Of course you could create more and more institutions which you could call universities and let them churn out more and more young people with what you could call degrees. All you had to do was lower standards. Figures pulled out of the air, like those Fulton used to demonstrate the ideal size for a new university, could be massaged and manipulated to produce whatever result you wanted.

Amis found evidence of how this kind of manipulation could work when he and his friend David Sims decided to boost their incomes by marking some 800 school exam papers in English for the Joint Welsh Board. They retreated to the home of Sims's mother in New Tredegar and settled down to a bout of nine-hour-a-day sessions. Reading through the sixteen-page papers was dispiriting enough, with their 'just-readable handwriting, spelling that was no worse than poor, sentences less than intolerably ugly or incoherent', not to mention their hackneyed opinions. The two did their best to be fair and Amis supposed they did that as well as most examiners. But, however conscientious, the most they could hope to do was rank candidates in order from best to worst. Standards were another matter. These were fixed by the examination board, which could do as it pleased. So if 70 per cent of candidates passed one year and 75 per cent the next it did not mean the candidates had become cleverer. It simply reflected a decision of the board. 'Intelligence, competence stay the same,' Amis wrote in his *Memoirs*; 'all you can do is monkey with percentages, usually for political reasons.'

In the 1960s Amis became notorious for what was commonly thought to be his personal slogan: 'more means worse'. What he actually wrote was 'more will mean worse', for he did not pretend to formulate some immutable Newtonian law of the educational universe, let alone the universe at large. Nor was he reacting to the expansion of the university system, since he wrote the words in 1960, before the expansion had properly got going. He was making a prediction based on his own experience as a lecturer at Swansea and in America. 'University graduates', he wrote, 'are like poems or bottles of hock, and unlike cars or tins of salmon, in that you cannot *decide* to have more good ones. All you can decide to have is more.' (Amis's phrase 'more will mean worse' was to haunt him all his life, getting twisted in this way and that as the years went by. By 1994, a review by John Banville in the *New York Review of Books* had claimed that Amis had 'formulated the neat if simplistic dictum that in life as in art, "more means less"', an utterance that Amis had never come near to making.)

But Amis's scepticism about the way universities achieve their results was already on show much earlier than 1960, in *Lucky Jim*, the typescript of which was finished by the spring of 1953. Jim Dixon shares digs with a lecturer from the English department, Alfred Beesley, who calls his professor Fred Karno but has at least one good thing to say about him. 'Fred's about the only prof. in the place who's resisting all this outside pressure to chuck Firsts around like teaching diplomas and push every bugger who can write his name through the Pass courses.' Dixon says that in his history department 'everyone gets through' because the person in charge of exams is 'a tender-hearted chap'. Beesley continues: 'go to most places and try and get someone turfed out merely because he's too stupid to pass his exams – it'd be easier to sack a prof.' Irritated as he is by the system, Beesley can see all too easily how pressures of one kind and another compel it to work the way it does. Most of the students are paid for by grants from education authorities which don't like to be told their students have failed because that means their money has been wasted. Failure could be prevented by entrance exams 'to keep out the ones who can't read or write'. But then the number of students would be halved, so staff numbers would have to be halved too. And then there was the demand for schoolteachers, for example. If two hundred were required, two hundred must be delivered, and that would be done, even if it meant lowering the pass mark to 20 per cent.

These are arguments in the mouths of fictional lecturers gossiping about their college and need not be taken as the literal truth about what Amis thought. But it was not so far from the truth either. Amis knew of a lecturer in the English department who had given pass marks to only two students on his first time out as an examiner. The prospect of such a poor set of results had caused so much upset that the marks were quietly readjusted to let most of the students pass through to the next year – for, if they hadn't, what would the teaching staff have been left to do? The lecturer had moved on but his legend lingered. When Amis went to arts faculty meetings most of the arguments he heard revolved around standards and how they should be set, and there always seemed to be a voice to argue for the soft option that would at least let students proceed to the next stage, however dismal their performance so far. Nor did Amis need exam results to convince him of low standards. In his very first year at Swansea, he came across students he believed should not have been there because they were 'not capable of benefiting from that kind of education. They had not wanted to be there and did not know what to do when they got there.' He also found a handful of good students that year and throughout his teaching career, their numbers never increasing but their proportion in an expanding student body continuously dwindling.

Amis's opinion of the dubious benefits of growth in higher education was confirmed by the year he spent in America in 1958–9. He taught at Princeton and lectured at a dozen other universities to supplement his stipend, regurgitating his thoughts on 'Problems of a Comic Novelist' at each one in return for a cheque for $300 or thereabouts. Being so much richer than Britain in the 1950s, America could afford to educate far more young people at universities, and did. But Amis found that, while quality was high at Princeton or Harvard, it fell away drastically at lesser campuses, the University of Illinois at Champaign-Urbana being the worst example he came across.

The Amis view of universities is undeniably lofty. He sees them as the guardians of the nation's culture. They should set the highest possible standards and be wholly autonomous, deciding for themselves what research and teaching to do. If, for instance, the philosophy department decided to take no new undergraduates for ten years while it rethought its programme then it should be allowed to do so, on the grounds that nobody could know better than the faculty. Science would be taught and studied but only at the highest level, while practical skills would be banished to technical colleges and the like. By these standards most of today's universities will inevitably seem third-rate. Right or wrong, however, Amis has been consistent over many years in his belief that the expansion of higher education has brought no increase in the numbers actually benefiting from it. In the 1990s he would add one further piece of bad news to his reading of the educational scene. If he had to choose between two graduates with identical qualifications, one of whom had acquired a degree in 1973 and the other in 1993, he would automatically choose the first of the two. The older graduate would have read more than the younger, who could be assumed to have spent more time with television and records and so would inevitably be the more ignorant of the two.

Amis once addressed Kenneth Baker, former Education Secretary in Mrs Thatcher's government, on the subject of higher education while both were descending the steps of the Garrick Club. Amis had three points to make. The government had not tried to use higher education for purposes of social engineering: good. Its expansion of vocational training was good too. But as for university education, he said, 'there your government has been at fault.' Baker's response was not quite what Amis had been hoping for. He shouted, 'Taxi!', dived into a black cab and disappeared. Mrs Thatcher's treatment of the universities was almost the only question of policy on which Amis dissented from her, and he thought Oxford University did exactly right by withholding from her the honorary degree normally given to any of its graduates who becomes prime minister.

It was Amis's views on education that were the first to earn him a wide reputation for political unsoundness (i.e. for not being unambiguously left-wing on every issue). During his time in Swansea, however, Amis continued his drift from the left across all fronts. He still voted Labour but with steadily diminishing enthusiasm. Beneath the Angry Young Man tag, which seemed such a hallmark of solid left-wing authenticity in the 1950s, he was changing.

Hope that there might, after all, be something worth salvaging in the Soviet variety of Communism sputtered briefly after the death of Stalin in 1953. Amis remembers a broadcast – one of the most moving he ever heard – in which a Soviet spokesman admitted that 'mistakes had been made'. But such hope as that roused was finally extinguished with the Soviet suppression of the Hungarian uprising in 1956.

The uprising coincided with another disaster, the Anglo-French-Israeli invasion of Suez, over which Amis took to the streets, though never much given to marching and banner-waving. The demonstration was organised by students (among them Paul Wilkinson, later professor of international relations at the University of St Andrews and director of the Research Institute for the Study of Conflict and Terrorism at St Andrews and in London). Most of the college staff were too cautious or hostile to join in. Grateful for Amis's support, the students invited him to the head of their march, hoping too that his presence there might act as a deterrent to the engineering students who, famously right-wing, were expected to cause trouble. As things developed it was not violence from those students Amis had to face but violence from an old lady, who abused him, tore the placard from his chest – it read, demurely, 'Eden – Abide By U.N. Decision' – and chased him down a side-street. Being faster on his feet, Amis escaped her wrath and managed to rejoin the column unharmed. He thought that Suez was not only wrong but a dreadful error because it made the West look as bad as the East. But it was the Soviet suppression of those supposed to be its allies that made the deepest impression. After 1956 he never shifted from a firm and total anti-Communism.

Amis made this perfectly clear in a letter to the Communist newspaper, the *Daily Worker*, in February 1957, after the newspaper's Arnold Kettle had seemed to be trying to enlist him as a sympathiser at heart, if not a party member. 'I have had Marxism – in both senses of the word,' Amis wrote. 'I have experienced the ailment and so am immune. And I have also utterly rejected it. No world-view, it seems to me, comes within light-years of being adequate to the world it professes to categorise. Each fact, each entity, each event is unique. To pretend otherwise is mere Victorian system-building. Marxism, I think, does just that. It repels me also by offering certainty instead of truth.' After scorning Kettle's claim that Communism is not violent, Amis concludes: 'I used to be able to say that some of my best friends are Communists. I can't any more.'

The piece that provoked this letter was a notice by Kettle of a pamphlet Amis wrote for the Fabian Society, *Socialism and the Intellectuals* (1957), in which he had speculated why it was that intellectuals like himself were politically apathetic. He distinguished between romantic and unromantic approaches to politics and concluded that the romantic type sought out causes while the unromantic were likely to be more pragmatic. A poet, say, would opt for Marxism, while a schoolmaster would stick with social democracy. But close reading of 1930s writers like Auden, Spender and Day Lewis had persuaded Amis that their politics went wide rather than deep. Their protest was personal as much as social; they were 'conducting in public a personal vendetta against their parents . . . and one or two unsympathetic headmasters'. Indeed, Amis was inclined to think that all middle-class Britain, not just the intellectuals, were pulled one way or another in politics by temperament alone. The family, for example, could 'seem a warm environment' or 'something to be fled from and rebelled against'. Here, surely, Amis had in mind his own phase of rebellion against his parents as much as Auden's or anybody else's.

Romanticism in politics Amis defined as 'an irrational capacity to become inflamed by interests and causes that are not one's own, that are outside oneself', and he said he was this kind of romantic himself. The trouble was that there were few causes on offer to the 'cruising rebel' – no Spain or Fascism or unemployment left to rekindle the zeal of the thirties. Cyprus, then engaged in its campaign for independence, Amis thought was the only issue around comparable to Spain, but nobody could be bothered even to go to protest meetings about it. All that was left to attract the contemporary romantic were non-political issues, 'or ones that are not in the first place political: the colour bar, horror-comics, juvenile delinquency, the abolition of capital punishment, the reform of the laws relating to divorce and homosexuality'. These, he thought, should be on Labour's programme, though he could see the difficulty of rousing the support of dockers, for instance, behind the homosexual cause.

Amis concluded that the apathy of the intellectuals was probably no bad thing, since the intellectual was as likely to be wrong as the next man, and probably more so. The reason for this was that the intellectual had no personal axe to grind. 'I think the best and most trustworthy political motive is self-interest. I share a widespread suspicion of the professional espouser of causes, the do-gooder, the archetypal social worker who knows better than I do what is good for me.' In his own secure job he really had nothing to complain about, so the only reason he could have for getting politically active would be 'a sense of guilt'. But this was not enough. How agreeable, he thinks, to have 'a respectable motive for being politically active'. Amis was writing only three weeks after the twin Suez–Hungary débâcles, but even these he thought unlikely to rouse the intelligentsia from its slumbers.

At this stage Amis still saw himself as firmly, if inactively, rooted in the left. In his 1957 Fabian pamphlet he says that 'any right-wing sentiment in the mouth of an intellectual (or anywhere else) is likely to annoy me' and he predicts that 'unless something very unexpected happens' he will vote Labour for the rest of his days. In an interview with *Life* magazine a year later he set out a political programme so extreme that it would have alarmed even Labour's own extremists at that time. 'I'd start by nationalizing everything in sight – we could use a great hurting swinge of socialism,' he said. 'I would also abolish the public schools, the aristocracy and, naturally, the House of Lords. As for the royal family, it serves a purpose as a sentimental glue for the Commonwealth and probably has to be preserved, but just as a personal feeling of my own I would like to get rid of them, queen and all.' This provoked an outburst of indignation from the *Daily Telegraph*'s Peter Simple at everything Amis represented – 'utter imbecility, sterility and vacuity . . . boundless ignorance and impertinence . . . a wasteland of cheap slogans and catcalls'. Perhaps the newspaper would have taken a more charitable line if it had known that Amis's interview had once again taken the form of a chat in a pub, this time with a *Life* reporter who had the embarrassing gift of total recall.

Soberly identifying self-interest as the only respectable motive in politics was, however, hardly evidence of socialist dedication to match this sort of talk, and Amis had had the candour to acknowledge it in his pamphlet, if only inside a pair of brackets: '(The only edge the Tories have over the socialists from my point of view is that they at least are not out to do anybody any good except themselves.)'. Clearly there was a contradiction here between ideas and feelings waiting to be resolved one way or the other.

Amis's pamphlet nettled some of the others who had been tagged with the same 'Angry' label as himself. Later in 1957 a collection of essays appeared in a book entitled *Declaration*. The contributors were Doris Lessing, Colin Wilson, John Osborne, John Wain, Kenneth Tynan, Bill Hopkins, Lindsay Anderson and Stuart Holroyd – as representative a group of supposed Angries as could be assembled – and each was invited to state his or her position towards society, in effect their political credos. Two took the opportunity to swipe at Amis's pamphlet. Doris Lessing denied his claim that self-interest was the only authentic political motive: 'everywhere in the world people with nothing to gain from being socialists (nothing to gain in the sense that Mr Amis uses) have become, are becoming, and will become socialists.' Amis, she concluded, was 'generalizing from an emotion which is current among a section of his generation now. It is a temporary mood of disillusion.' Lindsay Anderson was even more dismissive: 'Amis reveals himself as a coward, too scared to take up any stand at all . . . Socialism as a positive ideal, involving definable human values, apparently means nothing to him . . . One can only wonder why he continues to vote Left: through a lingering, irrational, shame-faced humanism, I suppose.' But Anderson did concede that 'Amis's human impulses are genuine; even if he cannot really bring himself to trust them.'

The voice most obviously missing from *Declaration* was Amis's own. The book's editor, Tom Maschler, invited him to contribute but Amis declined, writing to a somewhat surprised Maschler, 'I hate all this pharasaical twittering about the "state of our civilization" and I suspect anyone who wants to buttonhole me about "my role in society". This book is likely to prove a valuable addition to the cult of the Solemn Young Man: I predict a great success for it.'

Having said his piece in the pamphlet, Amis was not inclined to say it all over again. For one thing, he genuinely disliked theorising, whether about politics or about art. And then there was, after all, that apathy which to Amis was not only a political posture among the intelligentsia but a personal one too. The only issue besides Suez on which he felt strongly enough to demonstrate again was hanging – he turned out to picket Gwilym Lloyd George, the Home Secretary, when he visited Swansea. But by the late 1950s his apathy towards political activity was pretty much complete.

It was only during his brief romantic fling with Communism in his first terms at Oxford that Amis was ever enthusiastic enough to be counted an activist. When he turned up at a Labour Party meeting over Suez it was fifteen years since he had been to any meeting like it. He arrived so fired with indignation that he determined to join the party on the spot. Within fifteen minutes, however, 'I had silently released myself from that vow.' He had forgotten how boring such gatherings could be. Thereafter he reverted to his usual minimalist approach to political action – putting a poster in his window, casting his vote, lending his car at election time, demonstrating 'once in ten years' and 'chatting about politics' with friends.

In the 1950s it was fashionable for writers to be committed to the left and to stand up and be counted on its behalf. Amis did his bit to show where he stood in his Fabian pamphlet and that was enough for him. The fame or notoriety – or both – which he gained from that and from the counterblast to it he wrote a decade later for the Conservatives, *Lucky Jim's Politics*, were out of proportion to Amis's real interest in the political world. Throughout his life he has been far more concerned with, and given far more of his time and attention to, literary than political matters. But public reputations are not made by issuing a pamphlet arguing that, say, Housman and Graves were better poets than Eliot and Pound, as it is not difficult to imagine Amis doing. Reputations are made by having strong opinions on questions of the day and expressing them with pugnacity. And they become all the more controversial if, like Amis, you switch sides and can be handily caricatured as Angry Lefty turned Fascist Beast. Amis has been both a convinced left-winger and a convinced right-winger. Yet nothing about politics has mattered to him nearly as much as drink, women or literature.

*　　*　　*

Amis's early fictional characters closely reflect this balance of his own interests. Jim Dixon is certainly on the left, but *Lucky Jim* has to be read with care if the evidence of his politics is not to be missed. He gives an elementary, one-sentence lecture on moderate socialism – 'If one man's got ten buns and another's got two, and a bun has to be given up by one of them, then surely you take it from the man with ten buns' – and a few paragraphs later he predicts that there won't be rich people around much longer. But that is about all he has to say on matters of politics.

John Lewis in Amis's second novel, *That Uncertain Feeling*, is more of a class-warrior. His wife Jean calls him 'Karl Marx bloody Lewis'. He regards himself as a sworn enemy of the bourgeoisie, is glad that it doesn't take three generations to make a gentleman any more and thinks the new privileged classes have a lot to be said for them compared with the old. Lewis launches a diatribe on the Tories' treatment of the National Health Service but is interrupted in mid-sentence. Towards the end of the story he even finds himself becoming reconciled to the bourgeoisie on the ground that the members of it he has come across are 'amiable and light-hearted people' and there aren't so many of those around that he can afford to turn 'sour or sociological' on them. Lewis wears his ideological prejudices as a man might wear a tie to the office, not because he feels particularly comfortable in it but because it is expected of people like him, and anyway he's got used to dressing like that. When an attractive woman comes into the library where he works and turns out to be the wife of a local bigshot as well, Lewis has to remind himself that it is his 'political duty to seem unimpressed'.

Leftish politics in these novels are part of the background, the standard baggage that a young, poorly paid junior lecturer or librarian with origins as modest as his job might be expected to carry. It would have been wildly out of character for Dixon or Lewis to be militant Tories. In any case, politics do not take up much of their time or thought. What does is avoiding boredom, keeping jobs, propitiating the boss, paying bills, making the most of what pleasures are to be had and – most important and difficult of all – getting on somehow with the women in their lives. These were Amis's priorities too.

His short story 'Moral Fibre', published in 1958, is perhaps a better guide than the early novels to the way his mind was moving on political matters, though the observations in it are social rather than directly political. John Lewis is again the narrator (leading some to believe that the story is a stray chapter from *That Uncertain Feeling*, but it is quite separate, based on the incident of the generous whore already mentioned, John and Jean Lewis merely being re-enlisted as pegs on which to hang her tale, which had so intrigued Amis). Mair Webster, a social worker and friend of Jean Lewis, brings Betty, aged nineteen, to work as a domestic help in the Lewis household. She has two children, neither of them by her newly acquired husband, Bent Arnulfsen. Betty soon disappears from her job with the Lewises, though she shows up again a couple of months later when Jean is out. She offers herself to John, mentioning that she has 'got in with the business girls', by which she most definitely does not mean secretaries or typists. The story ends with Betty in jail, having been caught helping to burgle a café, and John regretting that he cannot be at the gate with a bunch of flowers on the day she is released.

For John Lewis's sympathies are all on Betty's side and against the social-working Mair Webster. Seeing Betty again, he reflects that 'being a prostitute was something you could be done a power of good by'. Seeing Mair prompts the contrary reflection: 'I put to myself the question whether the removal of all social workers, preferably by execution squad, wouldn't do everyone a power of good.' Mair thinks of herself and her colleagues as 'technicians in paternalism', while Lewis, watching her in action, thinks of her as more like a 'veteran colonial administrator' keeping the natives in order and patronising them too.

Lewis recognises that people like Betty are a problem but he can't see people like Mair as the solution. He concedes to himself that at least Mair, unlike him, is trying, but when he thinks harder about what she's trying to achieve 'the picture changed a bit' – 'just as things like the Labour Party looked better from some way away than close to', so did social work. That makes him think about all the other people you had to have besides such people as Mair: 'prison warders, local government officials, policemen, military policemen, nurses, parsons, scientists, mental-hospital attendants, politicians and – for the time being anyway, God forgive us all – hangmen'. You had to have such people but that didn't mean you should feel friendly towards any of them – 'bar the odd nurse'. The story ends with this question unresolved. It was unresolved in Amis's mind too. 'Moral Fibre' shows his mood moving against the constraints and restrictions of the welfare state and its meddling bureaucracies. But he had not yet made the connection to politics and shifted his support to the free market and the Tories. His ideas had not caught up with his feelings.

II

Portugal

Fame brought Amis invitations to travel, not always welcome. In his Swansea years he accepted two. One took him on an enforced holiday to Portugal, the other to teach at Princeton. He was tempted to stay on in America, though not in Portugal. A few years later Amis was also tempted to settle in Majorca and live the writer's life near Robert Graves, whom he much admired. But he resisted both temptations, settled for staying at home in Britain and never seriously regretted it. Amis had grave doubts about 'abroad' from the start – as did Garnet Bowen, the hero of his third novel *I Like It Here* (1958), which was closely based on Amis's Portuguese trip.

Bowen, a literary journalist, is more or less forced into going to Portugal by an irresistibly large commission for a travel article from an American magazine, made even more worthwhile by a second commission from a British publisher. In Amis's case, compulsion took the form of a literary prize which the winner could accept only on condition that he spent his prize-money abroad. Yet the mere idea of leaving Britain makes both uneasy. Bowen recognises that he suffers 'acute prejudice about abroad' and goes over in his mind some of the things he doesn't like about travelling: 'disinclination for change, dislike of fixing up complicated arrangements, and fear of making a fool of himself'.

What Bowen gloomily thinks of as his Deportation Order arrives in the post one clear bright morning early in April. Amis's came in a telegram from his publisher in late March 1955. 'BRAVO LUCKY JIM HAS WON SOMERSET MAUGHAM AWARD ALL OUR DELIGHTED CONGRATULATIONS . . . HILARY.' As recognition for his first novel this news was naturally welcome, the money too. The snag came in the condition described by Hilary Rubinstein in a letter he sent on the heels of his telegram. 'The Society of Authors man is writing to you today to remind you of the terms of the award – namely, that you agree to spend the sum of £400 on not less than three months travel or residence abroad.'

Abroad was bad enough, but being more or less forced into exile made it worse. Bowen protests to a publisher friend, Bennie Hyman (who lightly resembles Rubinstein): 'It's like being deprived of your citizenship.' When Hyman says he wishes he could get away abroad for a bit, Bowen replies, 'Ah, but I don't. I like it here, you see.' Amis's own reaction was similarly downbeat. A few days after he heard about the award he wrote to Larkin to complain: 'And now this deportation order from the Somerset Maugham Trust; forced to go abroad, bloody *forced* mun.' If it wasn't that Hilly would enjoy it and that it would look 'too eccentric' to turn the award down, he would have been tempted to do exactly that.

In May he wrote to Robert Conquest with news of his plans: 'We think we're going to Portugal after all now. What a fucking mess it all is. See England first, I say.' The day before he departed from Southampton on 23 June, he wrote in the same mood to Anthony Powell, '"Making the arrangements" has been very protracted and horrible. Well, see England first is what I always say.' 'When I write my travel-book, Book 1 (pp. 1–185) will be called GETTING FIXED UP,' Amis told Larkin. 'You really have no *idea* what the bloody business entails, or perhaps you can imagine.' There were passports to be organised, tickets, money, typhoid injections. But, he grudgingly conceded, it should 'be all right; paying guests with an English business type about a dozen miles from Lisbon and near the sea.'

The three months' enforced exile prescribed by the award could just about be fitted into Amis's summer vacation from the university. The bigger problem was deciding where to go. Not knowing much about abroad except what he had learned as a soldier and on one post-war trip to France with Hilly and his Oxford friend Christopher Tosswill, Amis consulted John Aeron-Thomas, a wealthy Swansea industrialist, owner of the *Newport Argus* and married to Margaret, Amis's social-working friend. John Aeron-Thomas put him on to a Portuguese family he knew, the Pinto Bastos, who promised to see that the Amises were looked after.

What this turned out to mean was that a Pinto Basto employee called Barley rented the Amises' part of his own home near Estoril. Barley, who by no very great leap of Amis's imagination becomes Oates in *I Like It Here*, had an English father but had grown up in Portugal and married a Portuguese woman. Soon after he arrived in Portugal Amis described him to Larkin: 'Billy Barley, alias Joe, has turned out to be very amiable in a childish way . . . He is pure English by birth, Portuguese by nationality, bilingual, and feels himself of no certain country, I mean he doesn't quite know which country he belongs to.' Barley drank coffee from a Coronation mug, had a picture of the Queen on the wall and a tea-caddy embellished with the Duke of Edinburgh's face. He was also a 'motor-bike maniac', and Amis imagined a scene in the Portuguese story he intended to write whose hero, Lucky Jim Dixon once more, would 'ride on the back of a motor-bike after dark, in a high wind, part of the way in very fast traffic through Estoril, with a Portuguese in front *practising his English*, turning his head to say "Would you like to take a cerp of cerffee later? Do you like Portuguese cerfee or is it too strerng?"' Garnet Bowen was offered such a ride but managed to avoid it.

The Barleys seemed pleasant enough to begin with. The food which they all ate together might look a bit different from the English dishes on which they were modelled but turned out not be so different after all – and, where different, 'not noticeably inferior or emetic', Amis reported to Larkin. Barley seemed generous with alcohol too. While claiming that he hardly touched the stuff, he poured out large drinks before, during and after meals for Amis and himself. 'All on the house, too,' Amis noted with satisfaction.

There were problems as well, though. One was the size of the house. As described by Bowen, the minuteness of the Oateses' establishment was its most striking feature. Its ground – and only – floor was about the size of half a badminton court and into it were crammed a lot of rooms: a dining-room, three bedrooms, a kitchen, a hall which doubled as maid's bedroom and a bathroom-cum-lavatory. The Amis family, like the Bowens, comprised two boys and a girl. In both families mother and father shared one bedroom with their daughter and the boys shared another, while their hosts' bedroom was converted during the day into what had been promised as their private sitting-room.

The lack of space might have been tolerable if only the household had been efficiently run. Or, as Amis complained to Larkin, 'if the place had been clean, or if the lavatory hadn't smelt, or if there hadn't been so many flies everywhere, or if, above all, the meals had been ready on time and had proved to consist when ready of food that Sally could eat.' Sally took to yelling for an hour before meals at the prospect of food which she didn't like or which made her ill, and nothing Hilly could say would persuade Mrs Barley – 'an amiable half-wit' – to change the menu.

Amis's tolerance frayed. Soon after he arrived he was able to tell Conquest: 'Yes, Portugal is where I am. Quite all right "considering it's abroad", as I always say. Wonderful hot days. Food a bit oily but v. palatable. Drink abundant. Women so fine that I have cultivated an efficient protective armour of nursing the baby, playing with the boys etc. and now hardly notice all the fine skins, imposing gaits, enormous busts etc. I bathe a lot, do crosswords, read Bloomfield's *Language*, and so on.' Ten days later he sent Conquest what he called 'a short verse on Portuguese culture', to be sung to the tune of 'She Was Poor but She Was Honest', which he later put into the mouth of Garnet Bowen in slightly amended form: 'See him gulping *vinho verde*, Scoffing filthy sheep's-milk cheese, Puffing fags of scent and mule-shit, While he searches for his fleas, But he tells them *Obrigado*, Matches all their courtly grace, Cause he's got his homeward ticket, Locked up safely in his case.' And he added, 'Feeling slightly sick – have just lit a Tip-Top (tobacco aromatico) esc. 3$30 for 20. Not like Players they aren't. Still nor's the price.'

But by the end of the month Amis had had enough. 'We've had a sort of muffled row with the chap who's putting us up here,' he wrote to Conquest on 31 July: 'he keeps getting more and more money out of us and giving us less and less in return. So we shall be getting out as soon as possible, in a week's time I hope, and going south to the Prince Henry country' – by which Amis meant: to the Algarve.

The 'muffled row' over money arose when Amis worked out that he could live both more cheaply and better in a *pensão*, or boarding house. He discovered this when he ran into a Swansea couple, the Browns, and visited them at the Pensão Continental in Estoril where they were staying. It would cost £2 10s. a day for the five Amises to live there, against the sum of 'about £3–7s' he was paying the Barleys. There would be '*slightly* better' food, there was 'a lounge with easy chairs' and 'the bathroom smells only of soap', Amis reported to Larkin. The Browns put Amis in the way of an even better deal when they introduced him to an Ulster couple, the Tyrrells, who had a 'mountain chalet' at Monchique in the Algarve. They offered to let the Amises have the chalet free, once a sister of Tyrrell's had moved out. Until then, the Amises could live in a *pensão* nearby.

Amis by now was anxious to move with all possible speed, so offered a fortnight's rent as compensation for shortening his stay with the Barleys. Billy Barley studied his accounts and declared that he would make a loss unless Amis paid him for a full month. So Amis did, in order to keep the peace until he could make final arrangements to move. Not content with this, Barley managed to chisel an extra £2 15s. out of Amis by getting him confused over the rate of exchange between the pound and escudo. It turned out that Barley had borrowed money from his firm in order to put in hand improvements that would make his house fit for the Amises to live in – including a hot-water geyser and a garage – and their defection meant he would not be able to pay his boss back. Amis finished up promising to send Barley yet another £40 when he got back to England. 'Oh the mean little bird-faced ignorant naturalised greaseball,' he expostulated to Larkin.

The Tyrrells, Amis told Larkin, 'are well worthy of fictional transcription', and they duly made their appearance in *I Like It Here* as the Bannions, Harry an Ulsterman (like Tyrrell) retired from an Indian bank, and his wife, Isabella, a Goanese (as was Tyrrell's wife). Tyrrell was notably, even embarrassingly, eccentric, given to 'periodical floor-shows' at the Pensão Continental in which he would address an audience of Frenchwomen in schoolboy patois ('La plume de ma tante est dans la poche du jardinier') or, at an Estoril café, recite 'The Charge of the Light Brigade' in four different voices, both of which performances Harry Bannion was to repeat in fictional form.

The Bannions 'extricated [the Bowens] from Oates's establishment and installed them rent-free in a chalet they owned in the mountains that overlooked the coastal plain', as Monchique overlooks the Algarve coast. The Bannions have an adopted son, a young priest, whom Bowen calls Afilhado, which is the Portuguese for affiliated, or adopted. The Tyrrells too had an adopted son whom they also called Afilhado. He, too, was a priest (and eventually rose through the hierarchy to the dizzy heights of the bishopric of Oporto).

The Amises had 'a tremendous drive down from Estoril, with an above-par dose of doubt, horror and despair at the Tagus ferry', Amis informed Larkin. The embarkation point had been changed, the officials did not speak English and did not anyway all know about the change, there was confusion over where to pay and how much, and Amis was down to his last pound. They stayed for three days at the Pensão Sol in Praia da Rocha – 'modest, quiet, clean, cheap, *excellent* food, polite staff, English-speaking proprietor; the lot', Amis wrote to Conquest. Then the family moved inland and uphill to Monchique. Amis told Conquest:

A lot of the Algarve *is* a ruddy desert, as you say: the part where we are (about 15 miles from Portimão, equidistant from Lagos and Silves) is up a mountain – getting on for 2000', I believe, but sodding hot – and greener than the remoter neighbourhood. While here I hope to draw as few sober breaths as possible, wine being 7s. a gallon, and not at all bad either, and there being 3 bottles of a kind of local applejack, made from a kind of arboreal strawberry [*medronha*], available by the courtesy of this C.H. Tyrrell character and absolutely FREE. There's not much to do here except drink, eat, sleep and beat down the stares of the natives, but I reckon I can get by on that for another four weeks.

Amis described a typical day at Monchique to Larkin. He had breakfast, then pumped the pump which supplied all their water. After that he wrote letters or worked on his novel. At noon he smoked and drank 'a couple of glasses of wine', ate a salad lunch and drank some *medronha*. In the afternoon he would rest or go to the beach at Praia, where a girl of about fourteen with 'a very deep-pitched laugh and the darkest fair hair I have ever seen' seemed to have taken a fancy to him. 'I think she knows what I'm thinking. So does Hilly.' In the evening Amis lit pressure lamps, ate dinner and drank a good deal before retiring to 'our hard but flealess couches'.

He described Praia at some length for the benefit of Conquest's mother, who was considering a holiday in the area: 'The beach is most impressive. The landscape otherwise is pretty arid now, but very attractive, I hear, in the autumn and winter. It's pretty quiet all round. No night life to speak of, though there are cafés. A small resident English population in Portimão and thereabouts . . . Hilly wants me to add (1) that she personally would go mad from boredom if left here on her own, and (2) that the town is an ugly little town. It is. They catch sardines, and then they tin sardines, and then they transport sardines . . .'

By the beginning of September the tedium predicted by Hilly as likely to consume someone on their own was beginning to take Amis over in spite of his not being alone. 'Ennui lies upon me like a hangover,' he told Conquest, 'or perhaps the other way round. It's the getting pissed every night so as to get to sleep on the hard bed which gets me down, if you get what I'm getting at. Bloody terrible, man. No more wine for me after I come back. I shall stick to splendid virile Nordic gin and healthy bracing tonic. There's something subtly vinous about the Lisbon gin I must have mentioned to you. I can't remember when I last voided a firm motion, if you'll pardon the expression.'

On the same day, 7 September, he wrote in more sombre mood to Victor Gollancz. 'The most important fact about Portugal as of now is, from my point of view, that I shall be leaving it soon and coming home. I thought I knew all the main categories of boredom one was ever likely to encounter, but I made the mistake of thinking it was only people who could be boring. Places can be as efficient in this way, if not more so. I am getting through the last few weeks by drinking a lot of local gin (11/– the bottle) and a kind of applejack-cum-Pernod that they go in for a lot hereabouts (and no wonder), also by having the DDT-spray within reach.'

The day of 7 September seems to have been an occasion on which Amis felt particularly low about the Portuguese version of abroad and particularly keen to complain about it. For he also sent a postcard that day to Anthony Powell, showing a massive, phallic-looking rock formation on the sea-shore. 'This is pretty impressive, no?' he wrote. 'And when you get tired of looking at it, you can always turn round and look at a hill. And when you get tired of that there's plenty of sport to be had going around the house with the DDT spray. And then you get your homeward vouchers out and read them attentively. Still, I've got nice and brown and that's something.'

But things were not all going badly. Amis's second novel, *That Uncertain Feeling*, was published while he was in Portugal and he had been anticipating its reception with some anxiety. 'As regards *That Uncertain Feeling*,' he wrote to Conquest towards the end of August, 'no doubt the blasts of critical hate are already beginning to poop off.' But he added, 'your mention of John Davenport's approval cheered me a lot, I can't think why.' (The reason for Amis's doubt about Davenport's approval was a recent letter in the *Spectator*: see shortly.) A few days later Hilary Rubinstein sent him copies of the book's early reviews and good news about sales. 'I am so immensely relieved that my neurotic forebodings about the book have proved to be unjustified,' Amis responded on 2 September. 'I really think the reviewers have been most generous, and the fact that about four of them have stigmatised the book in passing as "vulgar" has been almost as encouraging as their praise. I was afraid some of them might have praised it for being "sensitive". It seems I needn't have worried.'

In his letter to Gollancz five days afterwards, Amis repeated his gratitude for the way the book seemed to be going and his sense of relief: 'I'm mightily pleased that *That Uncertain Feeling* seems to be selling satisfactorily. I really was beginning to have kittens about it, in fact had been having mild ones for some time, and this was probably why the MS turned up on your desk some time after deadline. Every time I went through it there seemed to be more wrong with it. Apparently I needn't have worried over-much. A vast relief.'

Amis's gratitude that four of his reviewers should have found his new novel 'vulgar' rather than 'sensitive' may seem a bit odd. Surely a literary gent – even a straightforward literary chap – would find the second adjective preferable to the first when used about his own work? But even before he left for Portugal Amis was in a sensitive mood about 'sensitive' writers – especially writers who dedicated their sensitivity to that equivocal place, abroad.

In May 1955 Amis had shown what he thought about them when he pitched into the whole travel-writing genre in a *Spectator* review. Under the title 'Is the Travel-Book Dead?' he excoriated Laurie Lee's new book about Spain, *A Rose for Winter*, along with another book, by Peter Mayne, about the Pathan tribes of the Pakistan–Afghan border, *The Narrow Smile*. What Amis objected to in travel-writing were two things he found in unequal measure in both these books: their escapism and their style. It might be that the writers of books like these could justify their travels with such lofty-sounding motives as 'to get away from the exhausted sterilities of Western civilisation', though Amis was inclined to doubt it – all the more so when he found that what this usually boiled down to was nothing very different from the feeling that 'the other fellow's grass is greener, that the really good time, or good life, is going on somewhere else.' As for style, travel books leaned 'towards the more elaborate and unfashionable graces of prose . . . [but these] degenerate briskly into an empty and indecent poeticism, apparently based on a desire to get into the next edition of *The Oxford Book of English Prose*'.

Amis judged Laurie Lee guilty on both counts. On the stylistic charge he concluded: 'One way of summing up this book would be to call it a string of failed poems – failed not very good poems too.' But Amis found Lee even more at fault with the content of his book, with the way in which he portrayed the Spaniards he wrote about, 'so far as this can be debarnacled from rhetoric, generalisation and rhapsody'. According to Amis, where Lee 'seeks to show us gaiety, mere instability or hooliganism emerges, unselfconsciousness is detectable as coxcombry or self-pity, while the gift attributed to Andalusians of greeting others' misfortune with a shrug or a grin is neither mature nor admirable.' Garnet Bowen takes up the same theme of double standards. According to him, people when abroad behave as if they 'had invented the country instead of just living in it'. So they feel entitled to see things in ways they would never see them at home: the staring which he found the Portuguese went in for a lot was an example. If it happened in England, he surmised, it would be described as 'the mindless inquisitiveness of those whose greyly uniform lives were nourished on mere sensation'. But 'in the sunny South [it would be] a frank, free, healthy, open, uninhibited curiosity.'

In his review Amis turned with relief from 'Mr Lee's vulgar and sensational little book' to Mr Mayne's. Laurie Lee was not pleased about any of this, as was still evident nearly forty years later on the rare occasions when both he and Amis showed up in the Garrick Club bar at the same time. Lee would take care to ensure that Amis saw him, so that Amis would know that Lee was giving him the cold shoulder by appearing not to notice him in return.

Amis was kinder in his review to Peter Mayne, finding him 'agreeably unliterary' and not the kind of travel-writer 'who feels obliged to demonstrate his sensitivity at every turn'. But Mayne was still guilty of the escapist charge, of trying to demonstrate that Europeans could find some kind of answer to their spiritual problems in the lives of Pathan tribesmen. 'I for one cannot share Mr Mayne's thesis that violence tends to become morally neutral when pursued openly, in picturesque tribal costume, under conditions of physical discomfort, and against a background of boulders and precipices ... the Pathan avenger with his rifle and knife is as contemptible, however much more impressive looking, as our own species of gangster with his razor or bicycle chain.'

Whatever the separate defects of these two books in their style and their dubious morality, they seemed to Amis to share one further defect between them, a defect inherent in their form, the travelogue. This form depends on who the writer meets or what happens to happen to him on his journey. As a result, 'the encounters [Mayne] describes are for the most part indecisive, and the factual thread on which he strings them is inadequate to bear their weight; there is no principle of selection or emphasis.' Amis proposed an alternative. Writers like Lee and Mayne should choose another literary form in which to work out the kind of issues that interested them. They should write novels. And he mentioned E.M. Forster's *A Passage to India* as a possible model.

In the next week's *Spectator* John Davenport – the critic who was soon to give Amis's *That Uncertain Feeling* his approval – wrote a letter in defence of Lee and the travel book. 'Spain is hard and unsentimental. Mr Lee does not sentimentalise it. He respects the rock-like integrity individual Spaniards have retained.' Davenport was also unconvinced by the idea that the travel book should be replaced by the novel. Forster's *The Hill of Devi* might not be a travel book exactly but it wasn't a novel either, yet it was surely 'more successful than *A Passage to India* as a revelation of Indian character'. And there was A.J. Ackerley's *Hindoo Holiday* – could a novelist improve on that? Davenport concluded with a chirpy invitation to Amis: 'Is it an impertinence to hope that the talented winner of the Somerset Maugham Award may gain some tolerance from his coming experience of furrin parts? Who knows, perhaps he will bring back a travel-book with him?' Which was exactly what Amis had started to think of doing even before he left Swansea.

The following week Conquest came to his defence in the *Spectator*, admitting that Amis's proposal that travel books should become novels was rather eccentric but insisting that it was at least provocative and worth arguing about, and insisting even more strongly that, whatever might be thought about travel-books-as-novels, it would be a pity if that controversy distracted 'attention from the more important task of giving hell to "poetic" prose'. Amis got a letter of his own in the week after that – only the slowness of the posts between Portugal and home prevented him stepping in quicker – in which he defended his review and assured Davenport that 'I'm not being grumpy about furrin parts, only about people being silly about them, and that the furrin part I'm in now seems good-oh so far.'

Amis continued the argument privately in his correspondence with Conquest: 'What I think I feel about the business', he wrote to Conquest on the day his *Spectator* letter appeared, 'is first I'm all in favour of escapism – but people should realise that that is their motive (very often) for buzzing off down south, and not try to inflate their pleasure-etc. trips into a spiritual pilgrimage AND THEN COME AND TELL ME ABOUT IT in a travel-book; and second, the weather *is* fine and the girls *may well* fuck more (no personal experience), and that's all very fine and chaps are sensible to go after it, but again you get this frightful mystique-mongering which obscures the quite definite and concrete advantages of going abroad – not to mention the equally definite snags. Are you with me mate?' In his next letter to Conquest ten days later Amis returned to the subject.

I'm sure we agree about this abroad thing, as you say, and my only point this time is about the Pathans. I don't want to stop them being like they are, but I wish old Mayne would see (or say) that murder is murder wherever & however you do it . . . He can *still* admire them by all means *after* seeing & saying this, but I don't think he should gloss it over. Too many of these chaps think that if something happens abroad it's somehow all right, even if it's the kind of thing which over in the Old Country would send them (and me) running for a cop . . . L[ee]'s book was notable as much for bloody stupid *ideas* as for crappy style, and these ideas are shared by other craps (Davenport & his 'rock-like integrity' . . . etc). I really do mind the ideas more than the style, bloodiest as the latter undeniably is.

In the same letter Amis aired a mild worry to Conquest. He had attacked Laurie Lee in one *Spectator* piece and Evelyn Waugh's new novel, *Officers and Gentlemen*, in another and now he was about to attack a posthumous collection of Dylan Thomas's prose pieces, *A Prospect of the Sea*, in a third. 'Feel slightly nervous about this,' he wrote: 'the only honest way of doing it is to attack it, but after the Lee business and being rather less than cordial to old Evelyn it will seem as if I've gone all sour. Still *someone* ought to give Dylan a bouquet of old bogwort before long.' By the time he next wrote to Conquest thirteen days later, the Lee book was still on his mind, though now mixed up with the equally 'poetic' prose of Thomas.

To finish off this Lee business . . . I think the 2 classes (a) chaps who think abroad is mystically fine (b) chaps who think L.L. and D.T. [Laurie Lee and Dylan Thomas] are 'true poets' overlap quite a lot in practice, hard perhaps to distinguish, but after much consideration I will admit that it *is* a worse failure in *taste* etc to esteem L.L. & D.T. than to be one of (a). But literature not being very important I still think (a) worse, because it makes people *behave* stupidly, and that means that some other people are going to get hurt, which is undesirable. Liking D.T. is bloody silly but it doesn't do anyone much harm. I hate it all right though. My review has gone off to the *Spr* [*Spectator*]. Don't know whether you'll agree with it.

All this literary talk was mixed up with chat about Portugal and other matters in Amis's letters to Conquest, but even this much theorising was unusual for him and he evidently felt some awkwardness about it. "Fraid all this sounds a bit *pompous*,' he wrote down the margin of one letter. And in another he made a joke of the whole thing: 'It's nice to think that our exchange of views on the Laurie question will almost certainly appear, say about 1990 – you know, a book of about 50pp costing 8/6: *Spanish Fly*, a literary exchange: the Amis–Conquest correspondence edited by John Wain, O.M., published by Hartley & Tomlinson.'

Amis seemed to shy away from these arguments as if, by taking such literary questions too seriously, he might himself fall into the trap of pretentiousness, might even become the sort of self-conscious literary person who poeticises his prose, mistakes colourful foreigners for heroic souls and thinks abroad is mystically fine. But his articles and letters of that summer touched on issues he cared about very much, even if he was reluctant to theorise out loud about them – too much theory always being in his view the enemy of practice, apt always to clog the writer's mind with inhibiting second thoughts.

By reading travel books and by going abroad himself he had uncovered a whole new field where the aesthetic and social snobbery he so detested could mingle and thrive, just as they did at Professor Welch's cultural soirées. The bogusness of the travel mystique was different but no less real or offensive than the bogusness of the professor's notions of culture or Gareth Probert's notions of true Welshness. Travel-writers like Lee and Mayne were at fault because they deceived their readers into confusing mere enjoyable escapism with profound spiritual pilgrimage or into believing that gangsters were somehow all right if they dressed up in picturesque tribal costumes and carried knives instead of razors.

> See yonder peasant colourfully dressed
> By sterile Northern cultures unimpressed,
> Show his robustly independent heart,
> And kick the donkey in its tenderest part.

In these words Amis mocked the distorted vision and double standards of the writer dazzled by abroad in a verse pastiche he never published. But how exactly did you get the balance right – give furrin parts their due without getting absurdly reverential about them and knocking the old country at the expense of all things foreign? One way to find out might be for Amis to write a travel book himself. A new kind of travel book, if such could be devised, which would avoid the double pitfalls of overdoing the enchantments of abroad and overblowing the prose. He had already been turning over the possibility of doing something of this sort before he left home.

A few days before Amis sailed for Portugal Philip Larkin wrote to a friend: 'Kingsley sets out for Portugal this week, on the Maugham money, for three months. A travel book may result, like a space-thriller by John Hewitt [a poet], a verse play by Mickey Spillane [a writer of tough-guy thrillers], or a good poem by A. Alvarez [another poet of whom Larkin did not think much].' Larkin, in other words, did not expect Amis to come up with anything startling or innovatory in the travel-book line. And Larkin was right.

The book Amis started to write from his experience of Portugal was intended to be part fact, part fiction, with Jim Dixon returning to make a second appearance as its hero. *Lucky Jim* ended with Dixon taking a leap into an uncertain future when the wealthy Scotsman, Gore-Urquhart, offers him a job as his secretary. In the new book, Gore-Urquhart was to send him on a mission to Portugal where he would encounter the real Kingsley Amis – at least this was one possible strategy. But Amis could not make the book work in this hybrid part-fact-part-fiction form. Instead, he reshuffled his ideas into a more conventional novel, abandoning Jim Dixon and replacing him with Garnet Bowen and leaving out Kingsley Amis altogether. If this was the failure Larkin predicted, it can also be seen as Amis taking heed of his own advice to chroniclers of abroad like Laurie Lee: write a novel, not a travelogue.

While in Portugal, Bowen expends no effort on researching his travel article, though he thinks a lot about abroad in general and how writers cope with it in particular. Most of his energies are devoted to coping with his family and with Portugal, as well as making a not very whole-hearted attempt to get on with a play he is supposed to be writing – he tears it up on the boat home. The rest of his time he devotes to the second of his two commissions, the one given

211

him by his publisher friend, Bennie Hyman. This is to seek out Wulfstan Strether, a famous English writer who has lived for many years in Portugal. Strether was thought to have given up writing for good but a new novel has mysteriously arrived at Hyman's office where, on the evidence of the manuscript alone, opinion is divided over whether it is authentic Strether or fake. To make things more difficult, all personal contact has been lost between Strether himself and his publisher. Hyman asks Bowen to check out the man claiming to be Strether. 'Just the thing for you, this, isn't it?' Hyman says. 'You and your sham-detecting lark . . . A sort of test, in a way.' Sham-detecting was very much Amis's thing too.

The names he gave his exiled writer might have been and probably were designed to mark him down as a suspicious character from the start, the sort of names that would set anyone's sham-detecting antennae quivering: Wulfstan (Archbishop of York, died 1023, author of vernacular homilies and considered by Amis to be the most boring of all English writers, even of that gang of bore champions, writers in Old English); and Strether (Lewis Lambert, one of The Ambassadors in Henry James's novel of that name). And the early evidence – consisting mostly of Bowen's observations of the way the man calling himself Wulfstan Strether carries on – suggests that he is indeed the fake his name alone makes you suspect him of being.

After listening to him pronounce at some length about literary criticism, Bowen thinks, 'Could a man who had really written all those novels really be bounded by James and Conrad and Edith Wharton and Meredith, could he really not have noticed anything that had happened since? And wasn't [his] whole attitude that of a reader rather than a writer, wasn't his whole *persona* arranged on the lines of what a reader would expect a writer to be like?'

Bowen becomes even more suspicious when he and the dubious Strether visit the tomb of Henry Fielding in Lisbon. Fielding is one of those rare writers Bowen admits to thinking of very highly, 'the only non-contemporary novelist who could be read with unaffected and whole-hearted interest, the only one who never had to be apologised for or excused on the grounds of changing taste'. He is therefore flabbergasted when the man claiming to be Strether pronounces himself to be a cut above Fielding, on the grounds that Fielding is merely a comic writer while he, Strether, writes tragedy. 'In the field of the novel,' he sonorously declares, '[Fielding] is indeed the colossus of the eighteenth century, but I cannot feel that posterity

. . . will care to place him beside the colossus of the twentieth,' i.e. himself.

Eventually Bowen is convinced that Strether is who he says he is, though proof comes in evidence of a non-literary kind, including an old photograph. Once his identity is established, Strether's behaviour falls into place. He could not be a phoney because he would not have said he was better than Fielding if he was. He would have played safe by being humble and reverential at the Fielding tomb. The fact that he didn't 'meant he couldn't have been putting on an act'. And the reason that he behaved like a 'posturing, prancing phoney' was because it came naturally, Strether being old enough to belong to what Bowen calls 'the great writer period', which he defines as being 'roughly between *Roderick Hudson* and about 1930, death of Lawrence and the next bunch all just starting off – Greene, Waugh, Isherwood, Powell. Or perhaps 1939.' Being born when he was, Strether simply couldn't help being the sort of person he was.

Bowen goes to stay with Strether and gets to like the man, but nothing can make Bowen like his writing. After reading a passage of the novel under suspicion, *One Word More*, written with all the stately elaborations of Henry James, Bowen wants 'to put the man who had written that in the stocks and stand in front of him with a peck, or better a bushel, of ripe tomatoes and throw one at him for each time he failed to justify any phrase . . . on grounds of clarity, common sense, emotional decency and general morality.' Strether passes the test of being Strether. But as a writer he fails dismally, though partially excused on the grounds of his vintage and because Bowen has grown rather to like him.

In applying his sham-detecting faculties to Strether and his writing, Bowen incidentally sets out something of Amis's own literary position as a writer of the post-great-writer period, one who sees himself firmly attached to the Greene–Waugh–Isherwood–Powell line of descent. The Strether story is complete invention – there was no mystery for Amis to unravel while he was in Portugal – unlike the rest of the story, which follows in Amis's own footsteps more closely than does the plot of any other of his early novels. But Bowen's sham-detecting faculties are not reserved for literary matters alone. Like some ceaselessly watching radar system, Bowen is forever on the lookout for the bogus, including the bogus in his own reactions. While being driven to meet Strether, he inspects the passing peasantry 'as closely as he could for signs of instinctive wisdom and the rest of it, but all he could make out for certain was that they looked foreign, dowdy and on the whole rather amiable: some of them had even waved as they passed.' The natives are largely exonerated. It is the false expectations generated by travel-writers that are to blame for the traveller's overheated imaginings about them.

The attitudes Bowen displayed were to add a new note to Amis's growing reputation. Now he was not only a provincial philistine drunk, boor, Lefty and all-purpose Angry Young Man, he was an abroad-hater too. There was some truth in this. Amis did and does prefer his home turf to any imaginable green pasture across the sea. This is a matter partly of personal taste, partly of Amis's anxious dislike of the hazardous business of travel and dealing with day-to-day matters under foreign rules and in foreign languages. It is not a question of disliking foreign countries or people. 'I shouldn't like you to get the idea I'm trying to knock Portugal and the Portuguese,' Bowen says to Hyman when he is back in London. 'It's a very nice-looking place all round and if you exclude the Government and the upper classes the people are as decent as you'd find anywhere.' These are Amis's sentiments exactly, as are the feelings Bowen goes on to express: 'It's just that the place is located abroad and the people are foreigners, which . . . merely means that they and I belong to different nations, so we can't understand each other or get to know each other as well as chaps from the same nations can.'

For Amis as for Bowen, the essence of what is wrong with being abroad is that it cuts you off from your familiar habitat, the place where you can most naturally flourish. At least when the army took you abroad it 'imposed a purpose outside yourself and provided you with a job, a circle of friends, recognised modes of leisure and a coherent attitude to your surroundings'. For a civilian abroad, however, there is simply nothing to do. Away from what he knows, confused and bedazzled by unfamiliar surroundings, with no reason to be where he is more compelling than whim, the traveller misreads the behaviour of the locals and confers wholly unwarranted significance on things that happen to him which might just as easily have happened at home where, if they did, they would pass unregarded. The literary traveller is no less susceptible, in fact more so, because all too likely to disguise his ignorance about what is going on by calling it mystically fine and to further disguise his crappy opinions in overwrought prose.

In all this, Amis and Bowen are at one. Amis may have found it impossible to insert himself as a real-life character into his Portugal book, but he and his opinions are very much present in it all the same – so much so that at least one way of looking at the novel is to see it as a convenient structure, like a washing-line, on which to hang out his views about literature and abroad.

Amis never did quite find his own way through the problems of travel-writing. At ease in most literary genres from the novel to the restaurant review, writing about abroad is something he has mostly avoided. He salvaged only one piece of journalism from his Somerset Maugham excursion, published in the *Spectator*'s 1956 travel supplement under the title 'Lusitanian Liquors'. 'Among Portugal's many attractions for the tourist,' it begins, 'Portuguese drinking is paramount.' Here he was on safe ground. By sticking to the subject of booze, which he knew a good deal about and which, like music, speaks in an international language with local variations, he avoided the many snares that litter the path of the susceptible literary tourist. But thereafter he left travel-writing largely alone.

∞ 12 ∞

Princeton

Amis's second lengthy excursion abroad in the 1950s was a very different matter from Portugal. This time he went to America, as Visiting Fellow in Creative Writing at Princeton for the 1958–9 academic year. Here there was no risk of finding things mystically fine, or of being tempted to empurple his prose to describe what he saw. At Princeton he had a job with which to fill up his time, as he did not in Portugal – enforced idleness was always to be one of his principal objections to going abroad for no other purpose than to holiday. Better still, the American natives spoke English and were exceedingly hospitable. Amis's was the first generation of Britons to be brought up on a diet of Hollywood movies. He had acquired his own special liking for American science fiction and jazz. So going to America was not really like going to furrin parts at all, more like moving in for a spell with distant but congenial relatives.

There were, of course, the usual horrible preliminaries to be got through. 'Most of my creative energy for the last ten days has been siphoned off into the completion of the fourteen forms or copies of forms I must fill up,' Amis complained to Larkin in May 1958. 'You know the kind of thing. Mr., Mrs. or Miss. Then – Sex. How many female Mr's and male Misses do they get? Good mind to put "Yes" for Sex.' The mere thought of going to America aroused apprehensions in Amis. He felt 'panicky' about the trip, he told Larkin in August. Perhaps Americans wouldn't like him for being British. He was 'afraid of not being able to teach Creadive Wriding properly, or indeed at all'. Amis was not afraid of liking America so much that he might want to stay, but rather feared hating it so much he would want to come home 'by about Bonfire Night [5 November]'. He was worried about running out of money too.

Within hours of arrival in the *Queen Elizabeth*, however, all these fears were banished. After putting up with the sloth of New York dockers unloading his ship and being driven to Princeton along the New Jersey Turnpike with its 'wondrous multi-coloured lights', he lost any reservations he might have had, including those created by his lingering leftist doubts about American capitalism and power, and he made up his mind that this was his 'second country and always would be'.

At first four Amises moved into 271 Edgerstoune Road – Kingsley with father William and sons Philip and Martin, now aged ten and nine respectively – to be joined a week later by Hilly and four-year-old Sally, who had had to stay behind so that Sally could have an operation to remove a growth on her leg, which happily turned out to be benign. After only one day Amis told the *Princeton Packet* that he thought Princeton was a 'lovely community'. It reminded him of Sussex. The houses were as he had expected: 'white and boxy with neat, green shutters'. Having stopped off for a sandwich on the road from New York he was ready to denounce its British counterpart, in American style, as 'just out of it'.

Their house was rented from Murray Kempton, then as now a well-known New York journalist who was away for a year in Italy. Amis described it in his *Memoirs* as 'a residential bit laid out in the best posh-rural American style, with a great deal in the way of lawn, few flowers, no fences, and a patch of woodland at the back where the deer would sometimes wander, just a few minutes' drive from the centre of town'. The neighbours were helpful about everything, from cars to schools. They invited him in for drinks at all hours, in fact filling up every vacant slot in the day. Even brunch turned out to be less a merger of convenience between breakfast and lunch and more an excuse to make an early start on drinking. At the end of October, six weeks after his arrival, Amis wrote to Conquest: 'All very jolly here, settling in fine, with the smell of bourbon and King-size Chesterfields over all: cirrhosis and lung-cancer have moved into an altogether more proximate position relative to me.' The neighbours also found Amis a maid called May, a middle-aged black woman who mocked him with her bogus Southern accent and drove him on his first morning to the Firestone Library in a pink Cadillac, but had to be got rid of for, among other things, stealing the family's sweaters.

Some minor American habits Amis found hard to get used to. Everyone talks a lot, he told Conquest, 'though nobody ever says fuck or even shit and sex is regarded as a topic like any other, about as basic as horticulture or water-skiing. Never any jokes about it, even from N.Y. businessmen, etc.' He put the same point in a letter to Violet and Anthony Powell a few days later, though in less blunt terms: 'Princeton is fine, with crowds of hard-drinking academics who are good in the ways the ones at home are bad – e.g. drinking, amiability – but bad in the ways the others are good – these *never* swear (I had to wait for the arrival of another Englishman to do this, except to myself or in the family circle) and they like to talk of matters connected with the practice or appreciation of the arts.' Amis always preferred a conversation with a lot of no-nonsense swearing in it to one with a lot of nonsense about art.

He took a weekly class at the library on the principles of literature, or some similarly ponderous theme, but his main teaching duty was to go through with a couple of dozen students their fledgling attempts to write. He conducted these sessions in room 2–15–E–2 of the Firestone Library, tucked in between Philately and the Woodrow Wilson Collection. 'Creative writing is all right,' Amis told Conquest, 'especially with one Ed Hirsch, who turns in a semi-pornographic story every time. Students amiable, of immense physical stature, keen.' One would-be poet told him, 'Sir, I just don't take too much account of the reader,' and added that he didn't think he had 'anything too important to say'. Whatever the reader managed to get out of his poems was, it seemed, all right by him. Amis saw signs in this of the characteristically modern failing of careless obscurity. Since it was tolerated among the greats, it was perhaps only to be expected in beginners too. After all, if they turned out to be any good, someone would write a book explaining what they were talking about and so save them the trouble of making this clear from the start. Back home, such attitudes would probably have earned the student who expressed them a piece of the Amis mind, vigorously expressed. But in America the students were different. They spoke with a complete lack of arrogance that Amis found wholly disarming and therefore difficult to denounce.

He was genuinely impressed with the talent he saw, or thought he saw. More than thirty years later he could still remember 'Ochsner's accounts of life on the freight river-boats, Houghton on hunting (=shooting), and village feuds in Vermont, Collins on a father's ruthless social pushing of his son, Rose on the crew that dropped the Bomb'. Amis 'felt the excitement of somebody accidentally present at the first stirring of an outburst of new talent'. Writing in the *Observer* at the end of 1958, he compared the work he had seen at Princeton with what he had read in British student magazines and concluded: 'I am sorry to have to tell you this, gentlemen, but they – and you and I, conceivably – may look a little thin-blooded when the Williamsons and Bernsteins [not their real names] get going.' But they never did. Amis heard of one would-be playwright who managed to get a play produced, though not on Broadway, and a couple more students who published a single novel each. But none went on to any sort of literary glory, leading Amis to think that he 'had underrated the extent to which high intelligence and a sense of purpose find a real but incidental and passing literary expression in more than a few young people'. Teaching creative writing probably taught Amis more than his students, about America and about literature.

One other teaching obligation fell Amis's way. He was due to conduct the Christian Gauss Seminars in Criticism, taking over in the spring from the American critic Dwight Macdonald who had been teaching 'mass culture' for the first part of the academic year. R.P. Blackmur, the poet and critic who was responsible for Amis at Princeton, suggested he make science fiction the theme of his six lectures. Amis was willing enough but not well enough briefed. He had been a science-fiction fan ever since investigating a bin in his local Woolworth's labelled 'YANK MAGAZINES: *Interesting Reading*' when he was twelve or thereabouts. But he had never read the genre in any systematic way. Amis therefore sent Conquest, a fellow fan, an urgent message. 'If ever Amis did anything to win your support vouchsafe it now. Honestly, if you can send me any griff, or names of anything that might contain griff, any good stories especially stuff *about* s-f, let me know, eh?' Amis felt himself particularly weak on the early days of Jules Verne and even weaker about even earlier days, including an Alexandrian interplanetary romance he thought Conquest had told him about.

He soon found what he needed without Conquest's help. L. Sprague de Camp's *Science Fiction Handbook* gave him the necessary historical background and a trip with a large suitcase to Brentano's bookshop in New York yielded an abundance of book and magazine material. He gave his lectures in front of a notable audience – Macdonald, Mary McCarthy, Hannah Arendt and Robert Oppenheimer among them – and afterwards found that he could turn his script into a book without too much rewriting. *New Maps of Hell* was published in New York in 1960 and in London a year later, to more critical attention than anything he had previously published, most of it favourable.

This was gratifying, of course, but years later Amis came to think of his book as part of a regrettable change in the science-fiction climate, one that was to bring the genre out of its non-literary, non-academic popular ghetto into the inhospitable glare of respectability, where it suffered all the ills of modernism and experiment and went into what might be terminal decline. 'Science fiction has come from Chaucer to *Finnegans Wake* in less than fifty years,' Amis wrote in 1981 – in other words, from first bloom all the way to disintegration. Jazz, he thought, had gone the same way long before. The rise and fall of genres, even entire art-forms, he observed going on all the time. The difference with science fiction was that he feared he might, inadvertently, have given it a helping heave down the slope to destruction.

There was a good train service to New York, and Amis took advantage of it for frequent trips, usually making it back to Princeton to sleep. He found the city impressive as well as entertaining and wrote in his *Memoirs* that 'anyone who walks up Fifth Avenue (say) on a sunny morning without feeling his spirits lift is an asshole.' His first day there, Amis told the Powells, introduced him to

the English tourist doing the place in 12 hours, and included a conversation with a publisher (6'3", son of Middle West miner, Montenegrin ancestry), lunch with queer Broadway producer, party at magazine office (martinis at 4 p.m. – oogh ergh), dinner in Greenwich Village (mainly palate–eroding cheese), intellectual party (during which I heard one bearded man ask another if he really understood what intellectual integrity was – honest, now), jazz session with a lot of conscientious inter-racial camaraderie, and finally slumber in Auden's bed – all right, all right, WHA is in Rome.

The magazine office party was for *Esquire*'s twenty-first birthday issue, in which Amis by lucky chance had a story (it was 'Moral Fibre'). He was taken there by a man he knew who worked for *Esquire*, Gene Lichtenstein. Amis got drunk and 'instinctively' avoided Norman Mailer. The party led to a useful contract with the magazine to review foreign (i.e. non-American) films in a column with the title 'The Art of the Cinema'. The pay-rate of 'about 2/6 a word' was then so far beyond London rates it led Amis to tell Conquest that if he stopped writing for the *Spectator* or the *Observer* it was not out of 'laziness or contempt'. From the *Esquire* party he went on to hear his first American jazzman play live. This was the saxophonist Sonny Rollins, not Amis's kind of player, but he didn't let that stop him enjoying the experience. As it was too late to catch the train back to Princeton, Lichtenstein gave Amis a bed at his apartment in the Bowery which he had presumably borrowed or rented from W.H. Auden. 'Plucking down a copy of Crabbe's poems in the flat,' Amis told the Powells, 'I saw on the flyleaf "To Wystan, We must love one another AND die. Cyril." Didn't look at any others for fear of anti-climax.' This, Amis assumed, was a variation made by Cyril Connolly, the London literary critic, of Auden's famous line, 'We must love one another or die.'

In November, Amis went on another New York trip to a debate at the Young Men's Hebrew Association at Hunter College on the subject, 'Is There a Beat Generation?' Jack Kerouac, famous as the author of *On the Road*, was also on the platform, as were James Weschler, editor of the *New York Post*, and Ashley Montagu, a British anthropologist at Princeton. According to the version of what went on given by a reporter from the *Village Voice*, this was a pretty chaotic event, with Kerouac 'dashing off-stage a dozen times, clowning with a hat to the final stumble and wild dragging of poet Allen Ginsberg on stage'. According to Amis's own account in *Harper's Magazine*, Kerouac did his best to make them uneasy by greeting him with 'Hallo, my dear', and telling Montagu that he had seen him on the Jack Paar Show, but that Montagu had had nothing new to say. Then Kerouac went over to the backstage piano and began playing 'a version of the dear old Warsaw Concerto, but broke off every now and then to appear before the photographers.' Kerouac switched to left-handed boogie-woogie and when they finally went on stage he responded to the cheers of the audience (all directed at him) with 'weaves, bobs, and a chimpanzee-shuffle or two'. While Weschler was speaking, Kerouac wandered about the stage wearing his hat.

The Beats, Kerouac said in the *Village Voice* account, were 'a swinging group of new American boys intent on life . . . Being beat goes back to my ancestors, to the rebellious, the hungry, the weird, and the mad. To Laurel and Hardy, to Popeye, to Wimpy-looking wild-eyed over hamburgers . . . the size of which they make no more.' There were two types of Beat, he went on: 'the Cool:

bearded, sitting without moving in cafés, with their unfriendly girls dressed in black, who say nothing; and the Hot: crazy talkative, mad shining eyes, running from bar to bar only to be ignored by the cool subterraneans'. Amis, described as wearing a conservative light-brown suit, was reported to have been 'perplexed by the mad audience, but in a friendly way trying to understand the madness'. He poured a full bucket of cold water on the idea that there was any connection between Britain's Angry Young Men and the Beats, indeed on the belief that any such groups could be properly said to exist at all. The Angries were 'an invention of literary middlemen, desperate journalists who thrive on classifications and clichés, who put writers in pigeon holes and save people the trouble of reading'. Anyone under pensionable age was liable to be included in this mythical group, Amis said. 'Any day I expect to see Boris Pasternak [born 1890] so labeled.' John Osborne was angry, Amis conceded, but all the English writers put in this category along with him were only 'doing what writers have always done – they are going about the job of writing. There is no Angry Young Men movement. There may be a beat generation, but I doubt it.'

Weschler was puzzled – 'After listening to Kerouac I understand less about what [the Beats] stand for than before' – and unimpressed: 'There is no valor in their kind of flight and irresponsibility.' Montagu, described as 'white-haired, calm slightly sleepy-looking', was more sympathetic to Kerouac, if solemnly condescending too. 'The beats give personal testimony to the breakdown of Western values. These are the children who were failed by their parents . . . the beat writers are describing this generation.'

Afterwards, Amis told the *Village Voice* reporter: 'There I was sitting back, waiting for Mr Kerouac to be funny, but somehow he never quite got there, though he did a lot of shaping up to be funny . . . 50 minutes of it in fact' – compared with the ten minutes each that were allowed the other three. In his *Memoirs*, Amis wrote that his side 'had won the argument (argument?) but lost the performance'. The Beats came very firmly into Amis's no-good-AT-ALL category of writer, but he thought they had a malign influence all the same, especially on poetry, 'helping to reduce parts of it to spoken scripts or rock without the music'. He was dismayed when he spoke about them to Norman Mailer, whose 1949 novel *The Naked and the Dead* Amis valued highly, only to find himself getting 'a proclamation of their merits and significance', which on reflection Amis found boded worse for Mailer than it did for Kerouac.

Thinking over this literary skirmish in *Harper's*, Amis wondered why it was that people went to discussions or lectures by literary persons at all. For most of those present, he thought, the speakers might as well be talking Choctaw; 'the visual appeal is what counts. Mr Kerouac was shrewd enough to have grasped that.' Puzzled though Amis was by whatever 'system of inducement or threat' persuaded people to turn out for lectures, they did so in sufficient numbers 'to tickle even a writer's vanity', while greed was 'abundantly satisfied'. Since his earnings from Princeton were not going as far as expected, he willingly made himself available for lectures, usually for about $300 per performance, in any part of America reachable by a car or train journey of reasonable length. This took him among other places to Philadelphia, to Yale in New Haven, to Vassar at Poughkeepsie, New York State, and, via Chicago, to the University of Illinois at Champaign-Urbana.

On the whole he was well entertained and received on these trips. At Yale he was handed a dry Martini by his host, Archibald S. Foord, *before* he had time to remove his overcoat, a record of instant hospitality to remain for ever unbeaten. His audiences were at least 'polite and attentive'. The only man who made faces at him was, as Amis later discovered, 'an official of Her Majesty's Government, which I was mildly denouncing at the time'. But even the official's faces were mild ones. Amis only put a foot seriously wrong once, he believed, in Philadelphia, where he decided to abandon straightforward jokes in favour of 'a long, humourless, pseudo-academic diatribe on the comic spirit recited absolutely deadpan by a supposed comic writer'. He did not recall this incident too accurately later, being too busy keeping his own composure in the manner of someone who has had too much to drink, but his impression was that it went down badly, except for a reference to vomiting, 'which laid a single undergraduate in the aisle'.

Retribution for this lapse in humorous standards was swift. During the ensuing party, Amis's script was stolen. This was something of a setback, since he had been intending to recycle it for later use on some other campus. In *Harper's*, Amis reflected that he should have interpreted the theft as a 'blow against authority' to be welcomed as such, and also as a joke, 'a good deal more pointed and economical than my own'. He did not think this at the time, of course. But he did manage to recycle the incident in a different way. It became part of the plot of his American novel, *One Fat Englishman* (published in 1963), with the difference that its hero, Roger Micheldene, has his script stolen before he is due to give a lecture, thus giving him an excuse to display all his talents for bad manners, arrogance and anger by refusing to talk at all.

Experience of this lecture circuit helped confirm Amis's views on university education. There were places of the highest quality, like Princeton or Yale, but at the other end of the range you got the University of Illinois at Champaign-Urbana. As Amis travelled the campuses, the likelihood that he would soon come to the conclusion that 'more will mean worse' only grew stronger. Not that even Champaign-Urbana was all nightmare. On his second day there, the hangover Amis had built up the night before disappeared with his first lunchtime drink and he experienced a 'state of dazzling euphoria, as has happened to me only three or four times in my life, and never since'. At lunch Amis showed himself at the top of his form and later overheard one faculty wife say to another, 'Have you ever met a *reserved* Englishman?' A university that could produce this reaction to Amis's presence could surely not be bad all the way through.

In Portugal, the Amis bullshit-detecting antennae had tuned themselves to tremble at any signs, in himself or anybody else, of submission to the lure of peasant mystique and, worse still, of writing it up in overblown style, hiding a perverse ignorance beneath a froth of adjectives. But in America his antennae picked up a quite different sort of travelling ignoramus, what you might call the patronising Brit. This is a man – Amis does not seem to have come across any woman similarly afflicted – who, while quite prepared to pay 'a fuddled respect' to European civilisations because he has been told they are ancient and gracious, takes the opposite attitude in America, which he has been told is a naïve and materialistic place, and many other patronising things besides. Snobbery is stood on its head but remains snobbery even while upside-down.

A ripe example of this character type was Anthony Crosland, a rising young Labour MP and intellectual, later to hold a series of senior posts in Labour governments and dying in office in 1977 as Foreign Secretary. Amis had met him at a *Spectator* party in 1957. Taking a liking to each other, they proceeded from there to an *ad hoc* party at Crosland's Kensington home, at 19 The Boltons, where a quarrel broke out over the jazz records Amis had brought with him. In the version of this given in Crosland's biography by his wife Susan, the quarrel was over Amis's attempts to take charge of the Crosland gramophone and arrange his own records on its turntable. In Amis's version, Crosland was drunk in charge of the gramophone and heedlessly let the needle sweep across the surface of Amis's precious discs. Whatever the exact truth, it was essentially a dispute about the rights of the owner of the records against those of the owner of the gramophone. If Crosland had not been so big, Amis might have taken a swipe at him. Wisely, he decided to leave the party instead.

On the face of it, Crosland and Amis should have made natural allies. Crosland was no hairshirt socialist. Having fun was very much part of his agenda, and the brand of non-extreme social democracy he advocated in his highly influential book, *The Future of Socialism*, published in 1956, was of a kind that could be expected to appeal to the Amis of the late 1950s. But the trouble with Crosland, as Amis saw him, was that he was an upper-class snob and none the less of one for being supposedly on the left, or, when it came to America, for having an American girlfriend (later wife). Snobs like Crosland might be bright but they could be as prejudiced as anybody and just as likely to shove their prejudices down other people's throats when they got the chance (as Crosland did when he became Education Secretary in the 1960s, deploying what an infuriated Amis called the 'Crosland illiteracy commandos' throughout the schools and universities).

In an *Observer* piece early in January 1960, Amis reported an exchange he had heard about between an unnamed visiting Brit and an American student:

'And what do you happen to be working on?'
'Sir, I'm working on the "Aeneid" of Virgil.'
'Really? And tell me, what State do you come from?'
'I'm a Texan, sir.'
'Really. Ha, ha, I must say I find the notion of an Easterner working on Virgil interesting; a West Coaster working on Virgil would be amusing, but a TEXAN . . . it's simply GROTESQUE.'

Amis called this a 'little true-life chiller' and added that the villain, 'I record with a groan, was a promising young Labour politician'. He did not reveal that the villain in question was Crosland, but that's who it was. Amis came to think of meeting Britons in America as 'a major social hazard' of being there. Already, after just a few weeks, he was telling Conquest that 'the only unpleasant people I've met have been English'. The one variety of American who made Amis squirm almost as much as this kind of Briton was the American who fell for those Britons' line – who took British pretensions to maturity, ancientness and tradition at face value and were 'horrible Anglophiles', like the character Strode Atkins in *One Fat Englishman*. Americans like these were as capable of misreading England as Englishmen like Laurie Lee were capable of misreading Spain and just as much to be deplored and despised in their way as Lee in his.

Britons wandering in a state of Crosland-like insensitivity through America upset Amis on a number of counts. One was their habit of generalisation. No Briton thought of himself in caricature form, 'dressed in a bearskin and a choker and a frock-coat and kilt, with a cup of tea in one hand, a pint of warm beer in the other and a rolled umbrella over his arm'. Yet they persisted in fixing Americans in a similarly outlandish stereotype, as being, for example, 'naïve or brash or superficial or woman-dominated or shallowly bonhomous or hypnotised by gadgetry . . . or only interested in a fast buck and a fast car and a fast Martini-jag'. Well, yes, there was some of this materialism around, but where wasn't there? Then there was the British habit of carrying on like characters from the West End stage. One good reason for sending students across the Atlantic, Amis thought, was to let Americans know these were not the only kind of Briton.

The self-conscious Britishness that lay behind the theatrical presentation was a post-war phenomenon, a 'horrid index of a nation in decline'. A lot of it, of course, was to do with envy of American prosperity and power, especially anger at how Suez had exposed Britain's relative weakness. What particularly annoyed Amis when he got home was to hear about Britons who had been 'floated in free bourbon and not allowed to put their hand in their pocket for so much as a cigarette' and then went on about American 'immaturity'. Or the British academics, 'themselves barely capable of construing a Shakespeare sonnet', who patted American heads for their 'Teutonic thoroughness' instead of being 'healthily scared' by them.

There was one compensation in contemplating all this British awfulness. It gave Amis the idea for *One Fat Englishman* and its odious hero, Roger Micheldene, who displays all kinds of contempt for America on top of many personal horriblenesses of his own. He is fat, angry, gluttonous, drunken, vindictive, rude, aggressive, lecherous, arrogant – even to the Catholic God he occasionally makes his wishes known to. Apart from a fondness for snuff, Micheldene has no resemblance to Amis, although Amis was beginning to develop a mild tendency to stoutness which made him aware of what it might be like to be seriously fat. ('Outside every fat man there was an even fatter man trying to close in,' Amis wrote prophetically in the novel.) And of course Micheldene's enthusiasm for women is no greater than Amis's. America, Amis found, made an excellent hunting-ground, faculty wives and others turning out to be eager game. He 'really went to town' there, Hilly recalls. Philip Larkin reported to Conquest in August 1959, after at last getting a letter from Amis: 'He seems to have spent his time drinking and fucking, as if this should surprise me.' (What Amis actually told Larkin was that for the first half of his trip he was 'boozing and working harder than I have ever done since the Army, and that for the second half I was boozing and fucking harder than at any time at all. On the second count I found myself at it practically full-time.') One small incident in the novel was taken from Amis's amorous experience. When Micheldene is making love to the wife of the 'horrible Anglophile' Strode Atkins in the course of a picnic, he notices that they are being observed by tortoises, or turtles. The same thing happened to Amis.

More than any particular person, however, Micheldene is what any educated upper-class Englishman might turn out to be like if all his vices were given a free hand and all his virtues suppressed: an Anthony Crosland unbound, perhaps. His anti-Americanism takes the form of a generalised contempt, and being the egotist he is even the mildest counter-dig at the British becomes a vile insult directed personally at him. Playing a game in which one person has to act out an adverb and the other to guess what it is, Micheldene takes outraged offence when the woman he's after, Helene Bang, approaches a lamp standard like a Guardsman and gives it a single peck on its shade after being told to make love to it 'Britishly'. 'I'm afraid you'll have to excuse me from taking any further part in your sociological investigations,' Micheldene harrumphs, and so breaks up the party. After various adventures, Helene finally eludes him and he sets off for home, only to find the 'horrible Anglophile' Atkins is travelling on the same liner. The two are thus condemned to spend the journey together, an ending surely devised by Amis as fit punishment for the two types of people he most disliked in America.

Soon after he returned to Swansea, Amis tried to sum up his feelings about America in a letter to Larkin, to whom he had neglected to write all the time he was away. He had had 'a very fine time indeed,' he said. Both Hilly and he wanted to go back again as soon as they could. If England had been blown up while they were away Amis could 'happily' have made his home in America; 'but not as things are.' He had declined an offer of two more years at Princeton. Part of the reason was his wish to give his children an English education. Another part was his low literary output while in America. Swansea, being duller than Princeton, would at least let him get down to 'some sodding writing: 7pp of novel [*Take a Girl Like You*, published in 1960], 3 poems is not impressive for 10½ months.' Though he had also written the lectures that would become *New Maps of Hell*, that was not the work he thought he should be concentrating on.

Amis had a lot of good things to say about Americans: 'They have more energy than we have, and they are better at enjoying themselves. They are not complacent or woman-dominated or death-wishing or insecure or naive – especially not that. Mind you, you have to go there to see this: I can't make anybody here believe it quite. The women, with a few shining exceptions, are less attractive than ours.' Amis found it hard to come up with other reasons than its women for preferring Britain. The food tasted better? Yes, but Americans had clams and soft-shell crabs. There was less advertising in Britain? Yes, but there weren't 'so many nice things to buy'. English was Amis's 'own tongue', but he had had no trouble talking it in America. The balance of advantages was difficult to strike. 'Going to the land of the free gives one a kind of restlessness which may prove hard to eradicate.'

It turned out not to be so hard after all. Staying in America for two years would probably have meant Amis staying there for ever. He was thirty-seven – too old to become the citizen of another country? And perhaps too fond of his own to want to change, although residual 'leftiness' made it difficult for him yet to admit love of country as a powerful enough motive to determine his actions. America, however, had done its work on his politics too. He came to think of America as a 'critique-in-action' of Britain. Whatever leftish orthodoxy required you to think of the American system, the palpable fact was that the system delivered in all kinds of ways: efficiency, energy, conscientiousness, curiosity and social freedom were all on Amis's list. He remained a Labour supporter and was dismayed that the Tory Party responsible for the Suez fiasco should have been returned to office at the 1959 election. But Amis's left-wing loyalties were weakening as his sense of patriotism grew stronger.

∞ 13 ∞

Leaving Swansea

By the beginning of the 1960s Amis had made progress on all fronts except one, that of his own profession of university teacher.

His achievement as a writer was steadily advancing and consolidating. Already he had behind him four novels, three volumes of poetry, a science-fiction radio play, *Touch and Go* (about explorers on a hostile planet, snootily described in *The Times* of 7 November 1957 as 'an adroit anecdote, such as the Third Programme can afford to include without lowering of standards'), and a pioneering book about science fiction, *New Maps of Hell*. He had pronounced controversially on political matters in his Fabian pamphlet and written and broadcast numerous shorter pieces – learned, critical and amusing. (And sometimes all three, as when with his colleague David Sims he contrived to smuggle a smutty line into a 1954 broadcast about Dr Bowdler, famous for his cleaned-up *Family Shakespeare* and buried locally in Oystermouth churchyard. Dr Bowdler wrote travel journals as well and the Amis–Sims script had a woman ask him: 'Would you not devise some letters from France to cap your Dutch journal?' By the time the BBC's guardians of taste saw the joke it was too late to stop it going out.)

As an Angry Young Man, too, Amis was celebrated beyond any mere lecturer of his generation. But in terms of his own profession a mere lecturer is what he still was, earning the scale rate for his job as he had done since 1949. He had not climbed even a single rung on the career ladder to senior lecturer, let alone shinned up it as far as reader or professor. Nor had he shifted sideways to another, more prestigious university. Pleasant it may have been to work there but it

231

was not through Swansea's corridors that the great names of the Eng. Lit. faculties strode.

The money a shift in any direction might bring was not as important as it would once have been. Amis's university salary was of dwindling significance beside what he earned from his writing, perhaps now only a third or a quarter of his total income. A less cautious man might already have abandoned the university life and set himself up as a writer full-time. Amis thought of doing that soon after the success of *Lucky Jim*. If he had it would have been a ripe example of life imitating art, since such a leap in the dark was what Jim Dixon himself took when he gave up all prospect of a pensioned career in teaching for the insecurity of the job his patron, Gore-Urquhart, offered him in London. But the fictional Jim was a bachelor, the real Amis a married man with three children, for whom prudence prevailed. If the right opportunity for a step forward within his profession should come up, though, Amis was in the right frame of mind to take it.

As it happened, Peterhouse in Cambridge was thinking of appointing its first fellow in English, and as it further happened Amis had established an indirect connection with the college through his friend George Gale, who had been an undergraduate there and kept in touch with some of its fellows, among them the historians Brian Wormald and Michael Oakeshott. Gale was a journalist, at that time with the *Daily Express*, later editor of the *Spectator*, later still a columnist on the *Daily Mail*. He was a little toby jug of a man, with a growly voice that gave a natural edge of aggression to his manner, an edge which served him equally well as a gruffly opinionated broadcaster and a customer in a pub trying to get some attention from an idle barman. Gale was also a staunch right-winger, which in the fifties Amis was not. Nevertheless, the two of them hit it off when they met at a party in the *Spectator* offices in Gower Street some time in the middle of that decade.

Another connection emerged at that first encounter. Gale's wife Pat was a Swansea girl and her mother and two sisters still lived in the city. There was therefore every reason why Amis and Gale should meet again, and they soon did, in Amis's house at Glanmor Road, which Gale later pronounced as more 'awash with booze' than any house he had ever been in. The friendship grew around common interests – such as being pro-drink and anti-bullshit – and was cemented by reciprocal visits, Amis to Gale's house in Staines,

Gale to his in Swansea. Through Amis's friend Stuart Thomas they also rented Dylan Thomas's Boat House at Laugharne for holidays where others sometimes joined them, including Wormald from Peterhouse. (Stuart Thomas, though no relation of Dylan's, had been at school with him and when the poet died became solicitor to his estate.) When the Peterhouse appointment came up it was natural that Gale should mention it to Amis – or perhaps it was Pat who tipped off Hilly. However he learned about the vacancy, Amis let it be known he was interested.

Brian Wormald, encouraged by Pat Gale, put Amis's name forward for the new post of director of English studies at a meeting of Peterhouse fellows. Herbert Butterfield, the Master, went round the table soliciting opinions. Ted Kenney, later professor of Latin, remembers one fellow, a scientist, asking for a list of Amis's publications and being somewhat taken aback to be told that Amis wrote novels rather than learned papers. Another fellow, David Knowles, responded with a peculiarly donnish fastidiousness. According to Kenney, he said: 'Well, Master, I must confess to being a little disappointed in Mr Amis. Indeed, if I hadn't known who he was, I shouldn't have known who he was, so to say.' Knowles had started out as a monk at Downside before becoming a don and eventually Regius Professor of Modern History at Cambridge. In spite of his and other mild reservations, Amis's election passed its first hurdle.

The usual next step would have been to ask him to dinner so that he could meet and be met by his prospective colleagues, but this formality seems to have been omitted. It may have been because there was not much doubt that he would be offered the post. At any rate word got around that he was on his way to Cambridge. In March 1961 Philip Larkin mentioned it in a letter to Monica Jones: 'Isn't it a scream about Kingsley, a scream of laughter or rage as the case may be? . . . I must say I'm glad it's Cambridge & not Oxford who have done it . . . Queer College to have done it, though – all historians.' But the protocol of election had still to be observed. In May, Amis wrote to a friend at the BBC, Marguerite Cutforth, wife of the broadcaster Rene: 'You'll see that I don't mention my translation to Cambridge: this is because the final, utter, official rubber-stamping hasn't taken place yet, & until it does they've asked me to stay loftily silent on the subject in public.'

Soon afterwards, the last rubber-stamping was completed and Amis's election confirmed. He resigned from Swansea, took the family to Sitges in Spain for an August holiday and returned to look for a house in Cambridge.

Leaving Swansea was something of a wrench. Amis had grown very fond of Wales and the people he knew there.

When he first went to Swansea in 1949 Amis brought with him only a caricature notion of the country, thinking of it as 'a rugger ground with Cader Idris on one side and a spoil-heap on the other, populated by pit-dirty miners who only stopped singing "All Through the Night" to bawl "Look you to goodness whateffer" at one another, and subjected to continuous rain.'

About the rain, at least, he never found much need to change his mind; but about most other things he did. Swansea was English enough to be nothing like abroad and more like home than his caricature version of Wales allowed. Yet it was different all the same, and in ways Amis found almost entirely congenial. Chapel instead of church and the absence of an upper class or even much of a middle class made the place far less socially divided than England. Good relations were easier to achieve and social exchange with all kinds, from students to shopkeepers, more flexible and relaxed. Amis found the Welsh warmer, friendlier, affectionate towards children in a style more like Italy or Spain than England. They could be excessively nosey but, contrary to myth, they were neither more nor less devious or dishonest than the English. Somehow, though, they managed to be less offensive when they set about cheating you or prying into your affairs.

Wales had its own native hypocrisies and prejudices too, especially where women were concerned. When the wife of one of Amis's colleagues applied for a job in the college library it was to her husband that the library's first enquiry was directed – to see if he approved. A wife's place was in the home and to give a woman a job was to deprive a male breadwinner of the chance of it. Pub manners too were discriminatory in a distinctively Welsh way. Once, after walking the dog on the shore of Swansea bay, Amis took Hilly for a drink in the Woodman, ordering a pint of bitter for himself and a pint of shandy for her. Was the pint of shandy for a lady, enquired the woman behind the bar. When Amis said it was, the landlady declared that her regulars did not like to see ladies drinking pints. Would two half-pints be all right instead? There was a bar in a hotel called the Three Lamps in Swansea which Amis remembers showing off its upmarket status by serving drinks with cherries on sticks and by not serving pints of beer, even to men. It sold beer all right, but only in half-pint glasses. What would have been seen in England as the rather mean and narrow standards of the lower middle class generally prevailed, lightened by a Welsh talent for mildly hypocritical compromise.

The more extreme forms of Welshness grated on Amis but at that time did not much intrude. Sight of bilingual road-signs – 'Tacsi' alongside 'Taxi' – could still be relied on to rouse him to weary disgust in the 1990s but they were not yet to be seen in 1950s Swansea. Nationalism he regarded as unrepresentative of Welsh people, being largely the enthusiasm of a sect made up of clergymen, teachers, employees of BBC Wales and suchlike untypical persons. Dylan Thomas did more to earn Amis's approval by saying 'fuck Welsh nationalism' (or words to that effect) in an interview than by any other of his utterances. Thomas himself he regarded as a pernicious figure, a bad poet whose regrettable popularity did 'lasting harm' to both Wales and Welsh poetry, creating, as Amis wrote in his *Memoirs*, a picture that was 'false, sentimentalising, melodramatising, sensationalising, ingratiating'. Although he thought Thomas 'an outstandingly unpleasant man' he had some sympathy for him as a failed poet. Thomas's own name for himself – 'the Rimbaud of Cwmdonkin Drive' – was all too near the mark for comfort, since, like the real Rimbaud, Thomas was all but written out by the age of nineteen, an understandable reason for anyone to take to the bottle. But this did not excuse his overblown poetry and prose nor his promotion of a false image of Wales, and Amis was to caricature Thomas's kind of Welsh wind-baggery when he made Brydan, in his 1986 novel *The Old Devils*, resemble him in style not a little.

There had been some unhappy times at Swansea too. Amis's mother had died in 1957 in circumstances that were peculiarly upsetting. The three-year-old Sally Amis was staying at her grandparents' home in Berkhamsted when it happened. Grandfather William had scarcely left for work when his wife Peggy fell down in the bedroom while dressing, instantly dead from a stroke. Sally was left alone all day with her grandmother's body until William came home, and when she got back to Swansea family friends noticed how anxious the experience had made her. If she saw her father asleep, friends noticed, she would try to prise his eyes open for fear that he, too, had died. Sally had caused Amis and Hilly anxieties of their own the year before when she fell on her head in the garden and had to stay in hospital for weeks. Writing to Anthony Powell at the end of October, Amis said, 'Actually we have stopped worrying about her in all but a minimal sort of way, but that remains. You can imagine what a business it was at the time.' Fortunately there were no long-term effects.

With his mother dead, Amis felt obliged to bring his father, who retired soon afterwards, to live with him in Swansea. Filial duty rather than personal pleasure was the motive, for his feelings towards his father had long been a compound of boredom and indifference. Besides, William Amis took up bedroom space, which meant that it was difficult to have guests to stay unless the old boy could be got rid of. Amis wanted to invite Violet and Anthony Powell for a visit in 1958 before he set sail on the *Queen Elizabeth* in early September for his year at Princeton. 'I'm still hoping to drag the pair of you over here for a week-end before we sail,' he wrote to them in late July, 'but the chances of getting my father out of the way (a provision that really gives *sine qua non* its full meaning) don't look too rosy at the moment.' He added, 'A wan smile briefly lit my features on finding that my father is sharing a cabin with three strangers.' But a fortnight later Amis sent another letter: 'Alas, it looks as if my father (have I ever told you what I feel about him?) is going to be firmly anchored here until we sail.'

William Amis continued to irritate his son when the family returned from America, where it was briefly thought he might remain with his sister Gladys, who by then lived in Washington with her second husband. Instead, William came back and found a job in London and a room in a small flat belonging to a friend of Kingsley's. By June 1960 the friend was getting restless with this arrangement but didn't like to tell William Amis so to his face, preferring instead that Kingsley should deliver the bad news. This set Amis 'a very delicate problem of conscience', as he wrote to Powell. Might not William return to Swansea if told he was no longer welcome at the London flat? 'At the moment my friend's discomfort is torturing me frightfully, but I believe I can take it a little longer.'

In retrospect, Amis's Swansea years were to take on something of the glow of a golden age. The family was intact, the children growing, success accumulating and income increasing to the point where money was no longer a serious worry. Amis did not now have to ration his cigarette consumption (as did Lucky Jim) for financial reasons, though he began to make intermittent efforts to cut down for reasons of health, taking up snuff as a way of cutting down on cigarettes. Writing to David Sims's widow Joyce soon after her husband's death in 1989, Amis said: 'it was all very happy, wasn't it, on the whole, and terrifically *active*. How we got through all the beach trips, Jo Bartley's thrashes and the rest of it I can't imagine. Well it was the best time of my life and you and David probably did more than anyone else to make it so.'

In his *Memoirs* Amis said that he missed Swansea constantly and often wished he'd never left. But he also recognised that the Swansea of the 1950s had vanished. By the 1990s youth had gone, marriages were broken, children had grown up, old friends were dead. Amis kept in touch with Swansea and his friends there, though with a gap in the years when he was married to Elizabeth Jane Howard, she and Swansea seeming not to take to each other. He still contrives to spend a few weeks with Stuart and Eve Thomas in August and September when the Garrick Club is closed for the summer holidays. But once he left there was never any hope of recapturing the golden days and so no point in regretting them.

None of that, of course, was in Amis's mind when he left Swansea for Cambridge. In 1961 he was all optimism and expectation at what seemed a natural advance. Perhaps, in another ancient university, he could recapture the romantic excitement of his Oxford years, find a better class of student and a faculty of brilliant minds and literary enthusiasms, with a few new drinking chums thrown in. He was quickly disillusioned.

∞ 14 ∞

Cambridge

Hilly found the five Amises a place to live in at The Mill House in West Wratting a few miles from Cambridge from where she drove the children to school and her husband into town as necessary. They stayed for only two months before moving into Cambridge itself, close to the new Churchill College, at 9 Madingley Road – 'a rather posh house', Amis called it, one in a line of ample bourgeois residences, the only one to have been demolished since then. It had a large garden backing on to playing fields with a lawn, elms and rosebeds. There was a greenhouse and a shed big enough to accommodate Debbie, the new family donkey.

But Amis very soon began to feel uneasy about life at his new university. 'They all talk too loud in Cambridge. It's an arrogance. They are so bad mannered that you would almost think you were in Paris. The fact is that public school boys don't care.' This was Amis talking to Quentin Crewe of the *Daily Mail* in January 1962, after only one term at Cambridge.

What was it about the place that made him uncomfortable? There were little things. If you were walking down the street in Swansea, he told Crewe, and someone was coming towards you, then each person would sway a bit to the side so they missed one another. 'But not in Cambridge. Unless you get right out of the way they bump into you.'

At Cambridge, the students put on an act to disguise their ignorance. Ask a student at Swansea about Thorstein Veblen, the American sociologist, and the student would say, 'Who's he?', whereas a Cambridge student would say, 'Ah yes', even though he knew no more about Veblen than the Swansea student. However, Amis allowed, there was something to be said for the Cambridge man, who would probably go and look Veblen up, knowing he wouldn't get away with his evasion a second time.

Then there was the playboy factor. 'The upper classes are more entrenched [at Cambridge] than you would think. Many of them are there to fool around.' Nothing like that went on in Wales, of course. On the other hand, Welsh students were too likely to take the opposite extreme and think of their courses as little more than steps towards fulfilling their ambitions. But Amis liked this better than 'the fact that my best student at Cambridge thinks it slightly square to be enthusiastic'.

Summing up the differences between Swansea and Cambridge attitudes, Amis found all of them 'rather comic'; but class certainly weighed more heavily at Cambridge. It took up 'a fatiguingly disproportionate amount of a young student's time,' he told Crewe. And it infected his own subject: 'The view that Culture is Pall Mall is both prevalent and weakening.' The class structure, he thought, was 'tattered', yet still in good enough shape to last probably his lifetime. The biggest threat he saw to it was the Common Market. If that came off people would find out that class was 'not a law of nature. And this will be a real danger to those who own the class structure.' Approval of the Common Market as an instrument of positive change rings quaintly thirty years on when Amis has long been a stern opponent of all things 'European'. The need to shake off the negative effects of class has, however, been a preoccupation of his all his life, especially the effects of class on literature which tend always to turn it into the possession of a superior and exclusive minority intent on keeping the majority off their territory, as though literature were a private estate or a private Pall Mall.

And yet Amis acknowledged to Crewe that it was all more complicated in Cambridge than it had seemed in Swansea. Lucky Jim's targets had been obvious. 'The windmills to tilt at are bigger now.' But more complex and elusive too, somehow; for in a sense people made up class divisions as they went along. 'Rank pulling is in oneself. In the professor who is bumped into on the street . . .'

One example of this new class complexity was the way some dons tried to stand old class divisions on their head. Amis found his Cambridge colleagues were no longer exactly the upper-class port-and-walnuts men he had known at Oxford. Now they were likely to be professionals from any class. But undergraduates were still drawn in disproportionate numbers from the public schools, as they are now, and colleges made intermittent efforts to shift the balance of their intake towards state school pupils, as they do today. Amis was once at a college meeting held to decide a scholarship award for which there were two candidates of more or less equal merit, as judged by exam results and interviews. But one was from a well-known public school while the other was from a humble state-funded establishment. Someone proposed the latter candidate should get the award – at least that would dilute the public school dominance. There seems to have been relief round the table at finding a solution that was both simple and uplifting to what might otherwise have been vexingly tricky. Amis, however, asked the meeting to reconsider – surely a chap's having parents who could afford to pay school fees should not be held against him? Who the fellows finally chose is now forgotten. But here was the Labour-voting anti-snob Amis intervening on behalf of the better-off classes in the name of fairness. Things were indeed more socially complex than they had once appeared to him to be.

Amis hardly knew Cambridge when he arrived at Peterhouse. He had visited it only once before, in 1940, when he had unsuccessfully tried for a scholarship to St Catharine's, and he remembered that experience so little that in 1961 he found he recognised nothing whatever about St Catharine's when he went back to look it over.

Memory and nostalgia were part of the problem Amis had with Cambridge. He hoped living there would be like living in the Oxford he had first known twenty years earlier and that he would be able to recapture something of the pleasure and excitement Oxford had given him then and later. By the time he came to look back on his Swansea years for his *Memoirs* he knew better than to think he could recapture past times. But when he went to Cambridge he still thought he could relive something like his old Oxford experience. He was disappointed. His Cambridge turned out to be nothing like his Oxford, beginning with the town itself. Amis didn't feel at home in it. Its characteristic images, like King's College chapel at dusk in a thin mist, he wrote a few years later, 'seemed cold and lonely, a setting more appropriate to an unhappy love affair than to the bustling exchange of ideas that is supposed to go on'. He added: 'Arrival at that curious railway station, with its endless single platform like something out of Kafka or Chirico, ought to tip off the sensitive.' Peterhouse was the exception. He would have been happy with it, providing it could have been put on rollers and shifted elsewhere – along with the college members, the kitchen staff and the gardens – to the Cotswolds or upstate New York.

Amis's feelings about Cambridge society matched his feelings about the town. Each was as cold as the other. Cambridge he found 'inhospitable despite the ceaseless ceremonial parade of hospitality'. That was the trouble. It was all too formal. For Amis the essence of the enjoyable social occasion has always been that it should be fluid, relaxed, casual. Drinking with friends in the pub or at a cocktail party is fine. So is going on to eat afterwards. Having friends round to drink at home is all right too, and even feeding them there, though never more than three couples should be invited to sit down at table, in your own home or anybody else's. Anything on a larger scale gets into formal dinner-party territory and is to be strictly avoided. Dinner-parties inflict strangers on you, some of whom are sure to be disagreeable while at the same time impossible to escape. If compelled to enter a roomful of strangers they would all, in a perfect Amisian world, turn out to be easy-to-get-on-with Welsh people. But this was unlikely to happen in Cambridge.

The dinner-party, as Amis put it in his *Memoirs*, forbids all the things that informal arrangements allow, such as 'fairly late arrival, early departure without causing a stir, fairly manifest drunkenness, cancellation on sub-cosmic excuse', that is, something less than 'grave illness or domestic tragedy'. That is why the peak of social ease for Amis is a club like the Garrick, where members come and go as they please, eat and drink what and when they like, sit down with those people they find agreeable and steer clear of those they don't. He hated being forced to remain seated at the dining-table after the food had been consumed as a child and he continued to hate all restraints on his social freedom for the rest of his life.

If the common-or-garden domestic species of dinner-party is bad, the college function is worse. A college feast at that time typically excluded women, compelled the men to dress up in the dinner-jackets Amis hated, spun things out with toasts and other flummery. Although the food might be delicious and the wine excellent, there was sure to be too much of the first and not enough of the second. Besides, the hall in which the dinner was held was certain to be unheated and smoking forbidden until the whole assembly moved to another room for after-dinner drinks. These were first-class, but etiquette required drink to be circulated in decanters, passed leftwards from diner to diner, and an adequate supply depended on others remembering to keep them moving. (It is not difficult to imagine Amis keeping an anxious eye on this procedure, his face falling into scowls of dismay and despair when his preferred decanter got stuck in front of some thoughtless don or teetotal professor.) And if, by lucky chance, one found a neighbour at dinner worth talking to, one was bound to be put next to someone else after the move to post-dinner drinks. This, at any rate, was how these festivities struck Amis and it was to pass through his mind when he left Cambridge that burning his dinner-jacket in King's Parade might be the most appropriate farewell gesture he could make – better even than tearing up his notes on Jane Austen and dropping Coleridge's *Biographia Literaria* in the Cam.

'I dislike the excessive formality of faculty life,' he told *The Times* bluntly just before Christmas in 1962 when, in only his fourth term, he had already handed in his resignation from Peterhouse: 'too much dressing up and respectability. I do not like having to wait for a smoke after dinner until the port has gone round high table twice. Dining out in other colleges as a guest seems to me to involve an excessive amount of protocol.' And he spoke of escaping from the 'stuffiness' of Cambridge.

Struggling as best he could through this unsympathetic social terrain, Amis finally reached his ultimate dinner-party, the one which made him realise he could struggle no further. As he recalled in his *Memoirs*, he found himself on this clinching occasion among unfamiliar guests at one end of a college high table – it happened to be his own Peterhouse on a bad night, but it could just as well have been any other college. For the first half-hour or so the conversation was all about the paintings, drawings, engravings and other such that his neighbours had been buying. Since this was not an enthusiasm of Amis's, he said nothing. Noticing this, another guest tried to draw him into the conversation with, 'And what is *your* particular, er, line of country in this, er?' 'With Grand Guignol humility of tone and gesture,' Amis wrote, 'I said, "I'm afraid I don't go in for any of that kind of thing." The other man said "H'm" . . . Then he said, "I think that's a dreadful thing to say."' Amis remained silent after that, wishing 'for perhaps the hundredth time since arriving in Cambridge' that he was back in a Swansea pub like the Bryn-y-Mor with his old friends David Sims, Jo Bartley and Willie Smyth.

From all such charges as stuffiness Amis carefully excluded his own college, then and later. Though he was allocated some 'shitty' rooms in which to teach – first in what had once been a kind of bridge connecting parts of the college, later in a building some way from the college proper in Trumpington Street – he always liked the place and the people and continued to remember both with affection. 'The atmosphere in Peterhouse has always been most friendly,' he told *The Times*. 'My fellow dons were relaxed, unmalicious and never stooped to backbiting. I cannot say the same about the rest of the Cambridge academic world.' This academic world was yet another side of Cambridge life he found himself thoroughly disliking.

Amis had no university appointment. Under the double-barrelled Oxbridge system, which makes the colleges largely independent of the university, it was possible for him to be director of English studies at Peterhouse while not figuring at all on the strength of the university's English faculty. This spared him the need to give lectures – the only lectures he gave during his Cambridge years were to literary clubs or societies – and any direct involvement in the administration of his own subject, for which the one general meeting of the faculty he was invited to attend made him truly grateful. At the meeting, he wrote, he 'witnessed a first-rate clash of opposing juntas, perhaps staged expressly for my benefit'.

This sort of politicking within and between faculties – and probably in colleges too, though Amis found little or nothing of that in Peterhouse – seemed to him what interested his colleagues most. He had expected to find brighter, keener minds at Cambridge than at Swansea. After all, Cambridge was the top of the academic tree. Where were finer intellects to be found if not here? 'I sat back for a dose of original and well-grounded talk about English literature,' he wrote a year after he left Cambridge, but 'I did not get it.' What, to his dismay, he got instead was 'talk about intra-Faculty discord and personal quarrels, syllabus changes, the proportion of firsts to other classes, the attendance at old Joe Soap's lectures, etc.' In other words, gossip, and of a pretty high order of bitchiness too. But not a word was ever uttered about, say, Donne or Tennyson or Traherne. Now and again in Swansea somebody would mention a real writer in ordinary conversation and even if what was said was not especially profound the fact of talking about it showed the speaker took some interest in his subject for its own sake, not simply as a vehicle for politics and intrigue. There was no evidence to be had of similar interest in Cambridge. At least none came Amis's way.

One of the bitchiest of all remarks that did reach him was about himself, and it was reported to have been pronounced by Cambridge's leading literary critic of the day, Dr F.R. Leavis. According to frequent and well-authenticated accounts poured into Amis's ear, the fellows of Downing College where Leavis belonged were discussing a problem – it might have been car-parking – when somebody suggested following Peterhouse's example in solving it. Leavis is then alleged to have said: 'Peterhouse can't expect to be taken seriously about anything now that it's given a fellowship to a *pornographer*.' Leavis did not, it seems, mention Amis by name, but nobody was in any doubt that that was who he meant. His shaft seemed to Amis not only bitchy but wildly wide of the mark, since his novels were by most current standards uncommonly discreet and reticent about matters of sex. Deliberately so: not out of prudishness, but because Amis has always believed it impossible to write in detail about what people do to each other in bed without sounding laughable or absurd.

Amis had once been a moderate admirer of Leavis's and continued to admire a few of his pieces, particularly his essay on *Othello*. But he came to think that Leavis had done more harm than good to literature, though he did not get around to this judgement until after he had left Cambridge. Perhaps Leavis's description of him as a pornographer helped colour his mature view. But Leavis's published comments on Amis were certainly not of a kind to make Amis look kindly on him. In a sentence which yoked Amis curiously together with Ernest Hemingway, Leavis loftily declared that a period when Hemingway could be accepted as a great writer and an Amis novel described as 'a serious study in amorality' could only be one 'marked by a collapse of standards'. In a footnote in his book on Dickens, Leavis contrasts the two writers, claiming that Dickens 'does not use his remarkable powers of ridicule and satire irresponsibly – perhaps the true sign of great art', while Amis did. Amis never met Leavis, though he passed him once in the street, Leavis dressed in his usual open-necked shirt. Perhaps if they had sat down together they could have had that serious talk about literature which eluded Amis throughout his Cambridge days. Leavis might even have found Amis quite serious and well informed on the subject. Then again, he might not.

As with Leavis, so with the rest of the English fraternity. Amis did not hit it off with any of them. Donald Davie of Caius College accused him of thinking he, Davie, was 'square' – which Amis did think he was, though thinking somewhat better than that of his work.

George Steiner of Churchill College failed tests of possible friendship on two counts. First, though American by birth, he had acquired 'a heavy German-type accent', immersed himself in European culture and was 'not afraid of making statements about' the literature of most major European languages. This was bound to raise the Amis eyebrows, if not curl the Amis lip, since he is relentlessly sceptical of anybody's claims to understand literature in a foreign language unless he has immersed himself in that tongue for many years. Steiner also failed another standard Amis test. He did not keep the drink flowing at dinner-parties. Andrew Sinclair, the novelist and don, also at Churchill, fell at the drinks hurdle even more decisively, putting a seal on the improbability that he and Amis would ever become friends by turning up at the Merton Arms for a session only to discover that he had no money about his person. In one way or another all members of the English faculty that Amis came across failed to meet his standards of amiability. They were either unsympathetically strait-laced and mean or too zealously embroiled in faculty politics. Or they were both. Among fellow dons he came to find some of the scientists more congenial. They liked music and they liked contemporary writing and they did not know or care about the English tripos.

Failing to make friends in the obvious places, Amis looked elsewhere for companionship he could enjoy. George Gale was one person who, when available, could be counted on to fill that bill. During Amis's first summer at Cambridge the two of them appeared at a party given by Maurice Cowling, who was to become a fellow of Peterhouse after Amis left, not only having had a drink or two but dressed up in women's clothes as well. Their intention was to liven up, even shock, what was intended to be a fairly formal event and they succeeded. The following September they turned up tight again, this time at the front door of a Peterhouse flat where the Penneys lived. Gwyneth Penney was getting ready to give a lunch-party when the doorbell rang at five minutes to one, just when her guests were due. She opened the door to find Amis and Gale standing there, as she remembers them, 'swaying'. Gale said they wanted to take a look round the flat to see if it would do for Amis to have as rooms after the Penneys left – they were due to move elsewhere soon – and in spite of her protests they pushed their way in and had their look round. Only his considerable resources of charm allowed Amis to get away with this kind of drunken boorishness.

Other friends visited too. Stuart and Eve Thomas, Anthony Powell, Philip Larkin, John Davenport, the very large and very drunken literary journalist who aroused Amis's sense of pity by assuming that an invitation to spend the night meant that he was expected to sleep on the sofa, not in a proper bed in a real bedroom. Mervyn Brown, Amis's old tenor-sax-playing Oxford friend, dropped in to see him while playing tennis in the area – he had won an Oxford blue at the game – and found he was just in time to accompany Amis and Davenport to the pub. He clearly, or fairly clearly, remembers the three of them consuming four bottles of Beaujolais before returning to Amis's home for 'vodka soup'. Not being a drinking man in the same league as the others, Brown took the precaution of eating his lunch before making another call the following summer – in time, fortunately for him, to catch Amis and Davenport stepping out of a cab bringing them home from the pub. (The summer after that he called again. But this time Hilly was there alone. Amis was in Spain with Elizabeth Jane Howard.)

The one memorable encounter with a new face Amis had at Cambridge – the only one he thought worth writing up at the time – was with the Russian poet, Yevgeny Yevtushenko, who came to do a poetry reading and spent several hours with Amis first. They got on surprisingly well, though there was something of a gap in understanding between them. What was Amis to make of such gnomic utterances that came to him, through an interpreter, like this: 'Hitler's cigarettes were the chimneys of the extermination camps' and 'There is an old Russian proverb which says that some kinds of simplicity are worse than theft'? Amis coped as best he could and he was genuinely impressed by Yevtushenko's poetry reading, even though necessarily done in a foreign language. 'He is the first completely good reason I have met with for liking the U.S.S.R.,' he wrote a few weeks afterwards. When Yevtushenko was nominated in 1968 for the Oxford professorship of poetry, however, Amis was prominent in opposing him on the grounds that, behind a façade of liberalism, Yevtushenko was no more than a publicist for the Soviet regime which Amis by then so thoroughly detested.

The new acquaintances Amis made and most valued at Cambridge were the undergraduates he taught. One of these was Richard Eyre, later to become director of the Royal National Theatre. Eyre had come up to Cambridge against his military family's wishes. His father, being 'violently anti-academic', preferred the university of life to any other. Eyre was qualified in maths, physics and chemistry and had been accepted at Peterhouse to read science. But no objection was made when he proposed switching to English. On the contrary, having taken Amis on to teach that subject, Peterhouse had to provide him with a supply of undergraduates to justify his appointment – Eyre would do as well as anybody. His having no deep background in English literature or knowledge of the prevailing Leavis approach at Cambridge was no handicap with Amis. He was 'rather courteous and generous' when Eyre read out his essays.

Amis, Eyre soon discovered, was against adherence to any single approach or approved canon of critical thought and thoroughly disliked the Leavisite habit of dividing writers into those who were acceptable and those who were not. Eyre remembers Amis 'sneering' at Leavis. Whatever orthodoxy was in the ascendant, Amis was against its being swallowed whole. Eyre would produce an essay about, say, *Twelfth Night* 'marinated' in other critics' opinions. Having heard him through, Amis would insist on Eyre's own opinion. 'Do you think it's any good?' he would ask. Well, of course, it must be, it was by Shakespeare. 'Yes, but is it any *good*? I mean as a *play*. It says it's a comedy. Fine. But does it have any decent jokes?' Amis wanted his students to read for themselves and make up their own minds. There was some feeling among Amis's Peterhouse colleagues that his methods were too abrasive, too irreverent for some of his students to handle. Eyre felt a bit like that at the time but later changed his mind, coming to think that it was 'shocking, but healthy, for a young and impressionable man ripe to become a fundamentalist in matters of literary taste and ready to revere F.R. Leavis as the Ayatollah of "Cambridge English".'

Amis would also take Eyre for a drink at Miller's Wine Parlour opposite King's College, where he introduced him to John Davenport. 'They were the first two writers I'd ever met,' Eyre says. 'I thought I was in heaven.' Inviting his undergraduates out for a drink or to a party at his own home seemed to Amis natural things to do. They were normal in Swansea to the point where it would be abnormal in some pubs for there not to be circles of students and teachers drinking together. So why not Cambridge? But here it raised eyebrows and worse. Amis found he was acquiring the reputation of being homosexual. Whatever his practice in youth – before girls became available – by this stage in his life any such leanings, as he later wrote, 'must operate at a level inaccessible to introspection'. So rumours of his being queer were puzzling until he realised they must be prompted by his being seen in pubs with his undergraduate students, in those days inevitably at Peterhouse all male. That drinking with his students should appear perverse itself seemed perverse to Amis. He thought that contact outside the formal circumstances of lecture and tutorial were valuable because exchanges between student and teacher could take place on something nearer equal terms. Besides, shouldn't someone who taught young people actually enjoy their company?

Here was something else Amis found wrong with Cambridge – the gap between the teacher and the taught. He made a habit of asking undergraduates what they felt the lack of most in Cambridge life and they all told him the same: shortage of contact with senior members of the university. When he suggested that dons might be nervous of meetings beyond the odd official sherry-party, the undergraduates said the dons could at least try. 'The confusion and apathy of many students, the lack of sense of direction – not to speak of more painful troubles – so often commented upon, all derive in part, I suggest, from failure to bridge the gap.' Amis wrote this in 1964, well before the student unrest of the later sixties. But when these came he was not altogether surprised. Although he thought there were other factors involved – including too rapid expansion of the universities, bringing in too many young people unfitted for the academic life – he had already sensed an uneasiness in relations which, if not resolved, would inevitably become further soured and disintegrate in one way or another.

When Amis left Cambridge he gave up teaching for good, apart from a brief stint in Nashville, and so lost all contact with students. This was what he was to miss most about the academic world for the rest of his life. At university he had two sorts of relationship with his pupils: formal, in the lecture or the tutorial, and informal, in the pub or at a party. The first was necessary, not only for the student but for the teacher too. The teacher could try out his ideas on these occasions. But it was more than a matter of giving them a dry run in front of a captive audience. 'Teaching his ideas', Amis wrote, 'teaches him what these ideas are.' When he gave up teaching this stimulus was removed, leaving only the pub or the party. Such occasions, though, could only be a supplement to the formal side of teaching and you could not – or Amis could not – get down to sustained and serious talk about writing with a glass in your hand. Almost all that was to remain of that kind of discussion was a chat now and then with Philip Larkin or, later, with his son Martin.

However, if it was teaching students that Amis was to miss most, it was also what drove him out of his academic profession and out of Cambridge. He came to the conclusion that he could be a writer or he could be a don but he could not be both. This is one of those mysteries that baffle outsiders. How could anybody possibly claim that a mere dozen hours of tutorials a week could leave a man too exhausted to do anything else? To the outsider it is not enough to say that a certain amount of mugging up had to be done as well – Aristotle, Racine and Strindberg (to Amis, unwelcome intruders on an English course) as well as *Volpone*, *The Anatomy of Melancholy*, *Bleak House* and *Heart of Darkness* (also not much to Amis's taste, though at least written in English). Surely a Cambridge don could take all that in his stride? Some can and do, and find time to write as well. Amis could not. To him teaching had the same sort of draining effect as appearing on television – not as bad, perhaps, but bad enough. After his tutorial stints he felt fit to listen to records but not to write novels, which was what he wanted to do. Besides, he needed the money that writing brought in. The larger part of his income came from that source now and he could not possibly maintain his family on his Peterhouse earnings alone (which were not topped up by any payments he would have got if he had been on the university's payroll as well).

The burdens of teaching, social frigidity and faculty politics would perhaps not by themselves have been enough to drive Amis out of Cambridge. But they combined to make him feel uncomfortable and at the same time to frustrate his wish to write and make money by doing so, and Amis now accepted that these, rather than his academic career, were his priorities. By the beginning of his second Cambridge year he was ready to make another move. But where?

An idea had begun to take shape during the summer vacation, when Amis was offered a commission from the American magazine *Show* to interview the poet Robert Graves, who lived in Majorca. Amis had first written to Graves in 1954, pointing out that in an article in *Punch* Graves had said that he wrote just ten words a day, which according to Amis's calculation would give him only a 7,300-word novel every two years. 'Must stop now and write my word for the day,' Amis added. 'I plan to bring out a *conte* every twenty years.'

Graves acknowledged his postcard and Amis took this as a cue for a proper fan letter. 'I now have some excuse for saying what I have often wanted to say, that for my money you're the best poet now writing in English ... I know you know that all poets like your verse best, after their own of course, but I thought I'd just confirm it from my own corner of the battlefield.' In 1961 Amis was among the 123 MAs who nominated Graves for the Oxford poetry professorship, in succession to W.H. Auden. And in June that year, when Graves came over from Majorca to deliver the Creweian lecture, he lunched with Amis in Oxford. They got on well enough for Graves to be in receptive mood when the interview was proposed. Graves did not like journalists who used him to get a free holiday in Majorca, but he thought Amis was 'one of the few non-phoneys' and said it would be fun if he came.

Robert Graves put up Amis and his family for nine days in August in the Posada, or guesthouse, on his property at Deya, a tiny fishing village where he had lived since 1929, with a ten-year break from 1936 forced on him by wars. Looking round, Amis observed that Majorca seemed comfortable, the climate reliable, the cost of living low and the population 'scrupulously courteous'. There was no politics but lots of 'fruit, trees, wine, sea'. When he got home, Amis wrote to Graves saying that he had been 'fired by your example to attempt an 8am start to the writing day'. But Amis had been fired by more than that. Whatever reservations he might have had about living abroad, Graves was visible proof that exile need be no handicap to serious work. A move to Majorca would create an opportunity for undisturbed writing in congenial and affordable surroundings with at least one good writer for a neighbour.

Amis made up his mind to rent a house in Majorca for a year, move the family there and give exile a try. Exile was most often the resort of writers and artists who wanted to escape the English legal regime, which still made a crime out of homosexual behaviour. But nobody could accuse Graves of that. All in all, exile seemed a sound way out of Cambridge and worth at least a year's trial.

When he had made up his mind, Amis told the Master of Peterhouse, Herbert Butterfield: 'Sir, with great regret I want to resign my Fellowship at the college.' 'In heaven's name, why?' asked Butterfield. Amis gave a 'selective explanation' about the writer's need for solitude and so forth. 'Yes,' Butterfield said. 'Romanticism. That's what they used to call it in my day. The belief that you can help yourself to become an artist or be a better artist by going somewhere remote or in foreign parts.' Butterfield may have found this a quaint idea but he did not try to stop Amis from leaving.

If Amis did not care much for Cambridge, his feelings came to be reciprocated. At Peterhouse, Brian Wormald was particularly upset. He had invested much of his own credibility in putting Amis up for a fellowship, and now here was Amis planning to defect after barely four terms.

Other men have no doubt left the university with as much relief as Amis did, but they have mostly been tactful enough to keep their dark opinions of it to themselves. Not Amis. In interviews and in articles he said what he thought, about both Cambridge and the whole expanding university system. Philip Larkin surely reflected the resentful feelings of the university professionals when he wrote to Robert Conquest about Amis's departure: 'I think it wd have been more graceful of K. to glide out of Cambridge without all this public posturing. If he made a mistake in thinking he could write there that's his fault, not anybody else's: no need to put them in the wrong. I can't imagine Majorca will be any better.' Thirty years later, Amis's abrasive remarks still rankled, especially in Cambridge. And if one were to wonder why it is that Amis has never received an honorary degree – while Larkin, for example, got seven – Amis's outspokenness about the universities is surely a part of the answer.

Above: Hilly, Philip Larkin, Amis and Margaret Aeron-Thomas, Amis's social-working friend.

Below: Flat out on a Yugoslav beach, where an angry Hilly has written in lipstick on Amis's back, 'One Fat Englishman' (the title of an Amis novel) and below 'I fuck anything'. They split up shortly afterwards.

George Gale.

Philip Larkin.

Robert Conquest.

Anthony Powell.

Elizabeth Jane Howard.

Above: Amis and Jane flanked by their American friends, Mary and Mike Keeley, in Greece in the mid-1960s.

Below: Amis and Jane outside Lemmons in the 1970s.

Above: Sally Amis.

Martin Amis.   Philip Amis.

With washing in Regent's Park Road in the 1980s.

Amis with moustache and broken leg after an accident in 1982.

Home and away:
Above: the Welsh party – (*left to right*) Stuart Thomas, Eve Thomas, Paul Fussell, Mrs Fussell, Amis, Virginia Rush and Michael Rush.

Below: the London team. Saturday in the Queen's public house among friends and neighbours.

At St John's College,
Oxford, 1991.

With Sarah Snow the cat.

∞ 15 ∞

Elizabeth Jane Howard

Amis was due to depart at the end of the academic year, which meant that he would continue teaching at Peterhouse for two more terms after he had submitted his resignation. During the Easter vacation he went with Hilly to Majorca to look for a house, leaving Graves's address at Deya with Conquest as a point of contact if he needed to get in touch. They found a house that suited their purposes at Soller, within easy reach of Graves, and arranged to rent it for a year at £9 a week, beginning in the autumn. But before Amis could leave Cambridge – even before he had offered his resignation from Peterhouse – he had embarked on an adventure which was to disrupt the whole project of moving to Majorca and destroy his marriage to Hilly.

The adventure began with an invitation to the Cheltenham Literary Festival. There was nothing remarkable in such an invitation, though, in the light of what was to follow, a serious novelist would surely have hesitated to issue the same invitation to one of his characters, for fear of being seen to decline into farce by signalling the next twist in his plot with a leering lack of subtlety. For the seminar in which Amis was asked to take part was on 'Sex in Literature'.

The idea of inviting Amis came originally from the *Sunday Telegraph*, which had put up £1,500 to sponsor a day's activities at the festival and so acquired rights to determine that part of the programme. But it fell to the festival's artistic director that year, Elizabeth Jane Howard, to write the letter asking Amis to be one of the panel of four.

The seminar took place on 4 October 1962 in Cheltenham's

town hall before an audience of about a thousand people. On the platform alongside Amis were two Americans and a Frenchman with the *Sunday Telegraph*'s editor, Donald McLachlan, taking the chair. Accounts of the proceedings are sketchy. Amis remembers saying that Jane Austen did not write about sex but wrote instead what he called 'pecuniary pornography'. According to the *Daily Express*, he also declared that it was 'the novelist's duty not to be reticent about sex' but went on to add that sex scenes were 'difficult to handle properly' and that descriptions of 'physical love-making make me embarrassed – especially Lady Chatterley'. This was and is the standard Amis view: writers must not shirk sexual directness but are well advised to avoid blow-by-blow detail.

Romain Gary, the French writer and diplomat, took refuge in the ambivalent line that while he was 'dead against pornography' he wouldn't give any judge 'the power to draw the line'. Joseph Heller, author of the then recently published and enormously successful novel *Catch 22* (a 'largely undeserved' success, according to Amis), offered the view that the makers of mink coats had corrupted more girls than any literature had done. Amis thought little better of the fourth member of the panel, Carson McCullers, author of *The Heart is a Lonely Hunter* among others, whom he had already met in New York. She gave it as her somewhat question-begging opinion that 'so long as a book is true and beautiful' it could not be pornographic.

This was routine stuff, and the seminar appears to have been a fairly lacklustre event. According to Peregrine Worsthorne, then as now a journalist with the *Telegraph* and among the seminar audience, it suffered from the chairmanship of his editor, who didn't know much about literature nor much about sex either, and also from Carson McCullers being drunk. Amis remembers Heller livening things up a bit by responding to someone in the audience who wanted to know why sex always seemed more burden than pleasure in fiction. 'They had a hell of a good time in my book,' said Heller.

But it was away from the formal proceedings that Amis's adventure began, when he was introduced to the artistic director, Elizabeth Jane Howard, always to be known by her middle name, Jane. Had he been invited in any other year he would have missed her, for she was to be the director for 1962 only. She brightened things up a bit by spending £100 on fireworks and claims her year was a rare one in which the festival turned in a profit. But the pay was just £300 for a job that took eight months' work – not enough money to encourage a woman with a living to earn to take it on a second time round. She and Amis had run into each other a few times before but only by chance, at television studios where both happened to be making appearances, and once in the street when Amis came.across his old Oxford friend Kenneth Tynan with her on his arm. Amis was aware of her existence and her attractiveness but had not yet taken sufficient interest in her to go so far as to read even one of the novels for which she was already well known.

What happened when they met at Cheltenham is best described in Amis's own words. 'I sort of threw a pass at Jane which was sort of accepted.' This was not unusual behaviour for him. He was then, as he puts it, 'what I always was, on the lookout for a good time'. Amis was immediately attracted to Jane and she to him, but he had nothing more in mind than a possible fling, of no greater consequence than any other. As Hilly was with him, nothing could in any case develop there and then. Being familiar by now with his enthusiasms for women, Hilly noticed at once that there was something in the air between her husband and Jane. She thought Jane was 'tossing her hair' at him in a fairly blatant way. Still, though there might be another affair in the making for her to put up with, Hilly could not then have imagined it would go further than the others, any more than did her husband.

Elizabeth Jane Howard was undoubtedly something of a prize for a man on the lookout for a good time. Born a year after Amis, she was beautiful still at thirty-nine, with long, thick, blonde hair and attractively high cheekbones. When Evelyn Waugh insisted that he would only be interviewed for television by a beautiful young lady, she fitted the bill perfectly. She was talented too, already the author of three highly praised novels. And she was socially well connected.

Since people look for easily understood explanations, this combination was what most who knew Amis settled for when they later tried to account for his going off with Jane. Her glamour – a compelling mixture of beauty, talent and class – was thought to have been simply irresistible to the lower-middle-class Amis. It was 'a snob thing' in one view, and that view came to be widely held by Amis's friends. As often, the reality was more complicated.

For one thing, Jane's background was not aristocratic but rather – if one is looking for exactness among the intricate tracery of English class distinctions – well-off middle class. Her grandfather had started a timber business, Howard Brothers, which survived through four generations until it finally went broke in the 1980s. Grandfather bought a London house for his three sons to share, which meant that when Jane was a child there were a dozen or so cousins about in the school holidays, as well as her two younger brothers. The year before she was born her mother Katherine had given birth to a daughter who died. Jane came to believe that her mother disliked her, preferring her brothers to herself and later resenting the fact that Jane was able to pursue a career of her own as well as marry and have a child. Katherine had had to give up her own career as a dancer when she married – her husband insisted, as it was perhaps to be expected he would. The rule which then prevailed was that wives did not work. In this at least the Amis and Howard families resembled each other at their respective points either end of the middle-class spectrum. Neither mother had a job or career, though this does not appear to have troubled Peggy Amis as it did Katherine Howard.

Since women were not expected to work, it followed that there was no need to educate them very rigorously either. Jane went to school for only two terms. The rest of her education was undertaken by a governess, Miss Cobham, who had also undertaken her mother's instruction when she was a child. Jane remembers Miss Cobham as very old as well as being 'the ugliest woman in the world'. Yet she was not stupid, having written books on philosophy and mathematics, though her teaching curriculum was eccentric. Jane learned Greek and Latin, for example, but no French. With so many brothers and cousins around she did not miss the social contact that school would have provided. But later she came to regret her lack of a longer, more formal education. She read a lot for herself, but understood only about a third of what she read. School lessons ended once and for all when she was sixteen, and university was out of the question. Like work, it was not something gone in for by young women like herself. Besides, the Second World War had begun.

When she was nineteen in 1942 Jane married Peter Scott, son of the celebrated national hero Captain Robert Falcon Scott of the Antarctic, and later famous in his own right as a naturalist and painter and conserver of birds. Of course, Jane thought she was in love with this man fourteen years her senior, but later she believed it was more the case that she had married the first man who told her she was attractive. Perhaps, too, she wanted to escape from a loveless home – her parents were by now divorced and her father remarried – into one where she could hope to be loved. From Scott's side there was no doubt attraction too, but also a wish to have a child in these dangerous times before it was too late – he fought in destroyers through the Battle of the Atlantic, was mentioned in dispatches three times and won the Distinguished Service Cross twice. His own mother was anxious that he should become a father and in this the marriage did not let him down, Jane soon bearing him a daughter, Nicola.

Jane's youth and inexperience were put under extra strain by wartime conditions. She worried about her husband whenever he was at sea and found herself staying much of the time alone in hotels, sometimes the only woman for miles around among thousands of servicemen. If there had been no war she might never have married Scott. As it was the marriage only survived for a few post-war years. When Jane finally decided that her husband was more interested in natural life than in people – even his wife and daughter – she bolted, as she was to do from her second and third husbands too. 'Bolter' is the word she uses to describe herself, meaning that when she can put up with a relationship no longer she decamps, with little or no notice served on her partner.

In wartime conditions, the middle-class proprieties were set aside. Like everybody else, Jane worked for a living then and was to do so for the rest of her life. Before settling to become a writer, she went from job to job. Through the war years she was variously a model, a publisher's reader, a newsreader on the BBC and in 1947 secretary to the Inland Waterways Association. Then she turned to writing, and her first novel, *The Beautiful Visit*, was published in 1950, winning the John Llewellyn Rhys Memorial Prize.

Being highly attractive, she was not short of offers from men when she and Scott parted, including offers of marriage. At first she was more inclined to accept a proposal to make love than to marry, finding the men who wanted to become her husband unsuitable and being reluctant to take a second plunge into marriage too soon after the failure of her first. In time she changed her mind. She became tired of men wanting no more than to go to bed with her when they had known her only half an hour; and she wanted more children. Marriage was the best way of avoiding the first and achieving the second. But her second husband, whom she married in 1959, eight years after her divorce from Scott, turned out even less of a success than her first. James Douglas-Henry, an Australian broadcaster, was eager enough to make love to her before their wedding but stopped doing so afterwards. Jane came to think that she had been deceived by his charm and cleverness into failing to realise that he was only interested in her money and connections. It was a measure of the relative degrees of failure in her first two marriages that, while she maintained amiable relations with Scott, she preferred to draw a veil over her time with Douglas-Henry and never to mention him at all. Amis only ever caught sight of him once, from the rear, interviewing someone on television. When Amis came into Jane's life in 1962 she had bolted for a second time.

Unable to pursue matters with Jane at Cheltenham, Amis took advantage of his trips to London to pursue her there instead. This was how he had conducted his affairs in his Swansea days. But this time he allowed things to go further. As the months passed, the intensity of his feelings deepened. He enjoyed Jane's company more and more, in and out of bed. Still, though this may have been a more compelling affair than others, he had no thought-out intention of turning it into something more than an affair. He tried to keep his usual veil of discretion drawn. However, not for the first time, a flaw developed in his security system. Hilly discovered a compromising letter in his jacket pocket. In the past, and with other women, that might have been enough to end the affair or at least force Amis into taking greater care not to leave clues about what he was up to lying around for his wife to come across. Now, though, there were to be no promises of better behaviour. On the contrary, Amis announced that, come the summer, he intended to go to Spain for three weeks and to take Jane with him.

Meanwhile, ordinary life continued, more or less. In the early summer Amis and Hilly went with two other couples to a science-fiction film festival at Trieste. The other couples were the science-fiction writers Brian Aldiss and Harry Harrison, plus Aldiss's wife-to-be Margaret and Harrison's wife Joan; Amis and Hilly joined their ranks as a consequence of his pioneering study of the science-fiction genre, *New Maps of Hell*. One day the six of them slipped across the Yugoslav border to Koper, formerly Capodistria, where after a good lunch all six went down to the beach. Amis stripped off and fell asleep on his face in the sunshine. Overwhelmed with misery at the latest evidence of his infidelity, Hilly seized the lipstick from her bag, and on his bare back she wrote '1 Fat Englishman' (the title of the book he was working on). Underneath she wrote, 'I Fuck Anything'.

Amis had never before been so blatant about an infidelity. He had tried to avoid Hilly finding out about his women and, if she did, to make as little of it as he possibly could – to brush it aside as unimportant. Nor did he make a great fuss about Hilly's own affairs. This was part of the understanding that kept their half-open marriage going. And although their marriage had also been subjected to the inevitable wear and tear of the years – juggling budgets, managing three children, diminishing sexual excitement – Amis still loved his wife and children and still wanted very much to keep the family intact. He liked what was familiar as well as enjoying the thrill of what was different, and to a man of his anxious temperament the disruption involved in separation or divorce was a horror to be avoided. There was a sense, too, in which a holiday with Jane could be seen as no great break with the past. Hilly knew about the affair, after all. What was so very different about making it semi-official for a few weeks on the understanding, which Amis willingly gave, that he would return to the fold when the weeks were over? This was an adjustment to previous behaviour, not a radical departure from old ways. Or so Amis persuaded himself. In this mood, he saw no reason why he should not go away with Jane and no reason why he should not come back to Hilly either.

If this was his strategy, Amis might have been more tactful in his tactics. He left Cambridge for Jane on 21 July, Hilly's birthday, and on their bed he left behind as a birthday present a black, baby-doll style nightie which, to make matters worse, Hilly suspected had been chosen for her by Jane. Amis spent a night or two in London with Jane and then set off for Sitges, where he had taken the family for holidays before. For the rest of their three weeks he and Jane read, drank, ate, made love – though Amis was no more keen than he ever was on swimming or sitting in the sun and getting tanned, which Jane loved. Wind of their whereabouts had somehow reached Fleet Street and a reporter from the *Daily Express* came sniffing round their hotel. Luckily a sympathetic proprietor put him off the track, moving Amis and Jane to another set of rooms to make doubly sure.

Now, for the first time, Amis read one of Jane's novels, *The Sea Change*, taking two days on the beach to go through it. When he was finished he gave his verdict: it was good and he had enjoyed it. This was just as well. For a man as much given as Amis to speaking his mind about literature in the most categorical terms it would have been impossible for him to have had a long-term relationship with a woman novelist whose work he could not respect.

Yet Amis still did not anticipate having to sustain a relationship with Jane. His plan was to return to Cambridge, then take the whole family off to Majorca in time to get the children settled in and ready to start at the international school in Palma for the Christmas term. The first part of this plan Amis duly executed. He went back to Cambridge; but there he found the house in Madingley Road empty. There was no note but he soon realised that Hilly had forced the issue. From friends he learned that she had gone off to Majorca, taking the children with her to the house Amis had rented at Soller. Although Amis was surprised that she had gone, he was not surprised that, once she had decided to go, she should have chosen Majorca as refuge. His response was to get straight back on the train to London and head for Jane's flat in Maida Vale.

Hilly and Amis were never to live together again as husband and wife. It may be that their marriage unravelled more by miscalculation than by plan, certainly with no deliberate intention that it should on Amis's side. Perhaps, he thought, Hilly had gone off to Majorca as a way of calling his bluff, in the hope that he would rush after her full of contrition and eager to renounce Jane. If so she was mistaken. Amis took the line of least resistance. He had hoped to return to the bosom of his family but now he had another bosom to return to, Jane's. With her, he anticipated, there would be warmth and uncomplicated acceptance. There would be no rows to be gone through, no difficult reconciliations.

Soon after Amis moved into Jane's flat they decided to move out and rent another flat where for a while they could be on their own. Jane shared her flat in Blomfield Road with her brother Colin, and there was always the possibility that he and Amis might not get on. The two had met at Cheltenham and they were later to find that they could live in the same household perfectly well. For the time being, though, it seemed sensible for Amis and Jane to try out their new life together alone. They found a flat in Knightsbridge, in Basil Mansions, near Harrods, and moved in.

In Majorca, meanwhile, the three children were growing unhappy. They began to pine for their father. If he could not be there with them they wanted to hear from him at least. They took to going to the road at the end of the garden after breakfast to wait for the postman to arrive on his motor bike. Occasionally he brought with him what they hoped for – a card from Amis. But they wanted more than that and in the middle of November Hilly put fifteen-year-old Philip and fourteen-year-old Martin on to a plane for London, giving them Amis's address in Basil Mansions and telling them to head there from Heathrow.

The plane was late and it was around eleven at night when the two boys rang the bell at Amis and Jane's flat. Their father opened the door in his pyjamas, seemingly unsure how to handle their sudden appearance. He had issued a general invitation for them to come and see him but was not expecting them to show up at that precise moment and was uncertain how much they knew about what was going on. 'You know I'm not alone,' he said, and Jane loomed up dramatically in the background wearing a white towelling robe to prove it. They ate bacon and eggs and the boys tried to play it cool, as if meeting their father's new girlfriend was an everyday thing for them to be doing. Jane took care to see they had a good time for the few days they were in London and Amis settled down with them for a couple of lengthy afternoon chats. Philip impressed his younger brother by calling his father a cunt during one of these. But it became clear to them that their father was going to stay with Jane and not join the family in Majorca.

One other event during the boys' brief stay before they flew back to start the school term in Palma stayed in everybody's mind. George Gale had come for dinner and while they were having pre-dinner drinks Amis took a telephone call from the journalist Quentin Crewe. 'What?' the others heard Amis shout incredulously. When he repeated what Crewe had told him Gale promptly left for his newspaper office. This was how they all learned that President Kennedy had been shot on 22 November 1963.

If Amis's marriage to Hilly broke up through accident and miscalculation, like a ship in a sudden storm, the rock on which it came to pieces was solid enough and had been a lurking threat for years. This rock was a fundamental difference of view between Amis and Hilly about married life. He thought he should be free to have affairs and she did not, in spite of her own. Perhaps their attitudes were ultimately irreconcilable. But events and circumstances played their part. If Jane had not been without a partner in the one year she directed the Cheltenham Festival . . . If Amis had not been invited to take part in a seminar that year . . . If he had not been at that time fed up with his don's life and anxious to be on the move . . . If Hilly had not reacted to his brazen Spanish jaunt with Jane by rushing off to Majorca with the children . . . If Amis had followed Hilly to Majorca . . .

Later in life Amis was to speculate on how events might have worked out if each person in this little drama had played his or her hand differently. But it may have been written in their characters that, given the circumstances, each one should have behaved as he or she did. And given those characters – Amis's appetite for a good time with women, for instance – patching up the marriage might only have kept it temporarily together. If the break-up had not come then, it might have come later; if not over Jane, then over somebody else. These can only be speculations, as fruitless in the end as wondering how Amis's life might have been different if he had stayed in Swansea or stuck to Cambridge. What happened was what happened – although years afterwards Amis was bitterly to regret that he himself did not play his hand otherwise.

In his fourth novel, *Take a Girl Like You*, Amis had described something very like this clash of values between himself and Hilly. Patrick Standish is on the lookout at all times for girls to go to bed with. Jenny Bunn is on the lookout too, but for someone to love who will love her in return. Patrick is modern, knowing, devious; Jenny is old-fashioned, naïve, but sure of what she wants. She falls in love with Patrick but doesn't want him on his terms, which include sex before marriage. 'It's just your training,' Patrick tells her. 'Maybe it is, but that doesn't make any odds to me,' she replies. Eventually he forces himself on her during a drunken party. Afterwards they are reconciled and plan a life together but she knows it will not be what she had hoped. Patrick promises he will change but she hardly believes him. 'You know, Patrick Standish, I should really never have met you,' Jenny says. 'Or I should have got rid of you while I still had the chance.' Given the similar difference in values between herself and Amis, Hilly might have said, or thought, the same.

Take a Girl Like You was published in 1960 and it can be read as a remarkably prescient vision of the sexual turmoil that was to break out in the decade which had just begun. But it can also be read as an account of the differences between Amis and Hilly. Amis might have been on Patrick's side in the argument over sex but he knew there was something to respect in the argument on the other side too. Patrick tells Jenny that what has happened to her was 'bound to, you know, darling, with a girl like you'. Jenny thinks he may be right, though she adds, 'But I can't help feeling it's rather a pity.' A part of Amis felt the same. American friends of his thought the book was a love-letter from him to Hilly, and Amis later thought there was a good deal in that. Certainly Hilly continued to keep a powerful hold on Amis's imagination. There is at least a trace of her in all the nice, attractive girls in his fiction, from Christine in *Lucky Jim* onwards. The traces Jane left were to be very different.

Amis did not plan to leave Hilly and go off with Jane. Rather, he stumbled from one to the other at a time in his life when he was ripe for change and ready to make a move. But once the move was made he could see few reasons for regret. Jane was glamorous, intelligent, sexy and a writer whom he could respect. She paid Amis a lot of attention, which he enjoyed. She introduced him to new places and new people. She talked to him in his own terms, as one writer to another, in a way that Hilly had never been able or much wanted to do. For Amis and Jane these early years were happy, romantic, a new grand passion, and they slipped easily from living together into marriage when Amis's divorce came through in 1965.

After their first few weeks in Basil Mansions, they returned to Jane's flat at 16 Blomfield Road and her brother Colin, who designed and built sound systems and had been known since schooldays as Monkey. Amis liked and got on with him and was happy that he should be one of the household. Others were to come and go. Throughout their life together Jane and Amis were rarely alone in their own home. Colin stayed with them almost as long as their marriage lasted. He was joined by Sargy Mann, a jazz musician and painter friend of his, who moved in when Amis and Jane went away to Nashville in 1967 and stayed for eight years. Amis's boys, Philip and Martin, were in residence through most of their teens while Sally came for holidays. Jane's mother Katherine lived with them for three years until she died. Peregrine Worsthorne stayed for six months when his marriage went through a rocky phase. And there were often parties of weekend guests.

With his longstanding fears of solitude and his intense liking for company, big households like this suited Amis well. They were reassuring and protective. If Jane went away, Amis was not left on his own overnight, with all the potential for anxiety and panic that would mean, and if he wanted a drink and a chat he didn't even have to go as far as the pub to get them. If things got too much he could always escape into the privacy of his study. There was to be a literary dividend too. The ménage of himself and Jane, Mann and Monkey and Jane's mother presented him with the central situation for his 1974 novel *Ending Up*, which is about five people living together in the same house, with the difference that in the novel all five are old and in various ways at odds with each other. The Amis household was younger and calmer.

These arrangements may have suited Amis but they may also have worked against the long-term success of his relationship with Jane. She, at any rate, came to think so. Both had passed forty when they got together, and as a couple in middle age they should, she believed, have had more time on their own. Martin observed from the sidelines that when they had been on a trip by themselves they came back 'fortified' by the experience. For the first years, though, the size of their household did not stop Amis and Jane from being happy in a passionate way.

It did, however, prompt them to look for a larger establishment than Jane's flat, which was on two floors with only two bedrooms, one bathroom, a drawing-room and a dining-room, as well as a garden at the rear. By the end of 1964 Jane, Amis and Monkey were living in a flat at Keats Grove in Hampstead – 'very small,' Amis wrote to Conquest on New Year's Eve, 'and most of the time about a millimetre thick in coal-dust. To keep it warm (and thick in coal-dust) I have to go down and up 5 flights of stairs bearing first an empty, then a full coal-bucket, or "hod" as I believe they call the type in question.' Keats Grove was a stopgap while a house at 108 Maida Vale, on which they had acquired a ten-year lease for £2,000, was being got ready; it had 'a letter-box but practically fuck-all else yet', Amis said in his letter to Conquest. But within a couple of months Amis, Jane and Monkey were installed at their new address.

Amis had changed more in his life than his partner. He had left Cambridge as well as Hilly and so given up the teaching career that had paid at least his bread-and-butter bills since 1949. Now he had to depend on his writing to pay all his bills. And he had to devise a writer's way of life to replace the round of lectures and tutorials and administrative chores that had filled up much of his day at Swansea and Cambridge. For a man of Amis's temperament, this meant creating a regular routine which had to be stuck to no matter what – a shattering hangover, for instance. It had to be flexible too, though, to accommodate the fringe activities of a popular writer, such as visiting a broadcasting studio, and to allow time for social activities, especially drinks or lunch with friends, which to Amis were vitally necessary as both recreation and reward.

Mornings became the key to the Amis work routine. On a typical day with no outside engagements he would start slowly with breakfast and the newspapers, telling himself that it was his duty as a writer to keep up with events in the world while knowing perfectly well that what he was really doing was providing himself with an excuse for postponing the awful moment when he must sit down at the typewriter and start getting words on paper. Around 10.30, still unshaven and in a dressing-gown, he would force himself to his desk where he would stay until 1 p.m. or soon after. Then he would shower, shave and have a drink. After an hour or so's more work came lunch. The afternoon was always something of a black hole. Unless there was a piece to be finished against an inexorable deadline Amis would avoid writing until it was time for a cup of tea. He might read or take the dog for a walk for an hour or two until then. But whatever he did it was not meant to count. In the contract he had drawn up between himself and himself, the hours from lunch to tea were not expected to be productive – anything they did produce was a bonus on top of the regular daily output. After tea it was back to the typewriter where he stayed until 8.30, drinking more tea until he switched to alcohol at 6. Over the years Amis's routine has varied in detail, but always around this basic pattern: an intensive morning, followed by a fairly vacant afternoon and a final burst of energy in the evening before dinner.

Routine provided Amis with the necessary framework for his life. If he had not been a writer he would surely have had to find some other frame within which to cage and discipline himself. At times he has wondered if he would not have been happier as a journalist. He would still have written, but in a less lonely, more sociable atmosphere, in an office full of chums and with a pub around the corner. But it must be doubtful if he could have lived easily with the irregular rhythms of Fleet Street, typically made up of long bouts of idleness and uncertainty followed by sudden panic rushes to deadline. On one of the few occasions when he was subject to Fleet Street rules he was instructed to rewrite a column for the *Sunday Express* within an hour or so, and he hated having to do as he was told in so short and unyielding a period of time.

While Amis was following his writing routine in his study, Jane was following hers in another room, and when both were finished they would settle down and read aloud to each other what they had written during the day. As novelists they might have been jealous or competitive but they were not. Rivalry did not arise. In no other aspect of their lives did each understand and sympathise with the other's problems better. A partner who was not a writer of fiction would likely shrug off in irritable style worries about plot or character – about whether the personality of Amis's Mr X really was convincing or if too many words were not being wasted moving this or that bit of Jane's story along. But Amis and Jane knew that these were exactly the sort of seemingly trivial things that made writers anxious. They encouraged each other and offered suggestions, each becoming so much in tune with the other's writing that on one occasion they were both able to write a few pages of the other's novel-in-progress. Amis wrote a bit of Jane's *After Julius* and she wrote a bit of his *One Fat Englishman*. And they did it so plausibly that it was impossible for anyone else to tell who wrote what. (Monkey was the exception who spotted the differences.) Their co-operation as writers was to be the most durable part of their relationship, lasting to the very end when every other part was dissolving.

∾ 16 ∾

Maida Vale

During his first year with Jane, Amis worked on a novel, *The Egyptologists*, and a critical work, *The James Bond Dossier*, both of which were published in 1965. Here was evidence of his burgeoning professionalism and readiness to turn his hand to writing beyond his mainstream novels and poetry. They gave him a chance to earn some much-needed money too and to put behind him the more rigid austerities of university life. Amis had no colleagues now to call him a pornographer at the first hint of sexual impropriety. He could have some fun without worrying what effect it would have on his chances of promotion or his reputation as an academic of the required weight and substance.

The Egyptologists had been lurking around waiting to be completed since 1960. It lurked in the form of a rough draft by Robert Conquest, who had devised the idea for the story and given it a working title, 'Mummy Knows Best'. Conquest's plan was for Amis to knock his version into good enough shape for it to be published under both their names. The basic idea revolved around a club supposedly dedicated to the study of ancient Egypt but in fact a cover for its members to get away from their wives and pursue other women. As Amis remembers it, he invented new scenes and characters and generally 'put in the plot'. He took it with him in August 1964 when he and Jane went for a holiday in the house he had rented in Majorca, which had been empty since Hilly returned to England with the children before Christmas.

Amis reported progress to Conquest from there. 'Life proceeds in a full-of-fun way. With Cointreau at the equivalent of something like 30s. a *magnum* it can hardly do anything else. I am quite brown really. Graves seems a bit dejected but very ready to chat. Various creepy non-exhibiting painters, non-publishing writers etc. inhabit his ambience, but who cares a sod for them? Egyptology, given sun, wine, and other distractions, doesn't go forward very fast, but it is *steady*. Have just passed p. 100, which I reckon is half way. I think we stand to make a modest packet out of the bloody thing.' *The Egyptologists* was finished at the beginning of December. 'I ought to have told you earlier,' Amis wrote to Conquest at the end of that month, 'but general fucking about, including Xmas, and disinclination having finished typing out the Es to do anything whatever – this imposed a delay. The final scenes follow you on the whole . . . Anyway, I think you'll approve.' Amis's forecast of making a modest packet out of it came true. Richard [later Lord] Attenborough bought the film rights for £25,000, which Amis and Conquest divided between them, though Attenborough never got around to making the film.

A second Conquest draft novel was not so fruitful. 'Peach Key' – which sounds as if it must be a Caribbean island but when spoken aloud also sounds like the word for cunt in Bulgarian, a language Conquest knew – had been lurking alongside *The Egyptologists* since 1960, waiting its turn for Amis to get to grips with it. Over the next three decades it continued to lurk, until in October 1993 Amis got the typescript out once more. He had sent off his latest novel to his publisher the day before and wondered if now might not be the moment to get down to Conquest's project at last. But after looking through 'Peach Key' again he abandoned it once and for all. Like *The Egyptologists*, it too was about escaping from women but Amis decided there was not enough raw material in it for him to work on. The setting, for example, was the Caribbean, but there seemed no evidence in the text that Conquest had actually been there or knew anything much about it. So Amis put it away for the very last time and launched himself into another novel instead.

Before he got down to serious work on *The Egyptologists*, in fact by May 1964, Amis had already finished *The James Bond Dossier*. It took him five months. He had started out by carrying off on holiday all Ian Fleming's twelve Bond novels and one volume of Bond short stories, thinking he might be able to get an essay of perhaps 5,000 words out of them. But he found as he worked that there was more in them than he had realised and he finished up with a book of some

160 pages. The book has a serious purpose but in the context of its production it can also look like a cheeky two-fingered salute to the academic world, a farewell raspberry blown at all things pedantically donnish, in a manner Lucky Jim would surely have approved. For to Ian Fleming's *œuvre* Amis brought the anatomising and categorising zeal he never had devoted and never would devote to more elevated works of literature.

With almost parodic scholarly dedication he analysed the activities of secret agent 007 in the most minute detail. How many men did Bond kill? Answer: 38 ½ by his own hand, with another 70 killed by other people and 500 more blown up in one go at Fort Knox. How many girls did Bond go to bed with? Answer: one per foreign trip, though on a single occasion two, making him not much randier than the average travelling businessman or holiday-maker. What characteristics did Bond girls have in common? Answer: the majority were blondes and all but two had blue eyes. What did Bond villains have in common? Answer: not much, except that they were apt to be big and all had at least a glint of red in their villainous eyes. Dissecting Bond's boss M (otherwise Admiral Sir Miles Messervy, KCMG), Amis found him to be a 'peevish, priggish old monster', a description deduced in part from a count of the adjectives used to describe him; 'impatient', for example, showed up no fewer than seven times.

Nor could Amis resist a mild flourish of scholarly triumph at turning up for himself a few errors of fact in Fleming's work, which had a reputation for being particularly acute in its most sophisticated and worldly details. Most of Bond's early enemies were employed by an organisation called SMERSH. However, observed Amis, the Soviet body of that name had been redesignated OKR (for Otdely Kontrrazvedki) in 1946, several years before the first Bond adventure. And, he continued in learned vein, 'both SMERSH and O.K.R. were simply two of the various labels successively attached to what had originally (1921) been founded as Special Sections (Osobye Otdely) of the main U.S.S.R. Internal Affairs apparatus, the Cheka, later the G.P.U., later the O.G.P.U., later the N.K.V.D., later the M.V.D.' Rather than SMERSH, Bond's enemies should have been in the service of 'the I.N.U., the Second Chief Directorate (Foreign Intelligence) of the M.G.B., in particular . . . the I.N.U.'s Ninth Section for Terror and Diversion (Devyaty Otdel po Terroru i Diversii).'

The hand of Robert Conquest, Soviet scholar as well as drafter of novels for Amis to complete, in addition to pamphlets in the London Library, can be seen behind such recondite expertise. When it came to drink, though, Amis had his own expertise to draw on, with a little help from the wine-writer Cyril Ray. This enabled him to report that when Goldfinger gave Bond a Piesporter Goldtröpfchen '53 he was giving Bond a Moselle, not a hock as Goldfinger himself wrongly believed. Amis was even able to find a flaw in Bond's famous dry Martinis. According to the Bond recipe, these should be made with three measures of Gordon's gin and one of vodka (brand unspecified), plus a half-measure of Kina Lillet. But, Amis noted, Kina Lillet was in fact the name of a 'wine aperitif flavoured, I'm assured, with quinine'. What Bond should have used in his Martinis was a dry vermouth called simply Lillet – made, admittedly, by the same firm, Lillet Frères, but an altogether different concoction.

Amis even went so far as to provide a reference guide to the novels and short stories, breaking each down into ten component parts which he presented in columns across two pages at the end of the book, each column with headings like 'Places', 'Girl', 'Villain's Project'. In the last column, headed 'Remarks', Amis gave each a rating: 'Horrific and haunting in a way none of the others are, but travel-book material intrudes' was his verdict on *You Only Live Twice*.

If there seems something tongue-in-cheek about this performance, a turning of heavyweight critical guns on material Amis himself acknowledged could be called 'harmless tomfoolery', he had serious points to make too. He liked the Bond books, as he had always liked other forms of popular art, and he was happy to pay them scholarly tribute. Already, in *New Maps of Hell* (1960) he had written a pioneering study of science fiction. Here was another type of fiction unlikely to be much bothered with in the academy. But why not? Amis's investigation of Bond had proved two things to him: that the novels were 'just as complex and [had] just as much in them as more ambitious kinds of fiction'; and that they had been quite unjustly attacked for snobbery, sadism, anti-feminism, indeed for being 'a systematic onslaught on everything decent and sensible in modern life'. Amis turned most of these charges against the people who made them. Weren't they the real snobs? If only Ian Fleming had injected a bit of *portentousness* into his fiction, thrown in a few Freudian dreams and sprinkled around some knowing references 'to (say) Kokoschka, Schönberg, Reg Butler and Norman Douglas' – what a difference it would have made to the reviewers.

'That huge virtue of never stooping to any kind of pretentiousness, of never going in for any kind of arty or symbolical flannel, has cost Mr Fleming a formidable amount of critical acclaim,' Amis wrote, 'but it's done as much as anything to bring him readers.' The importance of attracting readers, of communicating with an audience, was something that had preoccupied Amis since the days of his Oxford thesis and continued to do so. The nine Bond books which had by this time been published in paperback had each sold more than a million copies and all these books had been bought for only one reason: pleasure. 'As a recently retired university teacher,' Amis wrote in his Preface, 'I can't help being drawn to any form of writing which (like science fiction) reaches no part of its audience through compulsion. One volunteer is worth twelve pressed men.'

Amis met Ian Fleming twice. The first time, at a *Spectator* party, he told Fleming he greatly admired his work. Fleming thanked him and said, 'They're all true, you know.' What? 'Yes, if you go to the Nievski Prospect in Moscow and you go to the seventh floor you will find the offices of SMERSH with Colonel General Grubozoaboyschikov sitting there.' Perhaps Fleming was pulling Amis's leg – at any rate Amis did not take him seriously. The second time they met was after Amis had written the *Dossier*. He sent Fleming a typescript and Fleming asked him to lunch at L'Etoile restaurant in Charlotte Street. Amis went with some misgivings. Although his book was more pro than anti, Amis had pulled Fleming up where he thought Fleming had gone wrong, complaining, for instance, when the Bond books slipped into 'the idiom of the novelette'. But Fleming wasn't worried about any of that. He had only two complaints to make. Oddjob had been sucked, not blown, out of the pressure cabin of an aircraft. And there was no St Andrews Golf Club; the club in question was the Royal and Ancient Golf Club. Apart from these, Fleming had no objections or corrections and no quarrel with Amis's critical judgements.

The Egyptologists and *The James Bond Dossier* not only provided Amis with some money and some intellectual light relief. They also marked an important stage in the decline of his relationship with Victor Gollancz, the publisher who had brought out Amis's books since *Lucky Jim* ten years before.

Early in May 1964 Amis wrote to Gollancz telling him that he was on the point of completing the *Dossier*. He hadn't mentioned it to him before because he didn't think it was 'at all your sort of book . . . I'm pretty sure in my own mind that you have very little time for Fleming and the Fleming cult.' Gollancz's heart would not be in it, and that was essential if the book was going to sell more than four or five thousand copies, and Amis felt he could not afford to do less well than that. The hearts of Jonathan Cape, on the other hand, would be 'very much in it'. Since they were Fleming's publisher they already had an 'immense' machine for distributing and publicising Bond. After 'some not altogether comfortable pondering' Amis had decided that Cape should publish his own contribution to the Bond cult; but he went out of his way to insist that he still thought of Gollancz as 'my publisher', that all his future novels and other books would be Gollancz's to publish, and that he had no intention of 'going over' to Cape.

Scarcely six months later, though, Amis found himself writing another apologetic letter, this time over *The Egyptologists*. 'You were so nice about the Bond book business that I don't like trespassing further on your good nature,' he told Victor Gollancz. But Conquest had already been in touch with Cape and they had agreed to publish the new book. Amis made light of the whole thing, saying that he had only given the book 'a quick whiz through, generally livening it up', and describing it as 'a pleasant vacation job' (he was writing from Majorca). Admittedly, it had meant putting off getting down to his next 'proper' novel, which was to be *The Anti-Death League*, but, said Amis, this delay would do it 'good rather than otherwise'. The same had happened with *Take a Girl Like You*, and 'the prolonging of the gestation period benefited the result substantially.' When he got home, Amis promised, he would settle down to writing and should be able to let Gollancz have his new novel by the following May.

Victor Gollancz responded to this with the best grace he could muster. He said Amis was entirely within his rights in having *The Egyptologists* published by Cape, since he had only collaborated in it and Cape had already accepted it from Conquest. For what it was worth, Gollancz gave the transaction his blessing. But, he went on, 'I cannot help feeling rather uneasy.' The fact that Amis had brought out two books with Cape before anything new with Gollancz was 'bound to cause widespread comment . . . a lot of people are not merely going to ask "Is Kingsley Amis leaving Gollancz for Cape?"

but are going to state it as a positive fact.' None of this would matter much in 'minor cases', Gollancz continued, 'but everyone knows that I think of you as much the most important (and I don't mean from the sales point of view) of my new, post-war authors.' Amis replied that he was prepared 'to go a long way towards actual violence if someone suggests that you and I are parting company'. More soberly, he was willing to drop a word in the ear of a friendly journalist on a publishing trade newspaper if Victor thought a public denial of his defection would be worthwhile.

There was a strong whiff of double-talk in all this on both sides of the argument. The truth was that Victor did not much like Amis's work for its own sake but was smart enough to know when a good thing had come his way. To keep Amis happy, rather than fervent enthusiasm, was the reason he had published Amis's 1956 volume of poetry, *A Case of Samples*, and his science-fiction survey, *New Maps of Hell*. (The latter, however, had done the company a favour. It made Hilary Rubinstein realise that there was potential for science fiction which had not been exploited in Britain and encouraged the company to start a regular and highly profitable SF list of its own.) Amis, for his part, was growing increasingly unhappy with the effort Gollancz was putting behind his work. Jane was published by Cape and a shift there was very much on the cards, as Gollancz may well have suspected. The one thing that kept Amis with Victor was his liking for the man.

Over the ten years since the publication of *Lucky Jim*, the relationship between the two had had its ups and downs. When Gollancz postponed publication of Amis's third novel, *I Like It Here*, without letting him know, Amis had exploded in synthetic fury: 'You are a perfect louse to have put off publication till 13th January,' he wrote in December 1957. 'Oh, curse you. A lesser man than I would have been on the blower to Collins within the hour. You are the most annoying old sod and I shall see that Graham Watson [Amis's agent] *bleeds you white* over the next book.' But Amis signed the letter off on a characteristic note of ambiguous exasperation: 'With renewed maledictions, threats, abuse, groans and affection, Yours ever, Kingsley.' He had reason to be grateful as well as annoyed, for the company had been consistently helpful with advances against Amis's earnings and in March 1962 Amis wrote to Victor to thank him for the latest example of this: 'Forgive me for not telling you long ago how sincerely grateful I am for your practical friendship.

You don't know how much you have done for us. If anything was needed to secure my allegiance to you and to V.G. Ltd., then this was it. But, as I hope you know, nothing was needed.'

Amis's allegiance was not as secure as that sounded. He liked Victor but recognised his flaws, of which his gigantic egotism was the most prominent. This egotism was as apt to show up in his political campaigns as it was in his publishing office. He liked to be in charge wherever he was. By the early 1960s his days as a crusader were largely over, though he was still chairman of the campaign against capital punishment. On this issue at least Amis could support him – he would by now have been unlikely to support him on Gollancz's old, left-wing, crusades – and joined him on the platform at a great Albert Hall rally in April 1961 alongside John Freeman, the *New Statesman* editor and future ambassador to Washington, and Lord Gardiner, the future Lord Chancellor. (Victor seized his chance to close the meeting with a rallying call for his own sub-campaign – to stop Adolf Eichmann being hanged in Israel.)

Victor's egotism had its good side. It made him capable of the big gesture which could mean more than cash advances. By the time Amis was ready to publish his sixth novel, *The Anti-Death League*, he had got fed up with the economical fashion in which Victor had so far produced his books. He wanted more than what he thought of as 'war-economy style' typesetting and he thought he was going to get it at last, as well as the artist of his choice for the cover. But when the proofs came in he found the pages had been set up in the same old penny-pinching way. Amis protested to Victor, who immediately ordered the book to be set up in print again from scratch. And he let Amis have the jacket illustrator he wanted too. These were expensive gestures but Victor had realised that no half-hearted titivating of the book's appearance would do – it had to be all or nothing. This was the kind of bold stroke Amis admired him for. On the other hand Victor could be petty and mean. It rankled mildly with Amis that he never gave a traditional publisher's party for any Amis book launch.

His egotism also made Victor a jealous man and a poor delegator. He might have been expected to be pleased with his nephew Hilary Rubinstein for finding Amis in the first place, but he saw that sort of initiative as a challenge to his pre-eminence in his own company and took over dealings with his most prestigious authors from Rubinstein or anyone else who brought them in. He seemed to lose no chance of putting Rubinstein in his place. On one occasion, lunching with an author, Victor declined to let Rubinstein choose his own food and pointedly failed to offer him a cigar at the end of the meal, while he and the author both lit up. In 1963 Rubinstein finally left the company and joined the *Observer* as a journalist. Soon afterwards, Victor said innocently to Amis, 'You know Hilary's joined the *Observer* – don't know what good he thinks he'll do himself there.' You shit, thought Amis, because you were the one who engineered Hilary's departure. This Victor had done by bringing in another man, James MacGibbon, while Rubinstein was in New York, leaving Rubinstein's position in the company ill defined, uncomfortable and ultimately intolerable.

The fact that Victor was bad at delegating made the firm of Gollancz too much of a one-man company and less good at its job of selling books than it should have been – or so Amis believed. So long as Victor was there to run it, Amis would stay with him. But what if Victor, who was getting on in years, were to be sidelined through ill-health or age? He might still be chairman according to the company's notepaper but he might not actually be running the shop. Victor was aware of Amis's feelings about this and one day summoned his daughter Livia to his office while Amis was there. She was instructed to change the option clause in Amis's contract which promised the company first chance to publish whatever his next book might be. The clause was to be effective only as long as Victor was in active control. This suited Victor's sense of his own importance as much as it did Amis's wish to be able to cut and run if for any reason Victor ceased to be in effective charge of his company.

When Victor died, Amis wasted no time in ending his connection with the Gollancz publishing house. On 24 February 1967 he wrote to Livia cancelling an impending lunch date. 'This is a painful letter to write,' he said, 'and I have been putting it off in a cowardly fashion. I had better just state the case bluntly. For some years I have been unhappy with Gollancz Ltd and would long ago have moved elsewhere but for my personal loyalty to Victor. Victor himself understood this perfectly well, and made no demur whatever about the get-out provision in the option clause of my last contract. Now, sadly, my loyalty is at an end, and I am off.' *The Anti-Death League* was the last of his novels to be published by Gollancz, in 1966. For the next dozen years, Cape was to be Amis's principal publisher.

* * *

Hilly and the children had moved on from Ovington Gardens soon after Christmas 1963. The flat there had only two bedrooms between four people. She and Sally slept in one, the two boys in the other. But something larger was needed for children now advancing towards and through their teens and a house was found in the Fulham Road at a rent of £12 a week. It was shabby and not very well furnished but, when Hilly went away for a while with Sally, Jane and Amis moved in, and over the next couple of weeks Jane saw to it that the house was cleaned up and a bit of carpet put down. Later Hilly took in boarders and got herself a job at the Battersea Park zoo which she liked very much, animals being her chief enthusiasm and joy throughout her life.

The family did not stay long in Fulham – only one Christmas. Then Hilly bought a house in Wivenhoe in Essex and moved there. Sally went off to boarding school, while the boys joined Jane and their father in Maida Vale, visiting their mother at weekends and holidays. At Wivenhoe the courtship began between Hilly and the man who was to become her second husband, David Roy Shackleton Bailey, then fellow and bursar at Gonville and Caius College, Cambridge, a classical scholar and for twenty years the university lecturer in Tibetan too. With a house, a car and a suitor, Hilly began to get over the break-up with Amis and to enjoy life again.

All this moving about was disruptive, particularly of the children's education. Martin's was not helped by his taking two months off school in order to go to Jamaica, Hilly and Sally going with him. He had been given a part in the film of *High Wind in Jamaica*, but this piece of glory cut no ice with his school, Sir Walter St John's in Battersea, which expelled him for his prolonged absence without leave. Martin came to think of this as something of an achievement, Sir Walter St John's being a fairly rough place at which it was hard to commit offences more worth being expelled for than everybody else's.

Jane took an interest in the boys' education, rather more so than their father. It was she who encouraged them to take their schooling seriously and to cram for exams. Amis himself took a back seat in such matters. Perhaps Jane, having had little school education and none at all at university, valued it more than Amis who, having had so good an education, took it for granted. As a father he was soft, generous, good fun, sweet, but not really a hands-on sort of parent, inclined to disappear into his study to avoid trouble. This did not stop there being some blazing rows. Amis, though tolerant enough of the girls the boys started to bring home, was much less easy-going about things like marijuana, then becoming fashionable among the rebellious 1960s young. In his hardening right-wing view, drugs were among the things that Communists used to undermine American soldiers in Vietnam, were thus part of the worldwide Communist conspiracy and so to be opposed on political grounds, never mind what harm they might do to health. Martin managed to live with this kind of argument but Philip was more rebellious and became the first of the two boys to leave home and set up on his own while still in his teens. To Jane, though, things looked rather different. There seemed to be something of an all-male alliance against her. She wanted the boys to be disciplined, to study and do well at school while the men in the household seemed to her united in their indifference, if not their hostility, to her good intentions.

Jane did nevertheless succeed in putting Martin on the right educational road. He went for a year to a crammer in Brighton, Sussex Tutors, and in that year did enough solid work to score a hat-trick of exam successes: five O-levels, three As and his Oxford entrance, which included getting his Latin up to O-level standard from scratch. (With Latin, his father played a larger role, claiming to have got Martin through that at least.) When the results of his Oxford entrance exams were announced Martin telephoned his father, who was then in Nashville, Tennessee. Amis heard Martin say he had got into Exeter. You mean Exeter University, Amis queried. No, no, Martin assured him – Exeter College, Oxford, a much greater coup. But if it had not been for Jane's encouragement Martin might not have made it to either university, and he was to remain always grateful for her having persuaded him to give schoolwork a serious try.

Amis retained his detestation of hanging and racism (which he was to find in peculiarly virulent form in Nashville in the late 1960s), attitudes more common on the political left than the right. In most other respects, however, he continued his move to the right, though by drift rather than jump. There was no moment of conversion, no dramatic recantation. His thinking evolved in fits and starts until at some point in the 1960s, impossible to say exactly when, the Amis who in 1940 would have had trouble seeing enemies to his left now found trouble seeing them to his right.

An *Observer* piece in July 1963 catches him in transition. Writing about patriotism, he said that its 'emphasis on hierarchy, order, responsibility and so on, could help to prop me up'. And then he added in a bracket, as though coyly admitting a small whisper of reproach from his conscience, or some imaginary left-wing friend's, '(You're moving Right anyhow, Amis.)'. But he was still left enough to end the piece by quoting George Orwell on England being a family 'with the wrong members in control' and concluding for himself that 'doing something about that would be another useful preliminary' to restoring a sense of proper patriotism. Wariness about the upper class and the establishment was something else that he never abandoned as he moved to the right.

Moving in that direction meant for Amis abandoning one set of ideas before it meant embracing another. He was repelled by the left rather than attracted by the alternative. What repelled him was the prevailing intellectual climate which was overwhelmingly left-wing and at the same time, he thought, overwhelmingly bogus and dishonest, a shoddy parade of fashionable nostrums and attitudes. As the years went by, Amis was to become more at ease with his newly acquired right-wing attitudes and, being unusually ready to say what he thought in the bluntest of terms, acquired the reputation of an enraged Tory clubman. But it is probably the case that, as he ceased to be left-wing he found himself knowing better what he disliked in politics than what he liked.

Europe was a case in point. He was an early and consistent opponent of British membership of the European Community. 'Nothing less than full political union, of course, is in store for us,' he wrote in his *Observer* piece, 'as is revealed by even the least expert attempt to peer through the fog of official reticence.' But union wouldn't work. He thought of each nation as a family, an organisation based on something more than self-interest; but a super-family of nations, a European family, was an impossibility. As a writer, he saw no cultural dividend to be drawn from a united Europe. Language was to him all-important, and Britain's eyes and ears should, he felt, be turned to North America and other places where English was spoken rather than to the Continent. If Britain had to join something, Amis would have preferred it to have become America's fifty-first state. But he recognised the impossibility of that even as he wrote it down. America wouldn't have it and there wasn't much enthusiasm in Britain either. That left only a revived patriotism to recommend as an alternative to Europe. Not perhaps much of a political platform but the best he could think of.

If there had been one pivotal point in Amis's political drift it was Hungary in 1956. From then on Communist tyranny could not be denied, became indeed the principal enemy. Any illusions he may have kept that things would improve after the death of Stalin were gone for good, along with any lingering idea that Soviet-style rule could ever be benign. In a letter to Victor Gollancz written in December 1957, he described what he thought would happen if the Soviets ever did come to Britain. 'The Russians . . . would take what loot they wanted, set up their bases and leave the population to rot. And they would simply suppress the B.B.C. and the Press, not to mention shooting half a million chaps out of hand.' The dreadful possibility of a Soviet takeover lingered in his mind for many years and eventually formed the background of his 1980 novel, *Russian Hide-and-Seek*, which describes a Britain desolate and inert after fifty years of Soviet occupation.

The Suez invasion was just as bad in its way as the occupation of Hungary and Amis had been shocked and dismayed by it to the point of demonstrating in the streets of Swansea. But Suez was different from Hungary too. This shameful mistake, having been made by democracies, could at least be put right, and so it was when the French, British and Israeli invasion forces withdrew from Egypt. The Soviet invaders, on the other hand, were totalitarians indifferent to public opinion, and therefore stayed put in Hungary. Yet this standing outrage seemed to be almost immediately forgotten by many on the left in Britain who chose to ignore it, even eradicate the whole event from their memory, the better to get on with the more agreeable business of finding fault with all things British and democratic. It was this capacity to put out of mind such bitter realities as did not fit a particular left-wing view of the world that stuck in Amis's craw and gradually stoked his bile against all things left and their representative champion, the Lefty.

Still, he could not easily bring himself to give up voting Labour. Ever since he had first done so in 1945 he saw himself as bound to keep on doing the same for the rest of his days. Habit was hard to break and the thought of voting for any other party made him frankly uncomfortable, almost scared. Besides, he still had some faith in the party of Ernest Bevin, Clem Attlee and Hugh Gaitskell – it could hardly, he thought, form worse governments than than those of Eden and Macmillan, who had given Britain Suez. In 1964 he voted Labour once more, for the last time as it was to turn out. At the next general election in 1966 he compromised and voted for the Anti-Common Market candidate in his constituency. It was at the local elections the following year that he voted Conservative for the first time, and not until 1970 did he vote Tory in a national election. Jane may have played a part in all this. Not that she preached her moderate Toryism at Amis. But living with someone who saw nothing wrong in being right-wing may have made it easier for him to give up the political habits of his adult life.

Other people certainly played a part in Amis's decision to switch. These were the people he saw in government after Labour's 1964 victory, among whom he took a particular dislike to Tony Benn. People like Benn, he thought, were very much not the same as the Labour stalwarts of 1945. Amis met Benn once, when he turned up unexpectedly on the Amis doorstep bringing his American wife Caroline and an American couple who were friends of hers and Amis's too. Amis offered them all a drink. Caroline took a gin and tonic but Tony, as Amis wrote in his *Memoirs*, 'reacted much as if I had said, "Glass of baby's blood? It's extra good today"', and would accept only a well-iced bitter lemon. It was characteristic of Amis that he should take an immediate dislike to Benn not just for his 'barmy' look or his unattractive policies but for his hostility to alcohol in any form.

Amis had already quarrelled with another prominent Labour politician, Tony Crosland, over gramophone records and Amis had taken a further dislike to Crosland for his superior attitudes to America. Later, when Crosland became Labour's Education Secretary, Amis had more substantial bones to pick with him and what he called the 'Crosland illiteracy commandos' who, Amis believed, were wrecking the new comprehensive schools even before they were properly established by doing away with streaming and traditional sixth-form preparation for university. Crosland was a ripe example of the Amisian Lefty: not by any means necessarily unintelligent, simply wrong-headed, and therefore a fucking fool no matter how clever. Although Crosland was an intellectual of considerable weight, he fitted into the Lefty mould all the same, with such remarks he made when Education Secretary as 'If it's the last thing I do, I'm going to destroy every fucking grammar school in England' or 'Let us now move away from our snobbish, caste-ridden, hierarchical obsession with university status.' Thus did the Lefty arrogantly mow down whole swathes of tried and tested institutions in the name of some half-baked theory of progress.

However bright, the Lefty as defined by Amis was moved primarily not by ideas but by feelings and emotions. He or she typically started out as a young person who had an 'unfocused dissatisfaction with the way things are', that is, with the way authority in the shape of employers or parents seemed to stand in the way of his or her freedom just for the sake of being obstructive. In stage two of the Lefty's development, the young man finds himself not getting on very far in a competitive world and, when he asks himself why, he finds a blissfully satisfying answer. Which is that it's not his fault for being lazy or stupid or anything else – it's all the fault of the system that holds him back. So from now on the system must be opposed. And even when the Lefty does manage to get on in the world he continues to oppose the system, out of habit or because the system has quite failed to take notice of his protests and adjust itself accordingly. The Lefty is someone who knows in his heart all about the evils of the Soviets but cannot let that knowledge interfere with his own beliefs. And because the Lefty hates 'the system' he must hate his country too. 'Affectionate portraits of England from a socialist standpoint are almost inconceivable,' Amis wrote in a footnote in *The James Bond Dossier*, and then tried his hand at a socialist essay in sentimental nostalgia: 'a world of soccer pitches and recreation grounds and Trade Union leaders, of Walsall, of the demonstrators being photographed with guitars across their knees in Trafalgar Square, of the pansies blooming outside Transport House . . .' It just couldn't be done, he concluded.

Amis knew all about his version of the Lefty from having been one himself. He was old enough now to see his early Communism as being much more about rebellion against his father than thought-through conviction. But he had grown up and out of self-deception. This too was an emotional experience. Amis the Righty was formed just as much by his feelings as Amis the Lefty had been. Experience – of the army, marriage, family, death, the first intimations of old age – had taught him a different set of feelings. Talking to the Communist *Morning Star* soon after his novel *The Anti-Death League* was published in 1966, Amis described its theme as being that 'what causes the most unhappiness and feelings of revolt is the inescapable condition of human life. That is why I am no longer a Left-Wing Socialist.' He had used to think, Amis went on, 'that human beings could be made significantly happier by transforming their outward lives'. Now he thought differently: 'Their real enemies are not landlords but cancer, not capitalism but death, not warmongering but the power to kill people.'

The next year he finally came out, not simply as not a Left-Wing Socialist but as no longer a socialist of any stripe. Now he was a Tory voter, and in a piece called 'Why Lucky Jim Turned Right' he explained his change of view. Experience had taught him that all-purpose blueprints for the brotherhood of man or the 'Just City' did not work because politics could not get rid of so many things that caused unhappiness and discontent – 'failure, loneliness, fear, boredom, inability to communicate'. And when politics did try to rescue people from these things the result was inevitable disaster. Amis was not made happy by giving up all his utopian hopes but he had come reluctantly to the conclusion that promoting the brotherhood of man led only to the enforcement of something very different. His own hopes had now dwindled to 'keeping things going', plus maybe 'an injustice righted here, an opportunity extended there'. The logic of this led him unenthusiastically towards the Tories. He supported them precisely because he saw the Tories not as system-builders or blueprint-creators but the opposite, as 'the party of non-politics, of resistance to politics', a party without a romantic programme, indeed with no programme at all. 'I like that,' he declared.

Having given up the ready-made ideology of the left, Amis was not cut out to have an alternative ideology fit for the 1960s. Then and later, that required above all a view of how economic affairs should be conducted and Amis had nothing whatever to say about that. 'I am no economist,' he wrote in 1967, and went on to prove the point by saying that if he strained himself he could 'nearly imagine that the unchecked rise in the cost of living is part of some frightfully subtle plan to send it rocketing down again'. This ignorance was not surprising in a man who left his own financial affairs to be conducted over the years by his father, his wives and his accountants. All such maters bored him stiff in his own life and in politics too. He stuck to issues he knew about, like education, or matters of broad principle, especially resistance to Communism in all its multifarious forms and disguises.

Once he had completed his shift to the right, Amis began to settle in and relish his new position, in public and in private, showing a strength of conviction he had not shown for Labour for a very long time. His letters to Conquest, up to now mostly about gossipy or literary or practical matters, had a new earnestness injected into them, above all about Communism and in particular about Communists in Vietnam. This became something of an obsession with Amis. Although American intervention there was widely opposed, in America and in Britain and by people with all kinds of non-lefty views on other issues, Amis was all in favour. Attitudes to the American presence in Vietnam he took as a test of political soundness. If you wanted the Americans out you were wrong and a fool, or a Communist dupe, or both. And whatever your reasons you were to be viewed with suspicion on other issues too. Vietnam was where Communism was currently making its most aggressive thrust so that was where it must be resisted and if you couldn't see that you couldn't be trusted on much. Even Martin felt the force of this argument, his father's insistence making him one of the exceptional young intellectuals of his time in not opposing American involvement in Vietnam.

Writing to Conquest from Nashville in November 1967, Amis describes someone in the university's Russian department there as a 'v. sound reactionary'. And he chides Conquest for not being more aggressively polemical:

I read in the T.L.S. [*Times Literary Supplement*] about Lenin's statesmanship and the balanced view which credits the Russian Revolution with its achievements. Aha! I think: Bob or Tibor [Szamuely, a Hungarian refugee from the Soviet Gulag and a friend of Amis's] will get going on the balanced view being that the R.R. was a disaster and *what* achievements? – and the following week – fuck all. Pull yourselves together, you lazy sods.

From Mexico the following February, Amis wrote again to Conquest:

I nearly shat myself with fury on hearing that the Pistener [i.e. the *Listener*] had put up that bastard Che Guevarra (?) as Man of the Year. And then, just yesterday, Mack [Reynolds, a science-fiction writer living in Mexico] told me that *Tony Richardson* was sending *Alan Sillitoe* to Mexico to write a film about *Che G.* AAAOOGHGH Did you know this? Where will it all end? We must plan a monster counter-offensive when I get back . . . Here, by the way, you get a rather different kind of Lefty, who seems genuinely anti-Commie, but can't or won't see that to be anti-US-in-Vietnam is to be pro-Commie. They all say there are no Commies in USA, no Commie component in draft-card burnings, etc. Anyway, fuck 'em.

Amis took up the same theme in another letter to Conquest in March, this time from Princeton.

Am sitting in front of the TV (though it's only 10.30 a.m.) watching the Senate Foreign Relations Committee doing its stuff: at the moment that pompous old poop Fulbright is sounding off, saying it's jolly understandable that the N. Viets are in Thailand, etc. [Dean] Rusk [US Secretary of State] dealing with him v. competently and unfussily. Not likely to go down well in P'ton. I met *one* chap who was sound on V Nam. His wife has enormous tits so he's clearly the sort of ally one wants. Otherwise it's all 'I'm fiercely anti-Communist but (or even and) the US should get out by next Tuesday.'

Amis had first become right-wing about education, calling attention to falling university standards as early as *Lucky Jim* and continuing to do so thereafter. In a 1960 piece he wrote of the 'pit of ignorance and incapacity into which British education has sunk since the war'. University students were already illiterate enough to have difficulties with 'hard words like "goes" and "its"', though this was happily ignored by people playing what Amis had heard called 'the university numbers racket'. So-called 'thinkers' cited figures to prove that Britain was 'falling behind' countries like Russia and America in its production of university graduates, so Britain would have to 'catch up'. But the only way of increasing the number of graduates, Amis insisted, would be by lowering the level of qualifications for admission, and that would quite simply 'wreck academic standards beyond repair'.

'*MORE* WILL MEAN *WORSE*' Amis resoundingly deduced from his own equation. The whole thrust of his piece was an attack on what he called the 'sociologising generalisation', or the kind of punditry that anathematised all things British on the slenderest of evidence. But that was exactly what he was now to find himself accused of doing. His '*MORE* WILL MEAN *WORSE*' was seized on, transformed and turned against him in the press and the public mind. It assumed the shorthand form of 'More means worse' and became an iron Amisian rule applicable to all after-school teaching or training. All Amis had meant to say was that what he understood as traditional university standards of excellence could not be maintained if the doors were to be opened to a flood of new students. There just weren't enough good ones around with the necessary qualifications to fill the places. He was not talking about colleges of further education or technical training or anything else of that sort, only the traditional university as he had known it at Oxford, Swansea and Cambridge. Nevertheless he was held to have condemned all forms of post-school education.

This kind of thing was irritating but harmless enough so long as it got no further than being chatter in newspapers and magazines. But if it got a firmer grip on politics and on life then it was much more worrying to Amis. He feared that the 'sociologising generalisation' would take over in a more sinister form, becoming a governing principle for ordinary life. 'What I dislike', he wrote in 1962, 'is the thought of having my life organized by those who believe that there is a scientific answer for everything, or everything that counts.' This wasn't true, but a society run on the belief that it was would, he thought, be 'hellish'. There was no need to fear that life actually could be put under scientific control; what should be feared were those who thought it could be: 'Efficiency consultants, statisticians, social engineers, market researchers, administrative planners, cyberneticists, technological advisers – here and there in the ranks of all these is the nucleus of a new priesthood.'

Amis may have started worrying about this by thinking about issues like education and the way in which foolish policies were determined by quasi-scientific formulae. But he found himself going far beyond that to imagine a whole world from which God, religion and art had all been expelled and replaced by a bogus scientism, directed by a corps of fake priestly experts. When Amis turned right he found the Tories less infected by such superstitions than Labour, but far from innocent of them all the same. They continued to expand university education, for example, with as much fervour as their opponents. In this sense at least the Tories were only the lesser evil. Dystopian fears of a world gone wrong far beyond any differences between two parties in a democracy were to form the basis of two of Amis's novels, *The Anti-Death League* and *Russian Hide-and-Seek*.

Prague and Nashville

Travel in the 1960s helped to confirm Amis in some of his political attitudes, as travel in the 1950s had illuminated some of his literary views and his feelings about the Briton abroad. He went with Jane to Prague in 1966 and found it fully as oppressive as he expected. The following year he went to teach for four months at Vanderbilt University in Nashville, Tennessee, and found it surprisingly oppressive there too – surprising because, after all, it was America. In Prague, Communism was the oppressor; in Nashville, racism. Tennessee may have been in the same United States as the Princeton where Amis had spent a year a decade earlier, but it was like being in a different world, one full of prejudice and hatred.

Amis had been interviewed for a job teaching English literature in Prague in 1948 but had not been taken on. Only a few months earlier the Communists had taken over in a coup, but the consequences had not yet become clear and, whatever they might be, they had not seemed more important to Amis at the time than getting work to support his wife and child. He was a very different political animal by the time he was invited to visit Prague for a week to deliver a lecture and talk to students.

The invitation, delivered in person by a Czech embassy official called Peter Pujman, was itself something of a surprise. Didn't they know that Amis was by this time a stout anti-Communist? But it seemed that the Angry Young Man reputation was the one which had stuck to his name in whatever files were kept on him in Prague. And when he proposed 'The Literature of Protest in Great Britain' as the topic of his talk Pujman was enthusiatic. 'Is good subject,' he pronounced. Perhaps he would have felt differently if he had been there to hear the lecture, which Amis used to pour 'brief doses of contempt' on sundry left-wing British writers to illustrate his theme that there really wasn't much in Britain for writers to protest about. When Amis was finished, a member of the audience came up to tell him that the plays of Arnold Wesker could no longer be performed in Prague. What did he mean? Amis went on guard, smelling a political trap. But all the young man wanted to say was that Wesker was too political for the local taste so nobody would turn up to hear his work. Which was not unlike how Amis thought of Wesker's plays himself.

Amis quite took to what he saw of the Czechs. They had after all been the only Communists to pay him in money he could use for their translation of *Lucky Jim* (*Státný Jim*), a decent three-figure sum too. And they employed a kind of 'straightfaced irony' in their dealings with officialdom which Amis could sympathise with. He had some fun in Prague as well. Jazz was available, and he and Jane ran into Johnny Dankworth and Cleo Laine, jazz musicians and friends of theirs. Briefly, Amis could even feel twinges of regret that he had not got that job in 1948 after all.

But only briefly. The overwhelming impression Prague made on Amis, as he described it in his *Memoirs*, was of a stereotypical Communist regime. The city was full of beautiful baroque churches and other buildings worth looking at, but they were all jumbled up with industrial works and dreadfully polluted as a result. People smelled but only, Amis surmised, because they preferred to spend their money on things like 'well-staged opera' instead of 'lousy' dry-cleaning. The absence of colour in the streets was puzzling until he realised, on spotting a decrepit old Martini ad, that the drabness was due to lack of advertising, which induced mild feelings of shame at how he had deplored the 'vile garish corrupting' advertisements to be seen at home and in America. Everywhere there was corruption and surveillance, and the ordinary citizen survived only by engaging in the first of these and learning to slip and slide his way around the second. An attaché at the British embassy told Amis that he could only talk freely to his wife on Sunday-afternoon walks in the park. His guide and interpreter, Karel, took Amis to see his apartment as an example of the rewards to be expected by a lecturer at Prague University. It consisted of one room of modest size. Karel pointed impassively to each corner, saying, 'this is the study . . . the living

room . . . the bedroom . . . the bathroom . . . the kitchen.' Here
was another of the ways in which Prague, with a few saving graces,
lived up to Amis's gloomy expectations, somehow turning out even
more as he had imagined by the more sinister characters being got
up to look like clichés in their own roles – players 'selected by a
Hollywood casting director . . . rather unimaginative in taste'. It was
with some relief that he left Czechoslovakia, even though the first
place his train came to in the West was Nuremberg.

Nashville aroused forebodings in Amis which were more than
fulfilled. When he received an invitation to teach there his first
instinct was to refuse. The reputation of the American South for
lynchings and the kind of popular music and writing (Truman
Capote, William Faulkner) that Amis particularly disliked was not
encouraging. Then he had second thoughts. The invitation came
from a reliable friend he had made on his earlier Princeton trip,
Russell Fraser, a New Yorker who had become head of the English
department at Vanderbilt University. Fraser assured Amis that things
were not as bad in Nashville as was popularly supposed. Besides, the
invitation was for only four months and the teaching burden would
not be excessive. Fraser added a point that carried particular weight
with Amis: Britons were rarely spotted in the area.

Amis had learned to keep clear of Britons in America – and of
Britain-loving Americans – during his Princeton year. In Nashville
he met a man called Tupper Saucey, whose wonderful name (which
would have been better still if put the other way round, Saucey
Tupper) Amis found no excuse at all for his being a 'flaming
Anglophile shag'. All such, he told Conquest, are 'v. horrible'.
They talked about 'graciousness, the Queen etc.' Amis proposed a
distinction between American lovers of Britain: 'Anglophiles (shits)
and pro-British (possibly okay)'.

Sticking to his rule of not flying anywhere, ever, Amis arrived
in New York with Jane on the *Queen Mary* liner in October
1967. He stayed at the Algonquin Hotel in the city for a few
days, making contact with science-fiction-writing friends like Fred
Pohl and Robert Sheckley, then spent a few more days in Princeton,
staying with Edmund and Mary Keeley, friends he had made on his
earlier visit, before embarking on a complicated zigzag train journey
that went by way of St Louis to Nashville. There they were picked
up at the station by Fraser's wife Phil and driven into town.

This turned out to be just as much of a cliché as Prague in its way. As Amis recalled it in his *Memoirs*, Nashville was 'a very lifelike version of what you see every time you put the TV on and are shown a main street in a standard US town'. No description was required to describe it, simply a list, including hamburger joint, gas station and '*nobody on foot*'. Amis later turned the list into a verse of his poem 'South': 'The usual: ranch-style, eat-o-mat, drive-in, / Headlight, tail-light, floodlight, neon, / And air-pollution —'. Nashville had some peculiarities of its own, besides the pollution: the Grand Ole Op'ry (unvisited by Amis), a Parthenon (with a roof, an improvement on the one in Athens) and a roadside Holy Family which changed colour from green to purple to yellow. But even these Amis regarded as more or less standard and therefore unremarkable features of American life. Nashville, though, had a few extra uncongenial idiosyncrasies in store, including a ban on buying drinks other than beer in bars — you brought your own and paid for glasses, ice, twists of lemon and so forth — and a climate which, while giving the appearance in October of being more or less like England in June, made you pour with sweat after a modest walk at no great speed. Once, undertaking one of those, Amis found a police car keeping pace alongside him — just as they did in the movies!

But the pervasive racism was the most uncongenial feature of all in the local landscape. Amis had hoped that, whatever ordinary people might feel about blacks, at least those at the university would be different. He was soon disillusioned. When he asked if there were any black students he was told there was one, a Mr Moore, who had seemed to enjoy Vanderbilt when he first arrived, but didn't much after two years there. Amis was invited to dinner by Walter Sullivan, the professor of English whose sabbatical had created the place Amis was in some not altogether clear sense taking. After a few drinks the conversation veered round to what Amis had come to recognise as the staple subject. What was said he refrained from recording in his *Memoirs* on the ground that it was too strong meat, if not actually against British race laws. But he quoted verbatim one sentence uttered by his host: 'I can't find it in my heart to give a negro [pron. nigra] or a Jew an A' (i.e. A-grade). For a moment, Amis thought Sullivan had switched to satirical mode and was taking off the local bigots. After all, Sullivan was a university teacher, a writer of novels. Surely he couldn't mean it? The chorus of support from the other dinner guests soon put Amis straight. Sullivan, he learned, was one of those who had given a party when President Kennedy was shot.

There were other dismaying encounters. A man in labour relations told Amis that scientific tests had proved the black man's brain was smaller than the white man's – and yet he'd heard the British were letting them come to their country from all over the place: had they gone mad? Amis's hostess at a party, wife of the professor of Iberian languages, could not understand why Laurence Olivier had acted Othello as a black man and, when Amis counter-argued (not very sincerely) that Olivier had done it rather well in that way, protested: 'But how could a real lady fall in love with a man like that?' Then there was the student called Julie Smith. She was one of the most beautiful girls Amis had ever seen and if Jane had not been there, well, he might have . . . well, done what he often did, or anyway try. Not only was Julie beautiful but bright, in fact the model student. And yet after a party Jane reported a conversation she had had with her in which Julie had revealed herself to be just like all the others under her attractive surface. Jane told her that there were reports of race troubles all over America but there seemed to be none around Nashville and Julie replied that that was because the Northern states 'didn't know how to keep 'em down'. Keep 'em down! Perhaps you couldn't expect the fiftyish Walter Sullivan to be unreconstructed but you might have hoped the twentyish Julie would be different. Maybe, Amis speculated in his *Memoirs*, her children now are.

Racism was the one issue that made Amis wobble on his right-wing plinth and see some surviving good reason for leftness. 'The nigra gets talked about in terms that recall the most tedious Lefty play,' he wrote to Conquest soon after he had settled in. 'Buggers haven't learnt a bloody thing. One can forgive a Lefty here, in that "conservative" opinion is so shitty. It would take some strength of mind not to say, in effect, "Anybody who thinks like that about Negroes must be wrong about everything. Stop US Murder in Vietnam."' Amis thought of various strategies for dealing with this racism when he came across it: Argue? Laugh? Fight? Walk out? Resign? What he actually did was 'kept my mouth shut, avoided assenting to what could not be assented to and made the most of the good bits, human and other'. Nor, in spite of his genuine hatred of this Southern racism, did he let it spoil his affection for America. Even though he was later to think of his spell in Nashville as the period other than his army service he would least wish to relive, he still very much liked the country. Amis found himself one night at a dinner where a Labour MP, Ivor (now Lord) Richard, made a speech saying that Britain was turning towards Europe and the old special relationship with America was on the way out. Called on to add his bit, Amis said Richards had talked balls and Britain was as attached to America as ever. No local displays of racism, however foul, could override Amis's feelings for America as a whole – the America of jazz, films and science fiction, of freedom, efficiency and generous hospitality.

As the weather grew colder, Amis and Jane settled into a new routine. 'Both of us working like bloody fools,' Amis told Conquest at the beginning of November: 'J. running the house – most unbelievably shagged and inconvenient south of the Mason-Dixon line – as well as novelising; good old K. doing 3 fucking lectures a week . . . and also doing his best to novelise.' Amis had the *Daily Telegraph* sent on – making him feel like 'a Maugham rubber-planter' when he went to pick it out of the mailbox. But he felt starved of gossip from home and asked Conquest for news.

The routine included, as always, a great deal of drinking, perhaps even more than usual. When he came to write his *Memoirs*, Amis felt able to state with complete confidence that he had a hangover the day after one incident because he had a hangover practically every day he was in Nashville. This was due partly to the local hospitality (some fifty social events were staged for or around the Amises), partly to the general strain of life in Nashville. On evenings in, Amis would sit drinking large dry Martinis in melancholy mood, chilly now because the central heating was inefficient and the logs in the fireplace damp, wondering if the ferocious dog from next door, Major, would finally get to him on his next trip to the mailbox, occasionally conscious of the 'dreadful' portrait of the woman who had rented him the house, Ivar-Lou Duncan, who was rapacious enough to have tried to make him pay rent on her husband's typewriter (though her husband vetoed the notion).

For Amis the best part of his time in Nashville was in his dealings with students, something he always enjoyed; there had been no such dealings since he left Cambridge and there were to be none again after he left Nashville – a matter for lasting regret. Jane drove him to the university for his thrice-weekly lectures and he was met at the gates by a graduate student called Don Schultz, assigned to him as a general guide and helper, who escorted him to the lecture theatre where he confronted a class of a hundred or more. These sessions – lecture plus questions – lasted an hour or so and Amis came to think of the class as 'about the best I ever taught: punctual, polite, attentive, ready but not overready with questions and objections, containing that ingredient essential for a decent course of lectures, however much the lecturer may squirm at its presence: a couple of students of whom he is very slightly afraid'.

Amis included in his course writers like Len Deighton, Ian Fleming and John le Carré as well as more 'literary' writers like Iris Murdoch, William Golding and Graham Greene. No doubt this was one reason for his popularity. But another was his lecturer's talent. A reporter described his style: 'He manages, as the best all do, to achieve balance between entertainment and pedagogy. His open, shoot-from-the-hip approach is different from that of other teachers at Vanderbilt; one of his students accurately observed that the chief difference was that Mr Amis' is a writer's approach to fiction. Another who seemed a bit overwhelmed described him as "anti-academic, anti-literary, almost anti-intellectual". But they loved him.'

By and large the Amises went down well too on the social circuit. The same reporter recorded the opinion of a woman who had seen a lot of them at parties. They were 'really an ideal couple'. Amis was 'virile, jolly, and a man who loves to entertain', while Jane was 'tall, mod, really good-looking, and very intense'. Amis's snuff-taking took them by surprise. (He found it impossible to get what he wanted in Nashville, where oral snuff was all that was available, and had to badger Conquest to send supplies of High Dry Toast from Fribourg & Treyer in the Haymarket.) 'It's the thumb in the nose that stops them every time,' the reporter wrote. Amis got into odd rows, as with a French Canadian on the question of separatism and a woman who took it for granted he was an atheist. '"Where do you get your ideas?" she asked him. "Why, from God, of course. Where else would you suppose they come from?"' Amis also shocked a large part of university opinion by his belligerence over Vietnam. American involvement was becoming a hot issue on campuses all over America, and Amis's enthusiasm for the cause – he even proposed that Britain should send a token force – was not at all welcome at Vanderbilt.

His reputation as a right-winger had preceded Amis to Nashville and, according to one version of what happened there, his hosts had played up to it, becoming 'stage-Southerners' for his benefit, especially hamming it up on the race issue. A friend Amis made at Vanderbilt, Richard Porter, a lecturer in German, later wrote that 'Racial talk in most academic circles [in Nashville] differed little from that in the North.' He claimed to have heard 'several people, wishing to please, make racial slurs around the Amises they would not have made otherwise'. Porter speculated that Amis might have been influenced by his sponsor, Russell Fraser, who ran into trouble with the university authorities over his teaching programme while Amis was in Nashville (the authorities wanted a greater emphasis on Southern writers than did Fraser) and came close to quitting. Amis described this to Conquest as 'a first-class Groves of Academe-type dust-up' and felt while it lasted that, if Fraser went, he would have to go too.

On this version, Amis became disillusioned with Nashville by the treatment dished out to Fraser and, with his sympathies twisted, went on to misread local attitudes because they were greatly exaggerated for him, so that the South should be made to appear full of exactly the kind of people he was thought to expect – indeed want – to find there. Was Amis misled by a charade staged just for him? It seems most unlikely, since it is difficult to believe that such a large cast of characters as he encountered in Nashville could have been so uniformly well drilled in simultaneous dissimulation; or that Amis, with his own gift for mimicry, would not have seen through any such attempt at mass deception. At any rate he left Nashville certain that his own reading of its deep and pervasive racism was the correct one. Writing to Conquest a month or so afterwards, he said that in retrospect his spell in Tennessee seemed 'rather like the days of the Captivity in the eyes of the children of Israel'. Later he summed up his bleak view of the place in his poem, 'South', in which he has an imaginary (or not so imaginary) Southerner say, 'You blind? Can't you see they're inferior? – / Our women's what they're really after – / You got to use fear – '. There was hope 'to north and west', in Mexico and even Africa, the poem went on. But, it finished, 'in the South, nothing now or ever. / For black and white, no future. / None. Not here.'

From Nashville in mid-January 1968 the Amises went by train, via St Louis again, to Mexico, where they met up with Edmund and Mary

Keeley. The four toured for a bit 'seeing things like the pyramids at Teotihuacan was it,' Amis wrote to Conquest, 'which impressed greatly, and Maximilian's palace in Cuernavaca, which was a fucking pain. Also Acapulco, where I had a suitcase stolen (nothing vital), and Mexico City, where we had a 40-second earthquake tremor.' Amis thought he was having a heart attack until Edmund Keeley put him straight on what was happening. At Acapulco Amis took his first and only dip in the Pacific but the plumbing was dreadful and the whole place struck him as 'fairly horrible'. It suggested to him a possible opening for a James Bond story: 'Bond had never liked Acapulco.'

Then at the beginning of February they went on to San Miguel de Allende, 'a bloody good spot: the first place we've really felt at home in since Princeton back at the beginning of September. 6,000 feet-plus altitude about 200 miles NW of Mexico City, weather like a hot English June.' Amis was stripped to the waist as he wrote his letter on the terrace of the Hotel Posada de las Monjas and drinking 'a kind of tequila Bloody Mary with a hell of a lot of tabasco and so on in the bloody part. Very sustaining.' He and Jane had gone to San Miguel for four weeks to write, aiming to finish the novels they had been working on in Nashville: he *I Want It Now*, she *Something in Disguise*. San Miguel was full of American tourists and 'artists', a class of person Amis might be expected not to want around him much, but they justified themselves in his eyes because their presence made things like shops, bars and doctors 'look up' and could be a source of evening conversation too. They were under the 'very solid wing' of Amis's new friend Mack Reynolds, an American science-fiction writer who lived in the town. To add to Amis's sense of well-being, two consignments of snuff turned up from Conquest within two days of each other. On top of which Amis avoided getting the shits for his entire six and a half weeks in Mexico – though a day after he crossed the border back into the United States he found himself 'hareing to the john'.

At the beginning of March the Amises started 'trickling northwards', again by train to stay with the Keeleys in Princeton for a few more days before boarding the *Queen Elizabeth* for home. They had a 'wonderful' journey, although the immigration authorities held them up at the US border, making them nearly miss their St Louis connection. Amis was now eager to be back in London. He was looking forward to lunching with Conquest and other friends the day after he reached Southampton on 20 March and was a little dashed when Conquest told him the date of the lunch had been moved forward, so he would miss it. But Conquest redeemed himself by arranging a party soon afterwards.

∞ *18* ∞

Barnet

'I heard just an hour ago that Hilly is getting married today,' Amis noted laconically in a letter to Robert Conquest from Nashville on 21 November 1967. 'Any news on this welcome.' Since Amis had by then been married to Jane for a couple of years, Hilly's marriage to her Cambridge don, Shackleton Bailey, looked like putting a firm and final end to their relationship, matters to do with the children apart. Any idea that they might one day get together again in any new set-up would have seemed outlandish, unimaginable. Amis would not have wanted it. He was happy as he was with Jane.

His feelings about her were not the kind of thing he would mention in his letters. Amis has been as discreet about the details of his own private life, even with friends, as he has been reticent about sex in his fiction. He could breezily recommend Mexico to Conquest for its charms in terms with which they were both at ease: 'Tequila is murder but local gin is good and about 15/- a bottle. Even Scotch hardly more than at home. Food excellent, wine awful piss, like tar-water, but beer drinkable.' But the charms of his life with Jane were not to be bandied about in any terms.

If not to be discussed in letters, Amis's feelings towards Jane do find their way, heavily camouflaged, into his fiction, as whatever is preoccupying him at the time he writes has a habit of doing. It is possible to make out the arc of his feelings about women in general and Jane in particular – first soaring upwards, then dismally sinking – through the long sequence of Amis's novels. These do not spell out exactly the story of his emotional life, but they do reflect it intermittently and sometimes closely.

The first two mainstream novels of Amis's years with Jane, *The Anti-Death League* and *I Want It Now*, are very different in tone and content from each other. The first is sombre and serious, ruminating on themes of life and death within the framework of a spy story; the second is social comedy about the adventures of an ambitious journalist and his rich, unhappy girlfriend. But in spite of their obvious differences the two novels have themes in common, themes which reveal a good deal about Amis and Jane.

In *The Anti-Death League*, published in 1966, Captain James Churchill falls in love with Catharine Casement (whose name Jane suggested to Amis). Catharine has had two husbands, as Jane had before she met Amis, marrying for the first time young at nineteen, as Jane did too. When Churchill asks why her first husband married her, she replies that 'it just must have been that he wanted to be married.' All his friends were, so he should be too. That marriage becomes monochrome, purposeless. Then Casement turns up and offers to extricate her from it. He is marvellous at first, but 'The moment we got married he started being different.' As for sex, with her first husband she didn't get 'a great deal out of it', at any rate 'early on'. And, as for her second husband, after they were married he took to hitting her. But now with Churchill it is love, strong and sure to last, virtually at first sight. 'I could never love anyone else in the way I love you,' he says. And she replies, 'I haven't done any loving before worth talking about.'

This is not, of course, an exact account of Jane's first two marriages or her third, but there are resemblances – her marrying young, for instance, and the style of her treatment by her first two husbands. Then, after two failures has come what appears to be a resounding success in love. It was as if Jane, like Catharine, has at last been fully aroused in the embrace of her lover. In *The Anti-Death League* this theme of arousal is only one of several. In *I Want It Now*, published in 1968, it becomes more central and more explicit.

The story and characters of *I Want It Now* are, if anything, even less close to Amis and Jane than those of *The Anti-Death League*. With all Amis novels the central situation is what comes into his mind first. *Lucky Jim*, for instance, sprang from the half-hour Amis spent in the Leicester University senior common room when Philip Larkin left him there in 1946. The central situation of *I Want It Now* arose from a chance remark of Jane's, when she wondered aloud what it might be like to be a daughter of her childless friend Dolly Burns, the rich socialite daughter of the art-dealer Lord Duveen. In response to this intriguing question Amis created Simon, the beautiful but sexually and otherwise bewildered daughter of the rich and domineering Lady Baldock. Amis also invented Ronnie Appleyard, a television presenter and journalist on-the-make, to be Simon's lover.

When the book was published people asked Amis if Appleyard was 'really' Robin Day or David Frost or some other real-life television star of the day. But Appleyard was nobody – he was pure invention, made a television personality mainly because Amis thought such people could be expected to possess the outsized egotism his character required. If anybody in the book resembled anybody in life it was Lady Baldock. At least Bobby Burns, Dolly's husband, saw a likeness in her to his wife and told Amis he would not be giving the novel to Dolly to read. Amis and Jane are not in the least like Ronnie and Simon. All the same, the fictitious story of the latter contains a version of the true romance of the former.

Ronnie Appleyard starts out as wholly dislikeable. He beats the anti-apartheid and ban-the-bomb drums, but only because they are intellectually in fashion and therefore expedient views for him to hold. His real interests are not in any kind of politics but in money, sex and fame, and at thirty-six his only interest in marriage is to promote his acquisition of all of these, particularly the first. When he meets Simon at a party he is instantly attracted by her beauty but her attraction is indefinitely multiplied when he discovers that she is the daughter of the wealthy Lady Baldock. Simon is in a mess, largely because of her mother. She first tried sex at fourteen and hated it but keeps energetically trying (she is the one who 'wants it now' from Ronnie within minutes of meeting him at a party) in the hope of its getting better and of finding a man who will love her. As Ronnie pursues her to Greece and America his feelings change. He falls genuinely in love. So does she. For her, sex with Ronnie becomes very nearly enjoyable, while he becomes willing, indeed determined, to marry her for herself, even without her money.

This story, stripped of its social and comic trappings, is a retelling of the Sleeping Beauty myth – Prince Ronnie awakens Princess Simon. Appleyard is aware that that is the part he is playing: 'over Scotch and water in the Duke of Marlborough . . . he thought he saw that he had indeed awakened the Sleeping Beauty.' To the cynical Appleyard, this is a mixed blessing: 'now here he was not only stuck with her, but, far more alarming, wanting to be stuck with her, wishing he were

299

stuck with her in a real sense instead of just being unable to stop thinking about her.' Falling in love diverts, even subverts, selfishness and ambition. Later, when he has been well and truly 'stuck' with Simon but without her money, he tells her: 'Very odd, this whole thing. I was a shit when I met you. I still am in lots of ways. But because of you I've had to give up being a dedicated, full-time shit.' If he has been a shit, Simon admits that she has been a 'terrible fool'. 'Perhaps we'll have to work on each other,' Appleyard says. 'Helping each other not to be as bad as we would be on our own,' she replies. Exactly so. Not only has Prince Ronnie awoken Princess Simon; her love has redeemed him, made him less self-absorbed, more like a proper prince of romance, a better man at least. In something like the same way, Prince Kingsley has awakened Princess Jane. And perhaps they, too, are living better lives together than they could hope to do separately.

This sense of helping one another to be better – of the couple adding up to more than the sum of two singles – comes across in an interview Amis and Jane gave in tandem to the *Daily Mail* soon after they came home from their honeymoon in Brighton in October 1965.

Kingsley: 'I remember you were very timid when I met you. She was, you know, shy and uncertain of herself.'

Jane: 'I agree.'

Kingsley: 'She was full of doubts. Whenever she proposed doing anything or cooking something, she'd ask two thousand times, "Is that all right?" Now she's sure . . . I've given her that. Women get sureness and confidence from being happy.'

Jane: 'You're more relaxed too.'

Kingsley: 'Yes, that's true. Is it that I'm older? No, I think Jane has changed me too.'

Jane: 'I admit it. I'm really dotty about Kingsley.'

This might be Ronnie and Simon speaking. Within the comedy and the cynicism of *I Want It Now* the novel's core is pure, uncynical romance. So, in those early years, did Amis and Jane's relationship seem to be. In this novel, Amis's romantic trajectory reached its highest point, as it did in his life while he was writing it. Here, and in his earlier novels, Amis's heroes had always liked women. Even when they only had sex on their minds, that was at least evidence that they liked women after their fashion. There were women too who thoroughly deserved to be liked, admired, loved, and often to be treated better than they were: Jean Lewis in *That Uncertain Feeling* or Jenny Bunn in *Take a Girl Like You*. After *I Want It Now*, romance went into decline. Women might still be desirable but they were no longer so often or so clearly admirable or nice. There had always been unpleasant women in the novels, from Margaret Peel in *Lucky Jim* to Lady Baldock. Now that unpleasant side of them began to take over, while some Amisian heroes could be seen less as men who loved women, more as misogynists. This decline in romance was not vertical or precipitate but gradual, matching, in fact, the spread of Amis's own disillusion about women as his relationship with Jane slowly and painfully disintegrated.

Soon after they returned to England from Nashville, Amis and Jane decided they must have a larger house. While they were in America, Sargy Mann had moved in at 108 Maida Vale to keep Monkey company and all now agreed they wanted him to stay on, putting further pressure on limited space. Jane was attracted to the idea of living in the countryside and Amis was too. Not that he had any great fondness for birds or flowers or landscapes. Jane once listed for *Vogue* his likes and dislikes and when she came to 'the country' she described it as something that was 'out' for him – 'except to go to a pub in, or to have lunch somewhere out there (indoors)'. What attracted him to country living was not the scenery but what he had heard years before about village life in Wales, where people were said to pop in and out of each other's houses and behave as neighbours are supposed to do. But Amis liked to be within reach of London too. A compromise was the answer and one was found. Someone spotted an advertisement in *Country Life* for a house called Gladsmuir, on Hadley Common in Barnet, and the whole household went out to take a look.

In front of the house was Hertfordshire, behind it London. The prospect was rural while the location was umbilically linked to the city by the London Tube system's Northern Line, which terminated a longish walk or short drive from Gladsmuir. Here was a country house with plenty of space that could be called a town house too. Philip, Martin and Monkey all liked what they saw and thought they should move there. So did Amis and Jane. The remainder of the lease on 108 Maida Vale was sold for £10,000 and the house bought for £47,000. Jane discovered that it had once been called

Lemmons, decided that that had a better ring to it than Gladsmuir and restored the old name.

When the decision to move was made she gave everyone a talking-to. Lemmons was not simply bigger than Maida Vale, it was very big indeed. The ground floor was sufficiently ample to provide them with a large hall, a study for Amis, a spacious drawing-room opening into the garden, a library which was a little smaller, a bedroom and bathroom *en suite*, a couple of 'tack' rooms and, down a few steps, a decent-sized kitchen. On the floor above there were five rooms (one of which became Jane's study) as well as bathrooms, and there were more rooms on the floor above that too. On top of all this, there was a nine-acre garden to cope with. They did not have money for a large staff, so if they were going to live there, Jane told them, they must all pull their weight. All agreed that they would, but none lived up to their promise – though in the summer of 1968, as reports of Soviet tanks rolling into Czechoslovakia could be heard on the radio, Monkey, Sargy Mann and Martin did rewire the house between them. Martin was rewarded with a bedroom on the first floor for his trouble, a cause of some later resentment on Philip's part.

Lemmons appears recognisably itself as the home of Sir Roy Vandervane in Amis's 1971 novel, *Girl, 20*. Opening the gates to his house, you 'walked into a paved courtyard adorned with small trees in a sickly or dead condition.' Entering the house you 'went through a glass porch of recent addition where there were a lot of very old coats on an old coat-stand and a lot of empty whisky- and wine-bottles. A further door gave on to a passage . . . [and] a near and distant staircase.' The passage led to 'a drawing-room with a large bow-window at the farther end'. Through this window could be seen 'a view of descending lawns, a sunlit wall with trees fastened to it, and some much bigger trees, cedars of different types, farther down. Much farther still were the roofs of the town, looking rather serious over the distant treetops.' Around the feet of a visitor to the Vandervanes' skittered a red cavalier spaniel bitch called Furry Barrel, 'a reactionary little dog: authoritarian, hierarchical, snobbish'. In fact, it was very like Jane's dog Rosie.

Some help around the house was affordable. Mr Mayhew worked in the garden, Mrs Lewsey helped with the vegetables and eventually Mrs Uniacke moved in – first to help look after Jane's ailing mother Katherine, later as housekeeper – occupying the cottage alongside the house proper. But that still left Jane with a lot of chores to handle. She took charge of the practical side of everything: money, decorators, staff, the kitchen, the garden. Since Amis, Monkey and Mann worked at home, all three were present for lunch as well as dinner. The boys were often there too, and Jane was expected to provide two meals a day for all of them.

Lemmons in Barnet was a long distance from Amis's first home at Buckingham Gardens in Norbury, not only in geographical strides across London but in relative splendour, in fewer than thirty years all the way from a little four-room box of a place to a house of nearly thirty rooms. This may have been a gratifying measure of professional success but the material or glamorous side of it was of no great importance to Amis. So long as he had drink, television, books, a room of his own in which to work and adequate access to chaps and conviviality, he was genuinely content to live without frills. When they had been together for fourteen years and Jane knew her husband thoroughly, she wrote in her *Vogue* article that he had 'less interest in money or his surroundings, or, indeed, in acquisitions of any kind than anyone I have ever met'. This had its advantages for her but it had its irritating side too: 'his total lack of interest in houses, for instance, has cut both ways. He finds out what he does *not* like too late for comfort.'

What Amis was ultimately to find most dislikeable about Lemmons was its location. It was too far from clubs, restaurants and chaps, especially for someone like him for whom travel by Tube was a real trial of nerves. And the compensating neighbourliness Amis had hoped to find in Barnet did not materialise. There were suitable people round about to cultivate: Robert Carr, later Home Secretary; Norman del Mar, the conductor; Ralph Harris of the Institute of Economic Affairs. In Amis's view, the cultivation of neighbours was down to the woman of the house. She should go next door for coffee, return the invitation in due course and generally get things off the ground. But Jane did not do enough of this, according to him. At any rate, not much Welsh-style popping in and out of houses took place.

In her *Vogue* article, Jane noted another rare attribute in her husband: 'He never pretends to like anything.' Book, music or film, he would not say he liked any single example of them merely to be fashionable or because not approving might reveal ignorance on his part. This too had its disadvantages because it also meant that there were 'whole areas of taste' in which he took no interest at all, not even simulated. Besides the country, other things that were 'out' for him included most architecture, the theatre (Shakespeare excepted), 'drama' on television or in films, foreign travel and games (indoor as well as outdoor). Some of these dislikes, such as the theatre, were longstanding. Others had built up over the years. Amis, for instance, continued to go on holidays with Jane through the sixties and seventies – to Greece, Turkey, France, even as far as Jamaica, where they were the guests of Dolly Burns – though with increasing reluctance.

Other irksome habits were there from the start too. Resistance to order, for instance. Books and papers accumulated in his study until there was nowhere to sit or even stand. He left his clothes where he took them off for the simple reason that he didn't like putting them away. He was 'extremely good at not being handy around the house', Jane wrote, and his incapacity extended to business matters like buying and selling property, all such transactions being left to her. Not being much interested in possessions himself he was bad at buying presents for other people but at the same time prone to grouse if his own presents were not up to scratch – 'bit *dull*, isn't it, being given pyjamas?'

Amis was fussy about his food too. The only things that he was certain to like were curries and onions. The HP Sauce bottle came to the aid of Jane's meals disconcertingly often, and dinner frequently came down to corned beef or grilled plaice, with liberal doses of sauce to make them palatable. Getting food down at all was a lifelong problem for Amis, dating back to his mother's insistence that he eat up. Infuriatingly to Jane, however, he seemed more tolerant when dining at other people's houses, although here too was something else where their tastes differed. Amis did not like dinner-parties while Jane did. When they had them at home, it was she who usually chose the guests, and he found them boring: chaps in office jobs, as he recollects those occasions, with wives who fluttered their eyes a lot and said that they supposed he must meet a lot of writers – to which he would reply that, no, as a matter of fact, he didn't. No doubt the wives were intimidated by him. For Amis, though, the domestic dinner-party was just a lesser version of the torture formal college dinners had been for him at Cambridge. As always he preferred the pub, the restaurant or the club.

Amis and Jane being thoroughly English and living an English life, it is impossible to ignore the push and pull of class in their differences. Both were middle class, but he from the bottom end of this ladder and she from the top. They might meet on common ground over matters like literature or sex that were of intense personal interest to them both and overrode any class barriers. But in matters of broad taste they reflected their origins. In the jargon of market research, which ranks the whole population according to income and status on a scale that runs downwards from A to E, Jane belonged up with the As, along with professional people and senior business executives, while the Amis background was definitely C1, down among the pharmacists, salesmen and clerks. Here is at least part of the reason why Jane felt more at home with dinner-parties and Amis more comfortable in the pub. Jane, too, liked to keep up standards. The house must be properly decorated, the garden maintained. Having had less of this in his own childhood, Amis expected less later. Besides, he was personally indifferent. He knew what did not interest him and his surroundings were one of those.

In Amis, class showed itself more in the way in which he preferred domestic affairs to be conducted. In lower-middle-class Norbury, horizons had necessarily been limited, and Amis's small family was particularly close and protective, verging on the claustrophobic. His mother not only saw to it that he ate up his food and weeded out friends thought unsuitable. She and his father brought Amis into everything they did and between them took care of every practical detail of his life. Even when he was at Oxford his father still looked after the bills, leaving Amis to handle only his pocket-money. Amis had long ago shaken off the restrictive side of all this, feeling himself entirely free to drink, chase girls and choose whatever friends he wanted, activities which his parents would certainly have interfered with if they had still been in a position to do so. But the supportive side of his old family life he clung on to. Amis liked to be looked after, to have all those same practical details of day-to-day living dealt with by other people, which now meant mainly Jane. In 1975 he wrote: 'Some men like being protected, even mothered a little, without being the less manly for it.' He was writing about Rudyard Kipling, but he might easily have been writing about himself.

This behaviour might seem selfish on Amis's part. Even persuading his own sons Philip and Martin to get themselves educated was left largely to Jane. By orthodox, modern, caring-and-sharing rules for middle-class, two-career families, it probably was selfish; but it made solid sense too. When the household moved to Lemmons it was short of money and Amis had by far the greatest potential to earn what was needed. Leaving him entirely free to get on with the job was the household's best bet for its own future. And nobody could complain that Amis was a slacker in this department of life, reluctant though he might be to wash up. From his daily routine there poured a steady stream of writing which paid most of the bills. All the same, the domestic routine she had to follow caused some resentment in Jane. She felt, rightly or wrongly, that she had sacrificed too much of her own writing career so that he could get on with his.

Money, class, tastes, lifestyles, habits, ambitions, geography – there was potential for disagreement and acrimony in all of them. But discordances existed between Amis and Jane from the start and it would be a mistake to exaggerate their importance. In retrospect, the nuances of difference between them can appear to take on malevolent shape and monstrous size. Besides, such nuances are easy to understand and describe and so make handy explanations for the mysteries of the eventual breakdown of their marriage. Undoubtedly the obvious differences were to play their part in that. In the early years, though, so long as passion and romance survived, they seemed perfectly manageable.

At Lemmons they shared a few sad occasions, especially when Jane's mother died there after moving with the family from Maida Vale. She had occupied a small suite of bedroom and bathroom conveniently located on the ground floor of the house. Her death meant that the suite was empty when an old friend of Jane's, the Poet Laureate Cecil Day Lewis, became ill. 'Poor old Cecil D-L is very ill, dying, in fact, and he will stay with us here until he dies,' Amis wrote to Larkin in April 1972. 'He's very weak but totally compos and cheerful (Christ) . . . Nobody can really tell, of course, but somewhere between a week and a month seems probable.' Day Lewis was strong enough to write his last poem there, dedicating it to the members of the household, 'Jane, Kingsley, Colin, Sargy', and calling it 'At Lemmons'. In the poem he describes 'the calm a loved house breeds'; around him, 'all is amenity, a bloom of / Magnolia uttering its requiems, / A climate of acceptance. Very well / I accept my weakness with my friends' / Good natures sweetening every day my sick room.' Day Lewis's fifteen-year-old son Daniel, later to be an Oscar-winning actor, arrived to hold his father's hand one last time before he died on 22 May.

But there was a lot of fun at Lemmons too. Monkey had never laughed so much in his life. Amis was still an outstanding turn as a mimic. His President Roosevelt, motorbike and truck-starting noises worked as well as ever. He even contrived a wartime thriller from end to end, complete with air raids and torture scenes, and at a party brought off a piece of fake Somerset Maugham with his journalist friend Colin Welch, one of them doing the dialogue, the other the jungle noises. After fourteen years Jane could write: 'I am lucky that [Amis] can still make me laugh now just as much as he did when I first met him.'

If Amis did not yet see Jane as full of unamiable defects, he was already developing a pretty clear-eyed view of his own, which he described at length in his 1969 novel *The Green Man*. Few people noticed the resemblance, since Amis disguised himself in the most un-Amisian figure of an innkeeper, Maurice Allington, proprietor of the Green Man in Hertfordshire, who finds himself plagued by a ghost. Only later, in his short story of 1972, 'Who or What Was It?', did Amis reveal that the likeness was there. In this story he wrote that all his heroes 'can't help having pretty fair chunks of me in them', and added, 'I'm more like [Allington] than I'm like most of the other [heroes].' Such a time-lag in revealing the likeness between himself and Allington is understandable, for the personality of Amis as reflected in that of Allington is not an attractive one.

As an innkeeper, Allington could hardly have had a job less like Amis's, and as someone involved in a hunt for a ghost Allington's adventures were remote from anything that had ever happened to his creator. The story's central situation had come to Amis through a remark by Philip Larkin about a ghostly character 'trying the doors and windows [of a house], trying to get in'. To this Amis added his own notion of the man inside the house being an alcoholic and so likely not to be believed when he claims there is a ghost about.

An alcoholic? This is a description of himself Amis has always rejected as being a term of abuse, not a diagnosis of clinical significance. Still, whatever name you put to it, Amis did drink a lot. Here was his first resemblance to Allington.

Allington is fifty-three, half a dozen years older than Amis at the time. When his doctor friend Jack Maybury asks how he is, Allington says he's all right, but Maybury is not deceived. He sees that Allington is sweating excessively, tells him he should get some exercise and asks how much he's had to drink that evening. 'Just a couple,' Allington answers. 'Huh,' says Maybury. 'I know your couples. Couple of trebles. You'll have another half a couple before we go up, and at least a couple and a half after dinner. That's well over half a bottle, plus three or four glasses of wine and whatever you had at midday. It's too much.' Allington admits that he's been feeling 'bloody awful' but when Maybury says that's because of his drinking Allington contradicts him: 'The only time I can be reasonably sure of not feeling bloody awful is a couple of hours or so at the end of a day's drinking.'

Yet at the very end of a day's drinking, and in some not very clear sense as a consequence of it, Allington suffers from jactitation and hallucinations, more precisely hypnagogic hallucinations. Here is a second resemblance to Amis. He, too, was afflicted by both of these and he devoted more than two full pages of the novel to describing them, pages of a vividness and detail that suggest first-hand knowledge. Most people, he explains, have some experience of jactitation: 'that convulsive straightening of the leg which is often accompanied by a short explanatory dream about stumbling, or missing the bottom stair'. The jerking movement may affect other muscles, including the face, and may be repeated a dozen or more times before the sufferer falls asleep or gives up trying. More serious cases, like Amis's own, find that jactitation is preceded by hypnagogic hallucinations which are not quite dreams but rather 'visions of no obvious meaning seen under poor conditions', something like what anyone sees when, after a long spell behind the wheel of a car, they close their eyes and find they can still see 'a kind of muted version' of what they have been looking at all day. The images are vague, unfamiliar, often human or a human part – face, buttock, thigh. They are not usually terrifying and on rare occasions 'something beautiful shows itself clearly, in a small flare of soft yellow light, before fading into nothing, into the state of a vanished fiction.' The worst thing about these hallucinations is that they herald the jerkings and twitchings of the jactitations that follow them.

When Amis first met Jane he cut back on his drinking but that phase did not last long. By the late 1960s he was drinking heavily again. As with Allington, alcohol did not interfere with his work. It did, though, help to bring on his night-time jerkings and visions. In the story of *The Green Man* the ghost that haunts the inn is finally exorcised and disappears for ever. Describing his own experiences for the novel turned out to have a similar consequence for Amis – his jactitations and hallucinations disappeared too, seemingly exorcised by the act of writing about them, making a curious parallel with the entirely fictitious ghost story he had invented. The same, however, did not happen to his drinking. It was not exorcised but continued and increased.

Allington and Amis share another affliction: hypochondria. Because hypochondriacs imagine they are ill when they are not and their imaginary illnesses respond to equally imaginary treatments they can seem figures of fun and Amis put comic descriptions of this process into Allington's head. Allington remembers how 'a cancer of the large intestine had begun retreating to dormancy as soon as I stopped eating the local greengages and plums, gave up drinking two bottles of claret a day, and curbed my fondness for raw onions, hot pickles and curry [these last three particular favourites of Amis]. A new, more powerful reading-light had cleared up a severe brain tumour inside a week.' And so on. Other people's hypochondria, Allington ruminates, is always good for a laugh and so is one's own when each case of non-existent disease and cure is over. The trouble with this way of looking at it, though, is that, while the hypochondriac may get it wrong a thousand times, he may not be wrong the thousand-and-first. He may laugh off a dozen groundless cancer scares only to find the thirteenth is the real thing. Since he must die of something, like everybody else, he is not altogether wrong to worry about his latest pain, for it might be the one that turns into the disease which finally carries him off. Allington was impaled wriggling on this hypochondriac's hook, and so, all his life, was Amis. Writing about it produced no exorcism of this anxiety, any more than did writing about an alcoholic Allington put him off drink.

Nor did work on the novel do anything to change Amis's grudging attitude to food, of which Allington is no great fan either. 'No other sensual activity must take place at a set time,' Allington complains, 'or comes up so inexorably and so often.' Some food, he concedes, is just about tolerable. 'Fruit slides down, bread soon goes to nothing, and all pungent swallowables have a value of their own that transcends mere food.' But 'chewing away at the vile texture of meat, pulling bones out of tasteless mouthfuls of fish or encompassing the sheer nullity of vegetables is not my idea of a treat.' After thinking a bit about this, Allington concocts his own lunch – ham and tongue covered with chutney and hot sauce, followed by a piece of Cheddar cheese made eatable by its accompanying onions and radishes. He gets through this with the help of a 'powerful' whisky and water and considers when he's done that he has 'made a good meal'. It is not difficult to imagine Amis, too, congratulating himself for having done his duty to his digestive system in this perfunctory fashion.

In a novel the hero behaves as the novel requires, and in *The Green Man* Allington spends most of his time doing things Amis has never done: running a hotel and chasing down a ghost. But Allington behaves towards the people close to him as Amis might have done if he had been Allington. Allington's wife, Margaret, has been killed in an accident after leaving him for another man. By the time the story opens, Allington is remarried to Joyce, and Amy, his thirteen-year-old daughter by his first wife, is living with the two of them at the inn. Allington treats both females badly, and badly in the same way. He ignores them, declines to spend any time talking to them, always has an excuse about having some footling task to do that takes priority. Allington has never talked to Amy about her mother's death or why she left him. 'Perhaps' he will make a start on doing that while they make a trip together next morning, he speculates. And then he adds another 'Perhaps', as if he already knows he won't be able to face up to it. In the event, he finds yet another excuse and cancels the trip. Amy is left to spend most of her days fretting alone in front of the television set.

Joyce manages to tolerate being ignored but then suddenly decides to leave Allington and go off, not with another man but with a woman. But it isn't sex that drives her out, though that is what she expects Allington to believe because, as she tells him, 'sex is all you know about.' She has tried to love him but he won't let her. He has his own ideas about what, how and when to do things and nobody can change them. What she misses and wants is 'being with someone. Who hasn't always got somewhere more important to be in the next two minutes.' Allington offers to make an effort to do better but she turns him down. She doesn't believe him.

Allington's adult son Nick berates his father for similar failings. 'You're just too lazy and arrogant and equal to everything (you think) to take the trouble to notice people like your son, and your wife, and deem them bloody well worthy of being let into the secret of how you feel and what you think about everything, in fact what you're like.'

Characters take it in turn to describe to Allington what he is already aware of about himself, that he is selfish, self-absorbed and indifferent to other people. Even at fifty-three, sex is the only form of human contact that he pursues with any serious enthusiasm. And the form in which he likes sex most is seduction. This he regards as 'the unique sensual pleasure', better than other pleasures, better even than sex itself, because all other such pleasures are 'mere activities, durative and repetitive'. Every seduction, by contrast, is 'a final and unchangeable thing, a part of history, like a century before lunch or a winning try'.

These thoughts come into Allington's head when he is fresh from the seduction of his doctor friend Maybury's wife Diana, a woman based on a real one with whom Amis himself had had a brief affair. Allington has pursued Diana without any sense of love or affection, planning his moves as though he were indeed taking part in a game like cricket or rugby. But then, as he reads her in his cold-eyed way, she is doing much the same, even if she is not playing quite the same game.

When they meet for their first assignation she is only five minutes late, 'quixotically early' by her standards, Allington thinks. Why has she come at all? Allington's father has died only a few hours earlier and Allington senses that she simply can't resist 'a meaty interrogation-session' with him, to do with how he could possibly think about sex at such a sensitive time. As he drives her to the spot he has picked out for them to make love in, he wonders if she will go through with it and then decides she will, and that he knows why she will. 'By opening her legs to me today of all days, she would be being strangely responsive to my strange need, finding herself strangely in tune with this strange man – in other words, she could represent herself as an interesting person.' Women adopting every sort of strategy to arouse interest or sympathy in men had been present in Amis's fiction from Margaret Peel's bogus suicide attempts in *Lucky Jim*, and it was to remain one of the ways in which he saw women all his life: that they had to work hard at being interesting, and would go to all sorts of lengths to do so. Overwhelming a conversation with talk, no matter how trivial or boring, he saw as a favoured strategy. This was one reason why he came to prefer men-only sessions in the pub or the club.

Drunken, hypochondriacal, victim of hallucinations and jactitations, selfish and self-absorbed, unconcerned about other people, keen on sex but largely from a Don Juan-like, points-scoring, fun-of-the-chase angle: the version of Amis contained in the character of Maurice Allington is bleak, even allowing for all the exaggerated colours of fiction. Nor does there seem to be any hope of change. Allington is trapped in his physical nature and his mental habits and all notions of reform are hopeless. At one point in the story the ghost offers, among other temptations, to teach him peace of mind, but Allington turns him down. Later he thinks over his reasons and wonders 'whether we are not all so firmly attached, in all senses, to what we are that any radical change, however unarguably for the better, is bound to seem a kind of self-destruction.' The life of Amis or Allington may be bleak but perhaps it is better than any other life for both of them. That leaves only one escape: death – for Allington 'my only means of getting away for good from this body and all its pseudo-symptoms of disease and fear, from the constant awareness of this body, from this person, with his ruthlessness and sentimentality and ineffective, insincere, impracticable notions of behaving better'.

Then again, that is not the whole story, of Amis or of Allington. The novel in which Amis draws this picture of himself is, after all, a comic tale and a ghost story too. There is more to life – to anybody's life – than can be found in even the most detailed scrutiny of a man's worst defects. Life must be got on with and made the most of while it lasts. As if to make the point, in the last sentence of *The Green Man* Allington puts on his dinner-jacket, swallows a large whisky and goes downstairs to see to his guests, as usual.

⚫ 19 ⚫

Hampstead

The prime object of the artist, according to Amis, is 'to portray human nature as it has always been, the permanent human passions of love, sorrow, ambition, fear, anger, frustration, joy and the rest'. He must also get his background right: he must 'describe his times'. Thus in art as in life there are things permanent and things that pass.

The permanent, though, moves on too. Love and fear are always about but they come and go in individual people. The times change around them as well. So do forms of expression in art. Some of these shifts are easy to chart, others less so. A period in life or in art may be changing without those involved being fully aware of what is going on until the change is complete. Both plain and mysterious forces are at work in such shifts. The decline of Amis's marriage to Jane was a bit like this: plain and mysterious, their relationship disintegrating even as it continued, like an art form moving slowly and imperceptibly to exhaustion.

Over the years, Amis developed a strong sense of passing eras. The poems he collected in his *Faber Popular Reciter* of 1978 were known to every schoolboy in his youth, but even while he was still at school, and unnoticed by him, 'this entire literary genre quite suddenly disappeared, never to return'. Science fiction, which Amis had celebrated in his 1960 book, *New Maps of Hell*, had by 1981 gone wrong in a different way: 'from Chaucer to Finnegans Wake in less than fifty years' – that is, from bright promise to near-terminal decay. Jazz in his view had begun to go the same way in the very year, 1941, that Amis had become seriously interested in it. The decline in other forms of music could be similarly dated. 'I do wish you'd try to fight

313

your way out of this box you're in about everything stopping when Brahms died [in 1897],' says Sir Roy Vandervane, the trendy-lefty conductor-composer of Amis's novel *Girl, 20*. 'No, it stopped with Schoenberg and serial technique,' replies Douglas Yandell, music critic and narrator of the story, here expressing Amis's own views.

Some eras were less precise, more fuzzy. When in 1977 he put together a collection of pieces from the *New Statesman* and the *Spectator* under the title *Harold's Years*, Amis was unsure whether the 'grubby eleven-odd years' of Harold Wilson's rule were worth calling an era at all, if only because they seemed indistinguishable to him from the 'interregnum' of another prime minister, Edward Heath.

By the late 1960s, evidence of changing times could be seen in Amis's own fiction. The 1950s world in which Margaret Peel of *Lucky Jim* and Jenny Bunn of *Take a Girl Like You* carefully preserved their virtue had given way to an utterly different world in which Simon Quick of *I Want It Now* meant exactly that when she used those words. The last sentence of Amis's 1971 novel *Girl, 20* is 'We're all free now.' There is an irony in this because the words are spoken by Sir Roy's daughter Penny, who has acquired all the freedom of a heroin addict. But there is truth too. The mood and manners of 1970 were much 'freer' than those of the 1950s, though both worlds seemed perfectly normal in their time. It is always easier to describe such change after it has happened than to account for its ever having happened at all. So it may be with a marriage like Amis's and Jane's.

One cause of change may be simple exhaustion, like an art form running out of steam. There are only so many ways in which a murder can be committed in a locked room (eighty-six, according to Amis's recollection of John Dickson Carr's count) and there has to be a limit to the number of things that can go wrong with a spaceship. When these limits are approached, Amis has noted, the country-house murder or the spaceship adventure is in trouble. Or a whole cultural context may crumble. The great themes of the popular verse Amis collected for Faber were the nation and the church. Yet these no longer commanded popular attention or support, probably because of the 'disintegrative shock of the Great War', so that even if a poet possessed the skills necessary to write more poems in this vein he would not find an audience to read them.

Another powerful cause of decay, in Amis's view, was fashion, in particular the scourge of modernism with its characteristic effects of 'mystification and outrage', which could undermine minor forms like science fiction just as successfully as major ones like poetry or painting. And then there was the dire influence of the academic and the critic, encouraging the writer to perform for their benefit rather than for an audience of readers. In time, Amis came to believe that the overall effect of making English a university subject had been malign. Had English literature got better or worse since it began to be taught at Oxford in 1894, he asked himself and answered: worse. And was its being taught at universities part of the reason for this decline? Yes, indeed.

It is not fanciful to see the decline in Amis's marriage to Jane in terms of disintegrative forces like those he saw altering or destroying forms in art and eras in life, the forces working simultaneously at different levels in their lives, some capable of being described, others not easily accessible to understanding – and, of course, as is always likely to be the case in the breakdown of a marriage, appearing differently to both parties involved.

Formally at least, dates can be put on the Jane era in Amis's life. They met in September 1962, started to live together the following July, were married in 1965, separated in 1980 and divorced three years later. Emotionally, the passage of their life together was not so neatly clear-cut. Exhaustion, like that of a literary genre running out of steam, played its part. Amis came to think that the marriage should have ended round about 1970. That it continued for another decade had a lot to do with them both not wanting it to fail. They might not be perfectly happy but the alternative, separation, might all too easily be much worse. Two previous marriages had gone wrong for Jane and one for Amis. Neither wanted it to happen again.

But there was more to this exhaustion than the ordinary wear and tear of married life, with passion diminishing and the excitingly new becoming the dully routine. That was part of it, naturally, but there was worse. For Amis, love wore off, then liking, then sex, though not necessarily in that precise and orderly sequence.

Sex, or the lack of it, was a reality about which something could perhaps be done and if sex could be put right then less tangible realities like love and sympathy might fall into place again. But what could be done? They could go to a therapist. In fact they went to two, one a doctor, the other not. Routines were prescribed for reviving sexual interest in each other of which Amis gave a version in comic form in his 1978 novel, *Jake's Thing*. Jake Richardson is fifty-nine and married, and has in his time been to bed with more than a hundred women. But lately his libido has fallen off, most notably when he re-encounters an old flame who happens to show up while his wife Brenda is away. In bed with her, Jake finds thoughts about his dinner and a James Bond film on television taking unexpected and unwanted precedence over thoughts about sex. Jake goes to see Dr Rosenberg, who questions him about fantasies, early-morning erections, masturbation – virtually his whole sexual history – then puts him on a regime he calls 'inceptive regrouping'. This involves studying pornography and masturbating in consequence, plus 'non-genital sensate focusing sessions' in which husband and wife fondle each other but do not have sex. Dr Rosenberg gives Jake a 'nocturnal mensurator', a complex device Jake is instructed to wear in bed so as to measure the rise and fall of his night-time erections. Jake dutifully follows the regime prescribed, and even goes to a workshop. Amis did similar things, though he invented regimes and names like nocturnal mensurator for his novel and filled out his own experience by talking to friends about theirs. But therapy doesn't do anything for Jake and it didn't do much for Amis's sex life either. At the end of September 1979, he told Larkin that he was suffering a 'total loss of sex-drive; I haven't had a fuck for more than a year and a wank for over a month. Don't tell anyone.'

Another thing that something could be done about was where they lived. Amis came to feel himself cut off in Barnet. He didn't like the Tube and increasingly travelled only by taxi if Jane was not available to drive him. Jane thought that moving back closer to town might help improve the general household climate and her relationship with Amis. By the mid-1970s she had already begun to think of leaving him. But even thinking about it terrified her. She had 'bolted', to use her own word, from two husbands and did not want to bolt a third time. If changing house would make a difference – make Amis more content and easier to live with – then it was worth a try.

After nearly eight years at Lemmons, they found a house worth moving to in Flask Walk in Hampstead. Gardnor House, a listed eighteenth-century building, was chosen for its size, its encircling garden (large by Hampstead standards, if small compared with Barnet) and its closeness to town. Amis told Jane he liked it for, among other reasons, its being handy for the Tube.

There were the usual alarms and hesitations over the buying-and-selling process. 'Our move has reached a very crappy stage,' Amis wrote to Conquest in May 1976. 'This morning, three days before the ultimate deadline for signing for the new house, the chap who's been going to buy ours for the last three months says he doesn't think he can now.' The buyer was short of cash and hoped to get a bridging loan out of a rich friend. A month later the move to Hampstead was still approaching 'in a sort of sidelong way, with the buyer of this place behaving like a shit and the mortgage company like a Central European bureacracy'. By the beginning of July these problems had been sorted out. London was immersed in a heatwave and Amis was 'getting terrified' of the upheaval of moving, now scheduled for the 21st of that month. 'I just can't believe that any human agency will move all these thousands of objects from here to there, or anywhere.' Amis took his own responsibilities in characteristic style. 'My main job so far', he told Conquest, 'had been drinking-up the nearly-empty bottles, horrible stuff like cherry vodka, Mavrodaphne, raki etc.' He warned Conquest, who was planning a visit, that the builders would be in the new house for several weeks, though a neighbour was going to lend him a room to work in, 'and there are plenty of pubs.' By October he was established. Doing the house up had been 'bloody expensive', he told Conquest, 'but not impossibly so. Am in my study now after nearly 3 months camping out in various bedrooms, and settled enough to start a longish poem.'

But moving house made no serious difference. Amis was happy to be nearer the heart of town, but that made him no happier with Jane.

The trouble was that Amis was coming seriously to dislike her. This was more than stopping fancying her or ceasing to be passionately in love. Or, rather, it was because he stopped liking her that love and passion finally died. He might still refer to her affectionately in his letters to Conquest ('poor old thing' when she was ill) but that was not how he was mostly feeling about her.

Amis later worked something of what he did feel about Jane into his 1984 novel *Stanley and the Women*, written after they had separated, giving her fictional form in the character of Nowell Hutchinson. Nowell is the first wife of the novel's hero, Stanley Duke. She is an actress, an appropriate choice of trade, since Amis thought there was a good deal of the actress in Jane. He remembers appreciatively an encounter between Jane and Victor Gollancz at a party. She complained to Amis afterwards that Gollancz could never remember who she was – 'You're an actress, Miss Howard?' he would say. Amis thought, but didn't dare to utter, 'That's because he's seen what you're like – affected.'

In the novel, Stanley Duke and Nowell have both remarried but are brought back into contact because of their son Steve, who goes mad. Stanley wonders whether Steve has scared her but then decides that 'all that scared her was the prospect of everybody not looking at her for five seconds. That was just as she plumped for being distracted rather than marvellous and began blinking a lot and making small sudden movements.' A doctor friend, meeting her again after ten years, observes to Stanley that she hardly looks a day older. 'But then egotists always do wear well. Like queers.'

Stanley struggles to describe Nowell to a psychiatrist. He picks on something the psychiatrist has said about people who can't make out what's going on and says, 'I don't think she can do that, not what's really going on . . . Nothing's what it is, it's always something else. Her sense of other people's not good. They can be sweet to her and they can be foul to her, and that's about as much scope as they've got . . . No use telling her to stow it or cheese it or come off it . . . That would just be everybody being foul to her.' Amis would ask Jane a question and instead of answering she would ask why he'd asked it, as if nothing could be straightforward, not even a simple question – nothing was what it was, it was always something else. Or so she seemed to Amis.

Nowell's world, Stanley thinks, is a place where 'the truth or untruth of a statement rated rather low.' She comes up with what he calls 'her own-brand facts'. Stanley tells the doctor who is looking after his mad son, 'In my experience nothing's what Nowell says to anybody . . . I mean whatever she said's got nothing to do with what happened.' Even if there are witnesses to prove it she will deny what she has said if it suits her. 'She makes up the past as she goes along. You know, like communists.' This notion of women being like Communists is brought up again by Stanley to account for Nowell's behaviour in relation to their son. Stanley remembers his doctor friend saying that women were like the Russians because 'if you did exactly what they wanted all the time you were being realistic and constructive and promoting the cause of peace, and if you ever stood up to them you were resorting to cold-war tactics and pursuing imperialistic designs and interfering in their internal affairs.' But if you went on promoting the cause of peace long enough you ended up 'Finlandized', or more or less under the woman's thumb.

Amis certainly thought of Jane in some of the same cold-war terms as Stanley considered Nowell. Early in 1982, after they had separated, Jane heard that Amis was in hospital following an accident and sent him a get-well note and a copy of her new book. Soon afterwards she gave an interview in the *Sunday Telegraph*. Amis's agent, Pat Kavanagh, reported that Jane had said she wasn't suited to living alone and had perhaps made a mistake leaving him – the reason she sent the book was because she wanted them to be friends again. In a letter to Conquest, Amis imagined Jane's apology: 'Sorry about that, Kingers; let's have a nice chat and forget it . . . Talk about women = the USSR. We regret the damage done to your homeland and the infliction of numerous casualties. We are now satisfied that the decision to rectify our common frontier was taken in error. Those responsible for it have been punished. There being no longer any cause for friction between our two peoples, we look forward to an era of common understanding and increased prosperity on a co-operative basis.'

A reason for this kind of behaviour occurs to Stanley Duke: insecurity. Nowell has accused him of neglecting their son Steve, whereas Stanley is pretty sure that it was the opposite way round. But Nowell's charge, he realises, is not a question of hostility, 'just self-protection, forestalment of the possible and well-founded charge that it was she who had done most of the neglecting. I had forgotten that her whole character was based on a gigantic sense of insecurity.'

Stanley finds himself in Soho getting matily drunk with Nowell's second husband, Bert, who has some additions of his own to make to the list of charges against Nowell. 'It's not much fun . . . living with somebody . . . you don't like much,' he declares. He has noticed something about Nowell that Stanley has missed. If you are in the same room as her while she's sticking safety-pins in or putting stamps on envelopes she will say you are watching her to see if she will be slow or clumsy or get things wrong. That's not what you're doing but how can you prove it?

Bert has developed a great refuge from this kind of thing. He is drunk all the time, or that's what he wants Nowell to think, even when he's sober. Being drunk excuses him from making love. As

Bert explains it, when you're young 'you're ready to fuck anything on two legs.' But when you get older you get choosy. It puts you off a bit if they 'chat to you about Harold Pinter while you're on the job' (a very Amisian choice of offensive subject, Pinter). And by the time you get to Bert's age – he is pushing fifty-three – you expect them to be a bit pleasant, to listen now and then, to be good company, 'A lot of unreasonable things like that.' 'If you don't like 'em, you don't want to fuck 'em,' Bert concludes. Of course, he could tell her that the reason he doesn't like her is that 'she's such a bloody horrible creature' but he would risk a knife in his guts if he did that and she would then talk her way out of any charges by some such excuse as unendurable provocation, which she would consider proved by the fact that she had been unable to endure it. QED.

This is, of course, a lurid, larger-than-life, Grand Guignol account of Amis's view of Jane's behaviour which he imagined for his novel rather than copied in exact detail from experience. Even more thoroughly imagined is the aspect of Jane which finds its way into Susan, Stanley's second wife. Unable to cope with Steve's madness and all that goes with it she finally walks out on Stanley. But before she goes she calls him a 'lower-class turd' and berates him for his 'gross table-manners and your boozing . . . and your mates and your whole ghastly south-of-the-river man's world'. This might have been a description of Amis as much as Stanley, both of them coming from south of the Thames – from an 'area', not a 'place' – and making it to Hampstead, bringing with them a strong appetite for drink and men's company. But it was not a charge Jane ever threw at Amis. On the other hand, was that what she sometimes thought about him? He suspected that it was.

Bert tells Stanley that Nowell only married him because he was earning a lot for his age and she married a second time so that Bert could get her parts in films. But she would never admit this – even to herself and even if filled full of a truth drug – any more than she could admit to herself that she wasn't a very good actress, nor very beautiful and anyway getting old, a combination of things which makes it impossible for her to leave Bert. So he's stuck with her. But why doesn't he leave Nowell, Stanley asks. Because he's been divorced once already, Bert answers, and 'it bloody near killed' me. Earlier in the story, Stanley thinks about his own divorce from Nowell: 'Stopping being married to someone is an incredibly violent thing to happen to you, not easy to take in completely, ever.' It was for reasons like these – like Bert's and Stanley's – that Amis did not want to leave Jane, however much he had come to dislike her. He preferred things to stay as they were, unsatisfactory though they might be, rather than endure the terrible violence of a divorce and face the prospect of being abandoned to that most dismal of fates – being left on his own.

Jane had no more wish to see her third marriage break up than Amis did his second. Yet his dislike for her became more and more obvious and increasingly hard to bear. She thought it was possible to live with someone who didn't love her but not with someone who didn't even like her or want to make love to her. This combination was insufferable. Still, she stayed put. Then in the late summer of 1980 she and Amis went on a two-week Mediterranean cruise with Violet and Anthony Powell and Amis's American friends Paul and Betty Fussell. From the cruise the Amises went on to stay with friends in the Dordogne and from there to Edinburgh for the Festival. This little expedition turned sour on Jane. It became the final straw that broke the back of her marriage.

The trip started off badly when she and Amis nearly missed their boat, getting off the train from Waterloo uncomfortably close to sailing-time and making it on board only with the help of a taxi-driver who happened to know whereabouts in Southampton docks the MTS *Orpheus* was tied up. In the piece he wrote about the trip afterwards – a rare venture into travel-writing – Amis glossed over his feelings about this incident. He put his anxiety down to 'travel angst' which he speculated might have something to do with 'the poor old naked ape' feeling particularly naked when the unfamiliar business of leaving home and group goes wrong. He congratulated himself 'for not ballocking Jane – she it was who had screwed the timetable.' If they had missed the boat the only way of catching up with it would probably have been by catching a plane to France, something Amis could not bear to contemplate. Jane was in charge of the travel arrangements, she should not have misread the documents, the anxiety it induced in Amis was unnecessary and entirely her fault. If she couldn't get that right, what could she get right?

The *Orpheus* took them to Brest, Nantes and La Rochelle, across the Bay of Biscay to La Coruña, then Vigo, Oporto, Lisbon and Portimão, on to Seville, Motril and Barcelona, finally dropping them off at Nice. After the poor start, Amis soon found he was enjoying himself. He had not been abroad for five years and anyway had never much cared for conventional holidays. The trouble with the usual hotel-and-beach vacation, he considered, was that it allowed no change of pace. You could put your feet up and drink all day, so there was never a moment to look forward to when you could permit yourself to do just that and feel you'd earned it. On a ship, though, there was a programme to be followed, rather like in the army, beginning with reveille and breakfast and going on deck for first parade in full marching order – which in Amis's case meant equipping himself with 'both pairs of glasses, passport, money, cheroots, lighter, pen, notebook, hat and Lomotil' (a pill that stopped 'any ordinary case of trots dead in its tracks'). From the deck you proceeded to a bus, trudged round a few sights and finally got back on board feeling you really had earned that drink. 'A very British arrangement,' Amis called it. Others might call it a very Amisian one, for with its routines and its regular changes of scene and pace the cruise programme provided him with a very tolerable alternative to the dreary liberty of a one-stop holiday in the Mediterranean sun.

Amis liked the Gulbenkian Museum in Lisbon because Calouste Gulbenkian had followed nothing but his own taste in making his collection, so had chosen well, at any rate when it came to sculpture or paintings – Amis was not interested in furniture or knick-knacks. He disliked the Picasso museum in Barcelona, having taken strongly against Picasso years before, on first seeing his work at the Tate Gallery as a schoolboy. Then he was coy of saying so, but not now. The paintings he saw in Barcelona were 'unappealing', 'really vile', 'grotesque, decadent'. Picasso's gravest weakness, he pronounced, was 'shortage of natural talent'. 'What a tiresome painter he is, to be sure,' he wrote to Conquest when he got home.

Travel did not now much broaden or change Amis's mind, as it had in America; it tended, rather, to confirm his opinions. 'Spanish, like English, nice people, nasty food; French, nasty people, nice food . . . Greeks, nice people, *terrifying* food . . . Germans, nasty people, nasty food.' He also found pleasing confirmation of his political views on encountering at the captain's table a 'rather pretty' Brazilian dentist called Mercedes Astle, married to an Englishman. It seemed that working in Africa had put her off Communism as a possible cure for her own country's problems. While in Botswana she had been called on to attend to the mouths of sundry Russians and 'what she had seen in this way of Soviet dentistry, especially the low quality of the cement used, had caused the scales to fall from her eyes,' Amis told Conquest. 'Truly there are innumerable roads from Moscow.' (Or, as he put it in his article, 'many paths from the Devil'.)

Amis as always was good at entertaining his friends. Visiting a church in Brittany he reduced Paul Fussell to uncontrollable laughter. They had been discussing how various writers – Henry James, William Faulkner, Ernest Hemingway – might handle the embarrassment of describing a fart. Fussell asked Amis how he thought Homer would have dealt with the problem and Amis came up with an answer at lightning speed: 'A great fart shook him.' Half an hour later Fussell was still convulsed in a paroxysm among the tombstones, 'shoulder-shaking, gasping and eye-watering'.

'Actually all in all the cruise was quite fun,' Amis told Conquest. It was when he and Jane got to La Tourette in the Dordogne that the temperature dropped again to something like Southampton level. 'Pretty enough, but filthy French food – you know, elaborate sauces to persuade you to swallow stuff you would otherwise have hesitated to give the dog – sod-all to do and no host.' Joy Law, a long-time friend of Jane's, was their hostess but her husband Richard had stayed in London. With nothing else to do and no males to talk to, Amis settled down to write the article about his trip, which was published in the *Sunday Times* the following July. Amis remembers rowing with Jane over a game. 'If you go on like this,' he told her, 'we won't be able to play Scrabble.' To which he recollects her replying, 'I think I'd better see my solicitor.' Jane, though, denies any talk of lawyers.

Matters did not improve when they reached Edinburgh for the Festival, where they had been invited to stay at a hotel owned by Amis's friend, Dickie Temple-Muir. The Festival programme was not to Amis's taste. 'Leftiness and trendiness have invaded it,' he told Conquest. 'There was a German shag whose art-form was chatting to you while he drew things on a blackboard. I didn't go to him. Nor to any of the concerts, all of which contained either a Mahler symphony or some jerk's first performance. Did let myself be taken to Lilian Hellman's play, *Watch on the Rhine*. Frightful piss, of course, but different from contemporary piss. She wasn't *trying* to write piss, like Beckett or Stoppard, she was just totally failing to be good. Vital point, or is it?' Amis wouldn't go to anything else and, according to Jane, wouldn't let her go to anything either. At a lunch with the conductor Claudio Abbado, Amis was, according to her, extremely rude. Amis has no recollection of the encounter. When they got home, she says, he accused her of losing or even destroying his travel article, a charge which appalled her, though it is one he denies making. Jane decided she must bolt once more.

Having made up her mind to leave, Jane planned her exit with some tact. She booked into Shrublands, a health farm she regularly visited, for ten days. That, she thought, would give Amis time to get used to her not being about the house, so that when she failed to return the immediate shock would not be too great. An amusing story later went the rounds that Jane announced her departure in a letter she left for Amis behind a teapot on the mantelpiece, where he failed to notice it for several days. This was not so. Jane arranged for a letter to be sent by hand to Gardnor House from her solicitors on the day she was due home. The letter was taken in by Mrs Uniacke, who had stayed on as housekeeper after the move from Barnet, and she passed it on to Amis. Jane duly returned to London at the end of her health-farm stint, not to Gardnor House, but to a room in the flat owned by her friend Ursula Vaughan-Williams.

Amis was taken aback by the arrival of Jane's letter, and yet not taken aback. Jane's defection was a complete surprise and at the same time exactly what he felt in his bones might be coming. His reactions were as mixed as his expectations. Writing to Larkin on 5 December 1980 he announced her departure: 'Not with anyone, just buggered off. She did it partly to punish me for stopping wanting to fuck her and partly because she realised I didn't like her much'. And yet 'not having her around is immeasurably more crappy than having her around'. A month later he wrote to Anthony Powell: 'It looks like a final parting of the ways, a thought that arouses in me, among many other things, a profound sense of relief.' Among the things it also aroused, at least to begin with, was a strong wish that Jane should come back, for reasons perhaps more negative than positive but no less compelling for that. As a man of habit and routine he liked stability and disliked change. In the middle of January, he told Larkin that 'the mere thought of the future terrifies me'. At least when he and Hilly had broken up he had Jane to go to; but now neither Jane nor he had another person to fall back on. Amis's anxieties, too, came into play. The last thing he wanted was to be left alone in an empty house with only Mrs Uniacke for company. Amis wrote to Jane saying that life would be even nastier if they were apart than it was when they were together, so why didn't they try to make things up between them?

A sort of negotiation followed, through the exchange of half a dozen notes. Jane said she would return, but only on certain conditions, the principal one being that Amis should give up drinking – not just moderate his intake or cut down a bit but stop completely and for ever. Drink, Jane argued, had been her husband's main problem, the chief reason why he had become unbearable and she could no longer live with him. Why else would he have accused her of destroying his travel piece during their Dordogne stay? Or pick quarrels with the likes of Claudio Abbado? Or even at times fall over, once so heavily that he had broken the humerus in his right arm near the shoulder? (This was an incident from which Jane, in Amis's view, emerged with even less credit than himself. She failed to accompany him in the taxi to hospital the next morning when his arm began to hurt, pleading business with a television producer which Amis thought could easily have been put off. He was obliged to type with his left hand only for several weeks, though this did not much slow him down.)

In singling out his drinking, Jane certainly chose an offence that was not implausible. Nobody could argue that he did not drink a lot, not even Amis himself. Monkey Howard remembers him going upstairs at Gardnor House on all fours because he could not make it standing up. Jane claims that he hallucinated, not just in bed like Maurice Allington in *The Green Man*, but while wide awake. She says they once had a row about people who had been present at a dinner-party which, she claims, had never taken place. Amis has no recollection of any such incident. From his point of view the important thing was that he could cope with his intake, however large. After some thirty-five years' acquaintance, his friend Paul Fussell reckoned that, though about half the time he had spent with Amis had been in pubs or other drinking-places, he had only seen him 'seriously under the influence' once – 'glazed eyes, silly smile – all the symptoms'. This happened when they were on holiday in Greece and attended a wine festival where they circulated from booth to booth sampling the vintages at no charge. Amis, Fussell considered, had 'remarkable powers of control', which he exercised over his drinking as much as anything else he did.

Amis knew he was vulnerable to charges of excess. He likes to tell a story against himself in which a taxi-driver drops him off after a heavy session. Finding that he has no cash, Amis is compelled to write a cheque. The driver watches him make heavy weather of getting out his cheque-book, signing his name and so forth, and observes: 'You seem like an educated man. Why d'you have to get yourself into this state?' The story can be seen as revealing the primness of the taxi-driving classes. But it also raises a question about Amis, as he well knows: why does someone like him drink so much? To which there is no real answer except the obvious ones of habit and pleasure.

Habit, even addiction, did not mean that, if he had wanted to badly enough, Amis could not have given up as Jane asked. Just over a year later, starting on 10 March 1982, he proved he could do exactly that, admittedly to begin with under the forced regime of a stay in hospital. He stumbled and broke his leg (the tibia and fibula) coming out of his bathroom, under the combined influence of a sleeping-pill and a day's drinking, begun at a lunch of the Society of Snuff Grinders, Blenders and Purveyors and continued off and on until bedtime. 'A drunken fall?' Amis wrote to Conquest a couple of weeks later, anticipating Conquest's inevitable question. 'Well yes, even if what I did was to break them *standing up*, trying to avoid a fall which I would have done much better to give in to.' The break needed an operation to put right and he was confined to the Royal Free Hospital in Hampstead for three weeks, during which alcohol and tobacco were hard to come by. The combination of being deprived of both while being pumped full of antibiotics, painkillers and sleeping-pills caused him some mild hallucinations to begin with: 'nothing spectacular, just a few voices and non-existent cats'. But he never smoked again and kept off drink for almost six months. Amis even managed to cope with the temptations of the Garrick Club bar and only succumbed when he went for his summer holiday to Swansea, where he found it impossible to stay on the wagon among his bibulous Welsh chums.

So giving up could be done. But why should he do it? It wasn't as if drink drove him to wild extremes. He did not hit his wife, embarrass her in public or default on his work and so fail to earn a living. Nobody could say that his drink affected his output of words, not even Jane. He was endowed with remarkable powers of recovery which saw him at his type-writer promptly every morning, no matter how much he had consumed the night before and how hungover his co-drinkers might be. Other wives put up with far more serious defects in their husbands, Amis thought. So why couldn't Jane live with his drinking?

In his exchange of notes with her, Amis did offer compromise: he would not drink spirits except at weekends – that sort of thing. It is very likely that he could and would have stuck to any bargain along those lines that Jane was willing to agree. If that would make her happy, then he had the determination to deliver. But she was adamant. Either he gave up completely or there was no deal. Amis declined to accept her idea of a bargain and negotiations were broken off. With that the marriage was effectively over, bar the practicalities of divorce and the associated division of property.

Why did Amis decline to go all the way to meet Jane's demands for total abstention? Not drinking turned out to be perfectly tolerable when he made the experiment. Surely it would have been worth trying a year earlier than it was forced on him in order to preserve his marriage? But Amis thought the price was too high, that Jane was being unreasonable to the point of vindictiveness. 'Can you imagine', he asked Conquest in a letter at the beginning of February 1981, 'the lifelong resentment I would feel?' Drinking and the sociability that went with it had been part of his life for forty years and stopping completely would deprive him of a fundamental pleasure. Besides, he did not accept Jane's analysis of the part it played in their relations. In his letter to Conquest he explained: 'Her scenario, you see, is that my drinking was what fucked up the marriage. And so it did in a way, in that when I'd had a few I would sometimes lose my prudence and tell her a little about her monstrous behaviour.'

Jane should have settled for compromise, in Amis's view, but wouldn't for reasons of her own which he speculated about in a letter to Anthony Powell a couple of weeks later. 'Jane said, wrote, that she was quite sure I wouldn't be able to cut down drink and my only chance was to give it up entirely. How did she know that? Because she can't cut down smoking but is going to give it up, or has, that's how.' In Jane's eyes, according to Amis, it had to be all or it would turn out to be nothing. That was certainly how it to turned out to be with Jane and smoking. A decade later she had still failed to give it up and was still puffing away at her Silk Cut cigarettes.

To choose the freedom to drink in preference to keeping a wife may seem a very strange order of priorities. But perhaps drink was a shadow issue as much as a real one, standing for a whole list of incompatibilities between Amis and his wife. If drink had not been the one on which Jane chose to make a stand, other differences might well have taken its place. There was certainly not much left of the usual ingredients of a marriage: love, sex, liking. Only their co-operation as writers had stayed largely intact.

When Jane left him Amis gave up on women, at any rate as sexual partners. At fifty-eight he was not looking for a new wife and he no longer pursued women in order to get them into bed as he had done thoughout his adult life. If he saw a beautiful girl through the window of his taxi he did not now desire to lay hands on her, though he could understand perfectly well why another man would. This made life less complicated but it left a hole in it too which was a matter for some nostalgic regret. But the simple fact was that he had lost the urge. Sex no longer interested him, a mixed new blessing which Amis celebrated in a poem, 'Senex'.

> To find his sexual drives had ceased
>   For Sophocles was no disaster;
> He said he felt like one released
>   From service with a cruel master.
>
> I envy him – I miss the lash
>   At which I used to snort and snivel;
> Oh that its unremitted slash
>   Were still what makes me drone and drivel!

Now that the slash of sex did not make Amis drone or drivel, if he needed a partner it was for company, not sex. Being left alone in the house at night was the most urgent problem created by Jane's departure. Fear of this was a phobia from which he had suffered since childhood. Conquest rallied round and offered to have Amis live with him in California but apart from any other objections another Amis phobia stood in the way: 'thanks all the same,' he wrote in March, 'but honestly the idea of climbing into a jet and being taken anywhere at all appeals to me about as much as that of walking a tightrope across Niagara Falls. My phobia about this is deep-seated and incurable.' His phobia about being left alone was 'milder but still bothersome'. 'I must say,' he told Conquest, 'it gives a booster shot to one's ordinary feelings over being deserted.' What he wanted was someone to live in the house, but 'Where the hell am I going to get the kind of companion/housekeeper I need?' This was a gap Mrs Uniacke could not fill.

A stopgap solution was found when the boys, Philip and Martin, also rallied round. As it happened, Philip was an enforced bachelor at the time, his own marriage having recently broken up. He came to sleep most nights at Gardnor House, with Martin filling in when he could not stay over. Amis's phobia about being alone may have been milder than his fear of flying but it was very real all the same. When neither of the boys was available he once asked a friend in whose house he was dining whether he might stay the night. Fortunately the friend agreed.

Meanwhile the disagreeable details of separation and divorce were quickly started. At the end of February 1981 Amis put Gardnor House on the market. 'Life goes on here undisturbed to outward view,' Amis wrote to Conquest early in March, 'though any moment flocks of shags will start being shown over the house.' By early May, Amis could report that 'a band of shags, potential buyers, can be heard creeping round' even as he wrote. He didn't like selling because he didn't want to leave his home. But he had no choice if he was to settle financial matters with Jane. In July a snag developed when the surveyor for a potential buyer discovered a fault in the wall which might cost as much as £100,000 to put right and therefore cause that much to be knocked off the sale price. By August things were on the move with an American buyer in prospect and by the end of the year Amis was out of Gardnor House and in a temporary flat around the corner, from where he was soon ready to move once again into 'a very decent little house in Kentish Town' early in January 1982. Gardnor House was sold for £265,000 to the American, and Amis was scarcely (though luckily) out of it when the suspect wall collapsed.

As he made arrangements to move, so divorce proceedings were begun, the grounds of which were to be Amis's 'unreasonable behaviour'. This was convenient but hardly in line with his own view of the case. 'Things like trying to stop her getting away with murder and bestial stuff like that' was how he described his unreasonable behaviour to Anthony Powell. 'You know, I've a bloody good mind to defend it,' he went on. 'It wouldn't half feed her up.' But the divorce went ahead undefended. Money was also a problem for which he blamed Jane. When he discovered the fault in the wall of Gardnor House he complained: 'I left all matters of house management etc. to Jane, because she's so good at it, you see.

My (new) accountant told me about my insurance-pension position the other day, another thing I left to Jane because, etc. He said that whereas Jane would be very well provided for if I died, my pension stands at £120 p.a. from age 65: literally that.' His new accountant had also told Amis that 'in his view Jane had lived very well off my income.'

A year after the break, when Amis had moved out of Gardnor House, he poured out some more complaints about Jane to Conquest. Jane had been behaving 'very shittily, an assertion which I'm sure will make you gasp and rub your eyes'. She had given interviews to the London *Evening Standard* and the *Sunday Express* in which she had blamed him for not going on holiday and stopping her writing by making her run the house – 'when we were two in family in the week, often but not always three at weekends, and had a living-in housekeeper' was his answer to that. Jane had visited Gardnor House in Amis's absence to take her share of the household goods but according to him she had done much more than that. She had 'rifled the place, taking all the best crockery, glassware, etc., including the remains of a decent case of claret a club-mate had given me for doing him a favour.'

That was not the end of Jane's attempt to get more than her fair share, according to Amis. She wanted not only half the sale price of Gardnor House, the proportion customarily awarded to each spouse in a divorce settlement, but another £40,000 on top, 'that being the value of the flat she sold in 1964 to help us buy 108 Maida Vale, adjusted for inflation', as he told Conquest. In Amis's pithy version, she got to keep things like the Victorian furniture, while he got to keep the paperback books. His sense of being stripped of his belongings was made especially acute when he found he had no suitcases to take with him on a weekend trip and had to borrow one from Mrs Uniacke. But, in Jane's version, she only got what was her due. She did not 'rifle' the house but took only things that had been hers before they moved to Lemmons, silver left her by her mother and 'a few pieces of good china' that had belonged to her grandparents. Nor did she take £40,000 but £14,000, which had 'bought the lease, paid for the extensive building repairs and also the furnishings' at Maida Vale. A version of the truth acceptable to them both seems unlikely ever to be computed.

∽ 20 ∾

Primrose Hill

Amis thought that Jane's attitude might be what he called 'purely routine female behaviour'. On the other hand it might be something with a more bitter, vindictive tang to it – 'her sense of injustice that I seem not only to have survived her departure but to be having an illegitimately uncrappy time.' One reason for his time being uncrappy was his developing habit of spending a lot of it at the Garrick Club. But there was another – far more important and far more likely to stir resentment in a departing wife. For by now Amis was comfortably installed in a new ménage with his first wife Hilly.

You'll have to hold on to your hat, Amis had warned Conquest in August 1981 as he prepared him for astonishing news. 'Hilly and her husband Ali (no, short for Alastair) are going to housekeep for me.' Ali was Alastair Boyd, Lord Kilmarnock, Hilly's third husband. Their young son Jaime was also to be part of the household. It sounds a 'very bizarre' sort of arrangement, Amis acknowledged – like something in an Iris Murdoch novel, as he would describe it later. The new arrangement began even before Amis left Gardnor House. Hilly and Ali moved in before Jane had finished moving her things out.

Bizarre it may have been but for Amis it made excellent sense. There was no other way he could 'get a bit of family round me, which I need above everything', as he told Conquest. Other options were considered. Amis thought of asking Susie Allison, the widow of his doctor, to become his housekeeper. She was John Betjeman's secretary and later became Amis's. Yet however useful and companionable a housekeeper might be, any outsider filling that role would have been a second-best solution. The influence of Amis's early years still made itself felt. He liked to live in the warmth of a family home as much as he had done when he was a child, even more so now that he had no youthful desire to break free.

But it was the boys, particularly Philip, who plotted the solution of reuniting their parents, not Amis himself. Philip first raised the idea, and tentative interest was expressed all round. A dinner was organised at Bertorellis where Hilly, Amis and the boys could discuss it. Kilmarnock, as the most recent adult arrival on the family scene, might have been expected to be the most lukewarm, but he was all for it too. Amis himself had moments of doubt. What would it be like to live in the same household as Hilly, for whom he still had strong feelings, not alone but with her third husband, whom Amis hardly knew? But doubts were suppressed and all agreed to give the new household a try. Philip and Martin were pleased at what they had helped to engineer but privately sceptical about its chances. They thought the new household might hang together for six months at the most before it inevitably fell apart. But they were wrong. It has lasted continuously from 1981 and shows no likelihood of falling apart, ever.

Hilly had come full circle back to living with Amis by a long and erratic route. In the nearly twenty years since they had last lived together, she had spent half her time in America and Spain and been married twice. She had first married her Cambridge suitor, David Roy Shackleton Bailey, in 1967. Shack, as he was known, was then senior bursar at Gonville and Caius but soon afterwards found a job as professor of Latin at the University of Michigan. The newly marrieds had moved to Ann Arbor in 1968, taking Sally with them. There Hilly opened a fish and chip shop in what had been an estate agent's office, giving it the name of Lucky Jim's. The relationship was not a success. Shack was nearly fifty when they married and had settled into being what many people thought of as the archetypal Cambridge don: crusty, rigid, eccentric, a confirmed bachelor – even a virgin – until he met Hilly. Although not badly off he seemed parsimonious to her; it was to make some spending money that she opened her chip shop. Hilly acquired a lover called Milton Cohen, who was also an academic. He soon retired from his job and went to live in Crete.

In the summer of 1970 Shack and Hilly decided to take a European holiday. They hired a car in Rome and drove to Ronda, some fifty kilometres north of the Costa del Sol in Andalusia, where Cohen had found a holiday apartment for them. Cohen wanted Hilly to join him in Crete and sent her a blank cheque for expenses. But Hilly decided she wanted to spend the summer in Spain. This did not go down well with Shack. He wanted her to go back to America. But she dug in her heels and said she was not going to go back there with him, then or ever.

Scouting around Ronda, the teenage Sally had come across a language school installed in the imposing Casa de Mondragon and run by Alastair Boyd. Alastair turned out to be a good reason for Hilly going neither to Crete with Cohen nor to America with Shack. She found a job as matron at the international school in Seville where Sally became a pupil. Hilly was at the school for a year and stayed most of the next six years at Ronda, where she and Ali made their living in a number of ways: running art courses and a bar and taking paying guests in to their house next to the Casa de Mondragon. In 1973 she returned to the home of her old Swansea friend, Margaret Aeron-Thomas, to give birth to Jaime.

The next year, 1974, she was divorced from Shack. The timing of these events – vindictively contrived by Shack, Hilly believed – meant that Jaime was born out of wedlock and therefore could not inherit his father's ancient title. (Although Ali is merely the seventh baron in a line stretching back only as far as 1831, the aristocratic pedigree reaches much further into history than that, at least to the thirteenth century. The Boyds are an old Scottish family, and Ali is the chief of the clan of that name. The barony he inherited had been revived after falling into abeyance for nearly a century, having been suspended after the Jacobite rebellion of 1745. For his part in this, the Lord Kilmarnock of the day was executed for treason at Tower Hill, where a plaque now marks the place at which he and Lord Balmerino died. But for this blot on the family escutcheon, Alastair Boyd would now be the fourteenth Earl of Kilmarnock.)

Ali had started out working in the City of London but he disliked that world, 'for whose ploys', as he wrote himself, he 'displayed a singular lack of talent'. He had moved to Ronda in 1957 with his first wife, Diana, and lived there at the Casa de Mondragon for the next twenty years, writing and running the language school. He married Hilly in 1977 and, following his father's death, decided to return to England the same year. They lived with Jaime in a cottage at Thornburgh in Buckinghamshire and Ali took his seat in the House of Lords. But the cottage was almost all he had and there was not much money to live on.

A ménage with Amis therefore made excellent sense from his point of view too. Amis had money, Hilly had the necessary household skills. As an ex-wife, she knew Amis's ways far better than any hired housekeeper. To Amis, her presence was a source of personal as well as domestic comfort and, since he had given up sex, there seemed little risk of awkwardness between competing males under the same roof. Living in North London gave Ali handy access to the Lords, where he became active. Through the mid-1980s he was successively chief whip and deputy leader of the small group of peers who joined up with the newly formed but short-lived Social Democratic Party. Ali did his bit around the house too. An establishment was thus created in which every adult put something in according to ability and got something out according to need – a sort of commune if you like, though this is not how Amis would naturally think of it. To an outside eye it all looked decidedly strange. But it worked, if for no other reason than that its members, seeing no better alternative for themselves, took care to moderate their own requirements to make sure they did not clash fatally with each other's. On this practical basis the Amis household has survived for longer and in better shape than most communes with only ideals for foundations.

After a few weeks in a temporary Hampstead flat, the new extended family moved early in 1982 to a house which Ali had found in Kentish Town at 186 Leighton Road. 'We're at the end of L[eighton] Rd away from Kentish T[own] tube, on the corner of Brecknock Road,' Amis told Conquest in mid-February, 'near a little down-market shopping centre where you can't get e.g. China tea or any malt whisky but Glenfiddich. All pubs so far visited have "music" in them but there are so many hereabouts that I still have hopes. The house itself is just right for the four of us, though not large. I have the top floor, small bedroom, fair-sized study. No bookshelves yet so I work half-surrounded by tea-chests. But work I do, and the telly operates and I am £1-worth nearer the Garrick by minicab.'

The house in Leighton Road eventually proved too small for comfort. When Philip moved in for a while there was no spare room in which to put up Conquest. In any case the four regular residents discovered that they needed more space. Tolerance did not thrive in confinement. A child, a writer, a politician and a housewife could not get on with their separate lives if they were falling over each other all the time. Having to share things like bathroom and kitchen had particularly disruptive potential. At times Hilly found herself standing on the stairs behaving rather like a policeman, directing the human traffic this way and that so as to make sure that people did not intrude on each other at awkward moments. The pop music which was played too late and too loud next door was an added incentive to moving.

By the middle of 1984 the search for 'somewhere more salubrious' was on. Amis put his archive up for sale in order to buy a bigger house in a better area. This caused him a slight twang of conscience. You look a bit of a shit, he told Conquest, putting out a catalogue that says: 'FOR SALE; 11 letters from John Betjeman to KA.' But there was no getting round it. The archive was eventually sold to the Huntington Library in California. Again it was Ali who found a suitable new house, this time at 194 Regent's Park Road in Primrose Hill. The family moved there in July 1985. 'Here we are then,' Amis wrote to Conquest, 'in just a week ago, study, drinks cupboard and telly (one on every floor) in operation, which is all that matters really.' Amis's requirements were as always simple enough. But there were still tension-provoking constraints, like there being only one kitchen. In time, this was sorted out too, with a kitchen being installed in the basement for Hilly and Kilmarnock as well as one on the first floor, where Amis's meals could be prepared and eaten in the adjacent dining-room, which was large enough for family meals as well.

The configuration of rooms finally arrived at represented a satisfactory carve-up of available space. All parties could lead their own lives, interact with each other as necessary and come together when they wanted. Hilly and Kilmarnock were self-contained in the basement; Amis had study, bedroom and bathroom on the ground floor; on the floor above there was sitting-room, dining-room and kitchen, which Amis had first rights in but which was big enough for family events too; and the floor above that could be occupied as required by Jaime or Philip or whoever else came to stay.

It was in this period of tumult and change that Amis suffered his only failure with a novel since his first aborted attempt at one, *The Legacy*, in the 1940s. All he managed to save from the new wreckage was the title, 'Difficulties with Girls', which he kept for a later and very different novel. The one with that name which he gave up was to be set in the early 1960s and told in the first person by a young homosexual. Among other things, the narrator was to be accused of interfering with her children by a woman whose advances he has rejected. When he protests his innocence to her husband – a character based on Amis's friend George Gale – the husband believes him but feels compelled to support his own wife's version for reasons of marital harmony.

Somehow, though, the story didn't come together. The fact that Amis's marriage to Jane was in a late state of disintegration while he was writing it surely undermined his confidence and his concentration. At any rate, he put aside the 130 pages he had written and never returned to them. Amis salvaged one slight crumb of consolation from this failure. By making his first-person hero a homosexual he risked arousing doubts about his own sexual inclinations in the suspicious minds of people like Judge Melford Stevenson, known to Amis for his impish sense of fun and his willingness to voice outrageous opinions at the Garrick Club bar, sometimes because he meant them and sometimes to provoke. By not finishing the novel, Amis avoided the possibility of a judicial dressing-down – mocking or real – for being a closet queer himself.

In 1980 Amis had published *Russian Hide-and-Seek*, a story of Britain in dismal condition after fifty years of Soviet rule. (As it happened, Amis was invited to a reception at 10 Downing Street on the day the novel was published and he brought a copy for Mrs Thatcher. When he told her its gist, she told him sternly: 'Get another crystal ball.') *Stanley and the Women* was published four years later in 1984. This was the longest gap between novels in Amis's writing career. Making apologies for delay in answering Conquest's letters, Amis wrote in November 1982: 'No excuse, just being frighteningly behind with work – no book out this year and none ready for next so far.' This falling off in output seemed to Martin a reliable indication that things were not going well in his father's private life.

Meanwhile, though, Amis picked up a couple of enjoyable sidelines. In 1982, through George Gale, he started a column on drink in the *Daily Express* from which he was able to spin off a book, *Every Day Drinking*. Two years later he was invited by the *Daily Mirror*'s editor Mike Malloy to become his newspaper's poetry editor. 'Yes, a nice spot next to the Old Codgers, where Andy Capp is now,' Amis told Conquest. His job was to pick a poem five days a week, which would be fun to do and not badly paid either. But 'the real satisfaction is the thought of how cross it'll make them all [Amis meant the literary establishment]. Dosing the honest masses with stuff about patriotism and religion and the countryside and out-of-date ways of thought. *You* know.' Six months later Robert Maxwell bought the *Mirror* and Amis feared this new broom might sweep him out, though he had a one-year contract and thought Maxwell should find others more urgently in need of being got rid of than him, like the left-wing journalists John Pilger and Paul Foot. 'I should be sorry to stop my column, not least because of the thought of the rage engendered in Al[varez]-et-al circles. There I am feeding them (I quote from the last few weeks) Browning, Tennyson, Housman, Kipling, Edward Thomas and Isaac Watts when I could be giving them Adrian Mitchell, Tony Harrison and John Ashbery.' His contract was terminated as feared, but Amis later salvaged another book from his editorial stint, *The Pleasure of Poetry* (1990).

Soon after this Amis began to write about food, first in the *Illustrated London News*, then *Harpers & Queen*, most recently in the *Oldie*. In each case he was eventually moved on for one reason or another, as is common enough with magazine columns, which come and go according to editorial whim. The *Oldie's* patience proved shortest. Amis had time to write only three pieces – about Simpson's-in-the-Strand, the Ritz and his neighbourhood restaurant in Primrose Hill, Odette's – before being summarily fired by the editor, Richard Ingrams, in mid-1994. Ingrams's grievance was that Amis went to places which cost more money than his readers could afford – or, perhaps, than his magazine wanted to pay for. Whichever it was, Amis himself felt aggrieved, since Ingrams could not have had time to read his latest piece, which had singled out Odette's for giving value for money and would therefore not be costing the *Oldie* much. But as always there was nothing to be done when an editor displays his whim of iron.

For someone who relished eating in general as little as Amis did, he was not obvious restaurant-critic material. A drinks column was more naturally his line. But he liked lunching out, even if he did not much care about lunch itself. Hilly came too, which was pleasant for its own sake. She also made notes, advised him about what had gone into what he was eating and carried home any leftovers in a doggy bag as possible ingredients for Ali's supper. Apart from this, writing about food made a change, and was an extra string to Amis's all-round professional writer's bow. It also gave him one more chance to take a crack at the so-called experts, the 'Alvarez-et-al circles' of the food world. The rule of the expert, the sort of person who preferred Tony

Harrison to Tennyson, also infested this world and deceived people into putting trendy rubbish above good, no-nonsense fare. These bogus experts were everywhere to be found, in Amis's view, and nowhere with more harmful consequence for the public at large than the restaurant trade.

Amis made the comparison between food and literature explicit in a 1987 piece about the much-esteemed Chelsea restaurant, La Tante Claire. 'Contemporary poetry is written to impress other poets or would-be poets, not to please the ordinary reader . . . And restaurant catering is designed less and less for those who simply want a good meal somewhere nice and more and more for food specialists of various kinds. And snobs, of course.' The fact that the restaurant had been recommended to Amis by a world-famous expert on food and drink should have warned him what he was in for.

Characteristically, though, little as he may have enjoyed his lunch, he gave credit where he thought it was due. The restaurant was an extremely pleasant place to eat, 'light, airy, uncrowded'. The clientele was reassuringly 'unbusinessy' and 'unarty' too, and included a well-fed old fellow eating on his own, another good sign. There was, however, hardly any space set aside for pre-lunch drinks, always a bad omen in the Amis book. (If invited out to lunch, he is fond of saying, the very worst start is when your host asks, 'Shall we go straight in?' – meaning there will be no detour to the bar. The second worst thing for an Amis host to say is: 'Shall we go straight on to wine?' And the third is: 'Shall we stick to the house wines? I always think they're the best value.' At each stage the host reinforces his determination to be stingy.) Another black mark against La Tante Claire was its having a menu in 'advanced-level' French which required translation. All restaurants which do not have their menus written down in plain English but rely on recitation or translation by a waiter automatically earn the Amis seal of disapproval, since they impose on their customers the unnecessary burden of trying to remember what they have been told when they should be able to ruminate at leisure over an intelligible written list.

When it came to the food, Amis really dished out low marks. The starters 'rose as high as being patchy'. The lobster was 'flabby, tasteless, dubious sauce'. The lamb 'looked fine but completely tasteless'. Dover sole 'tasted of what thin strips of leather may well taste of'. And so on. The desserts were good. So were the bread, the 'sensible' large wine-glasses and the service. But nearly everything else was bad.

'And it was ferociously expensive. Perhaps nobody would go there if it were not.'

Contrast Amis's discriminating diatribe with, for instance, the unqualified approval of La Tante Claire's entry in the Egon Ronay guide for the same year, in which it was given the top, three-star rating: 'cooking of rare refinement and balance . . . everything from the wonderful nibbles to the classic main dishes is a total delight to both eye and palate . . . marvellous wines and service that allies friendliness with discreet, polished attentiveness'. This kind of licking the boots of fashion Amis detested, wherever he found it. And he found it everywhere. Arbiters of taste, in restaurants as in poetry, deliberately cultivated narrow and exclusive idioms in their fields, he believed, the better to set themselves apart from and above ordinary people. They promoted cooks and poets in their own image, and the cooks and poets then foisted their stuff on a bewildered public. Charging a lot, as at La Tante Claire, was a sure way of keeping things exclusive. When you boiled it all down this was no different from old-fashioned snobbery, insiderdom in food or poetry standing in for insiderdom in class. At all signs of this sort of pretentiousness Amis continued to have a go, using restaurant piece or poetry column or any other means of attack he could lay hands on.

Novels like *Jake's Thing* and *Stanley and the Women* gave Amis an increasing reputation for misogyny. It was not much help for him to argue that women like Brenda, Jake's wife, behaved better than men like Jake, or that Stanley was no more of a saint than the women in his life. Amis had started making his unattractive heroes attractive in spite of themselves as early as Roger Micheldene of *One Fat Englishman* and he did not lose the habit. Besides, there was some substance in the misogynist charge. Amis certainly did not hate women but his opinion of them dimmed. Perhaps it was his experience of Jane that cast a shadow across his feelings about the whole of her sex. Stanley Duke thought that women only wanted one thing: 'for men to want to fuck them'. Then they could fuck up their men. But Amis did not want to fuck women any more and so could see them more clearly, his vision unclouded by lust. Or so he seemed to believe.

Amis's variety of misogyny was partial and qualified. He was sympathetic to women as a class, believing they got the raw end of most deals in life, as wives or mothers or in work. They had to bear the children, they were poorer financially, weaker physically, had more to put up with, were more at risk generally than men, led less entertaining lives. The looks on which they so heavily relied wore off earlier than those of men, who did not need them so much. These sympathies of Amis's were real. In his 1990 novel *The Folks that Live on the Hill*, he wrote with compassionate irony of 'the superiority and seniority of the sex that has always had to do the rough work, the real work, the clearing up of the sick and the shit and the afterbirth and the dead bodies while the men think and create art'. But his sympathies were also detached, like the feelings anyone might have for a clearly disadvantaged group of people seen suffering on television in some distant spot on the globe. In his own life Amis wanted less to do with women and as the years went by became less inclined to seek out their company.

Women's main practical offence in his eyes was no great moral outrage or sin but their social behaviour: they talked too much. Conversation was important to Amis, and women, he believed, monopolised it whenever they could and so spoiled the flow. By talking a lot they made themselves seem important and by overwhelming the conversation with chatter they concealed the poverty of what they had to say. There was an inverse relation between quantity and quality; more, in this case, really did mean worse. Some men were just as bad, obviously. Not so many, though; and there was a better chance that any man would have more to say that was of interest, if only because men led lives that were more interesting – at any rate, more interesting to Amis. That was one reason why the male-member Garrick Club was such a good place for him to spend time.

To any distaste for women's company that Amis might feel there was always one exception: Hilly. She did not talk too much and her presence was always a delight. Amis took her with him whenever he could, out to supper with friends or to lunches at restaurants he had to review. As the years went by his marriage to Jane came to seem more and more of a puzzle. He did not attempt to deny that he must once have loved her but he found it difficult to remember exactly how or why. In a letter to Conquest in May 1988 he tapped out an impromptu verse, after the Cockney song 'My Old Dutch': 'We've been apart for over seven year / And it don't seem a day too much, / For there ain't a lady living in the land / I'd as soon wallop in the crotch. / I called her Jane / Her proper name is Lizzie / She's a real bleeding pain / And she'll get you in a tizzie / It's many years since

340

first we met / And thoughts of her still make me sweat / So pop her in the oubliette / Old E—J—H.' As verse it needed working on, Amis admitted, but he was happy enough with the sentiments it expressed.

Three years later he returned to the subject in slightly less aggressive, more reflective vein. 'I continue to bear a charmed life and never set eyes on the bag. Almost unbelievable that it's now *8 years last Nov* that I last did . . . It amazes me now that for several months after she went I was v cut up about it, wanted her back, contemplated a *poem* on the subj[ect] if you don't fucking well mind. Now I wish it [the break-up] had happened – well I suppose about 1970 would be right. Well it's all experience, though it's a pity there had to be so much of it.' Amis and Jane did later run into each other on a couple of occasions. The first was at Odette's restaurant, where both happened to be dining in different parties – Amis turned his back on her, according to Jane, but Amis does not remember seeing her there at all. The second time was at a birthday party for William Cooper in the Savile Club when Amis discovered reasons for making a rapid exit on catching sight of Jane across the room.

But these peevish outbursts to Conquest were exceptional. As the years went by Amis found himself hardly thinking of Jane at all, except when compelled to by being reminded of her. This was one of the things that puzzled him about the whole Jane episode in his life. How could it be that the eighteen years he spent with her had made so little impression, left so light a trace in his conscious mind? With Hilly it was the opposite. He thought about her a great deal, took her with him whenever he could, hated her to be away from Regent's Park Road or to leave her behind there when he went off to Swansea. Amis constructed a life that was as much like being married to Hilly as he could contrive to make it without going so far as to destroy the harmony of Regent's Park Road.

Amis continued to make his opinions heard on matters of the day, directly in articles as well as more obliquely in his fiction. His targets were many: sloppy language; education – Steve in *Stanley and the Women* 'belonged to one of the generations which had never been taught anything about anything'; the Arts Council – 'Subsidy damages art by tending to foster irresponsibility, showiness, cliquism and self-indulgence in the artist,' he wrote to *The Times* in 1985; restaurants (see above); explicit sexual description, or ESD;

the condition of the Church of England, which he thought by 1987 might achieve what even Communism had failed to do and rob England of religion once and for all; 'sod-the-public' tendencies, meaning roughly that the customer is always wrong, to be found in everything from architecture to typography; and, of course, leftyism in all of its many forms.

Britain turned formally to the right with the election of Mrs Thatcher in 1979. But the dominant intellectual-literary tone was still left-wing, shading into 'political correctness' when the collapse of Communism and the Berlin Wall made traditional socialism unfashionable except to a handful of diehards. By these prevailing standards Amis was undoubtedly a reactionary. Such a simple and dismissive label, though, contains only enough likeness to him for a caricature but too little to make a portrait.

Amis discovered many of his opinions by testing fashion or trend against reality as he understood it. Feminism, for instance, was fashionable, and one of its beliefs was that women's talents had been consistently undervalued. But was this true? One obvious test for Amis to make was in English poetry. A don called Smith in *Jake's Thing* ticks off the record of women's achievement in this art, as he – and Amis – sees it. There is Katherine Philips, otherwise the Matchless Orinda, who lived from 1631–64; after her, Anne Finch, Countess of Winchilsea, 1661–1720; which left only one more, Christina Rossetti, 1830–94. Not much to show for six hundred years of English literature, and no other contenders in sight – certainly not Sylvia Plath, the most widely admired woman poet of recent times. It 'shook' Amis when Philip Larkin chose her collected poems as his book-of-the-year in 1981. Surely she was 'no good really', as he put it to Conquest. Yet there was no obvious reason why women should not write poetry. Being a poet is not a full-time occupation, ideally suited, in fact, to the odd hours and half-hours of spare time that crop up regularly in the course of women's lives. These were just what a poet needed to get out a poem in draft and put in a bit of work on it, as Amis himself did. Women had performed much better as novelists, though why was mysterious. The point was that they had failed in poetry, which must indicate some defect or deficiency. This failure was just as mysterious as women's success in novels, but there it undeniably was.

A judgement like this was typical of Amis. It did not involve sweeping generalisations but covered only poetry, which Amis knew about. His list of good women poets was exceedingly short. But then his list of good male ones was not very long either. He was no more likely to include Ted Hughes, Sylvia Plath's former husband, in his tally of good male poets than he was Plath in his tally of good female ones just because both were fashionable. 'Saw another shortest-book tother day,' Amis once wrote to Conquest: 'The Achievement of Ted Hughes.' Nor did Philip Larkin's endorsement of Plath have the slightest effect on Amis's opinion of her. Amis might be deeply fond of Larkin and admire him as a poet second only to A.E. Housman, but that did not mean Amis accepted his judgements. 'Old Philip was given to literary passions that I have never shared at all,' Amis wrote to Anthony Powell after Larkin died, 'as for the poetry of Hardy, Wm Barnes & Stevie Smith, the novels of B Pym, etc.' Lastly, Amis did not offer any explanation or theory about why women had performed so poorly as poets because he did not have any. He merely observed what he considered to be the facts and argued his case as far as he rigorously could take it. A serious Amis judgement had to be defensible in argument and it had to be one that he had arrived at strictly for himself. If the opinion he reached in this way turned out to be 'reactionary', so be it.

It was their readiness to follow fashionable opinion that Amis found contemptible in far too many supposed opinion-formers, especially literary folk. They didn't make up their own minds but followed the pack. Talking about the writer Blake Morrison, Amis said in a letter to Conquest: 'The trouble with chaps like that is they have no taste – I don't mean bad taste, just the mental organ that makes you say This is bloody good and This is piss is simply missing, and they have to orientate themselves by things like "importance" and "seriousness" and "depth" and "originality" and "consensus" (= trend).' It was this lack of taste that let writers like William Golding, John Fowles and John le Carré (and many others too, including Anthony Burgess, Thomas Pynchon, Patrick White and Samuel Beckett) earn reputations they didn't deserve. Once a reputation began to take shape, literary folk were afraid to question it for fear of being out of step with fashion. They didn't trust their own judgement. Amis trusted his. You might disagree with his opinions but you were not in doubt as to whose opinions they were.

Something more than taste or preference went into Amis's opinions. There were moral factors too. If women talked too much they were wasting your time and so robbing you of the most valuable of possessions. 'Sod-the-public' architects and many other suppliers of goods and services were simply cheating clients and consumers: over-priced, over-fancy restaurants, for example. Coteries of self-appointed experts who foisted unreadable poetry or non-musical music or meaningless paintings on the public were the same or worse. Sloppy language was laziness. Critics and pundits who took their opinions from others were lazy too. They were simply not working hard enough at their jobs if they could not make up their own minds but had to take in their views at second hand. The ethics of his father, especially the virtue of hard work, were never far from Amis's mind, consciously or not.

Many of his sharpest opinions Amis kept for his letters to Robert Conquest. By the 1980s Conquest was firmly established in America, but he liked to be kept in touch with what was going on back in Britain and was an enthusiastic letter-writer. Amis was less so, often responding to two or three of Conquest's letters with only a single one of his own. When he did get around to writing he set out to be entertaining as well as informative. A certain amount of creative embroidery went into embellishing the gossip and the opinions. But the authentic Amis sounds through too, often the more authentic-sounding for being heard in a private letter rather than a book or an article.

Amis could be amusing and abrasive even about his son Martin. In February 1981 he reported to Conquest that Harold Evans had been switched from the editorship of the *Sunday Times* to that of *The Times* and repeated the crack he had been making in the Garrick Club: 'Things have come to a pretty pass when the editor of the Times's girl friend is one of my son's cast-offs.' (Amis meant Tina Brown, who later became Evans's wife as well as editor of *Vanity Fair* and the *New Yorker*.) Five years later he reported with dismay that Martin had 'gone all lefty and of the crappiest neutralist kind, challenging me to guess how many times over the world can destroy itself . . . in fact going on like H Pinter does these days I hear.' Amis asked Conquest to recommend a book that might change Martin's mind. 'He's bright, you see, but a fucking fool, and the worse, far worse, for having come to it late in life, aetat. [aged] nearly 37, not 17.'

Amis was no kinder about Martin's novels. 'Between ourselves I only read about half,' Amis said of one published in 1981; 'too boring. Little sod said on TV you had to read it twice. Well then HE's FAILED hasn't he?' But Martin knew more or less what his father thought of his work, could indeed make a reasonable guess at what page of any given novel of his Amis *père* would send the book spinning across the room in exasperation. Martin continued to read his father's work while knowing his father did not read his own – or not all the way through. The difference between their novels was something of a puzzle to Amis senior, since their methods seemed to have so much in common. Both agreed on the importance of traditional elements like plot and character, and yet the results were so unalike. Martin was inclined to think that the novel had simply moved on into postmodern forms, leaving his father behind stuck in old-fashioned realism. Any suggestions of that kind were apt to rouse snorts of derision from his father. The two of them tended to keep off such delicate matters so as not to spoil their regular weekly dinners together or the weekends when Martin brought round his children to Regent's Park Road.

Other writers came in for similarly uninhibited treatment in Amis's letters. Tom Stoppard, the playwright, was 'amiable enough – pity his work is no good.' Bernard Levin, the journalist, 'has become a pompous buffoon'. Tom Paulin, the poet, was a 'little twit'; another poet, Craig Raine, was 'that fool'. George Hartley, Philip Larkin's Hull publisher, was 'that little turd'. But Amis was even more uninhibited in his opinions about events or trends. In his letters he thought the unthinkable – or at any rate the not easily printable.

One persistent theme was what he called 'the fallacy of impartiality', the assumption that in every story there are two sides meriting equal attention and respect. For example, the IRA and the British army in Ulster or urban rioters and the police or 'the two giant Powers', America and the Soviet Union. 'That little bastard Jonathan Dimbleby is doing a series on that,' Amis wrote in September 1981, 'called the Eagle and the Bear, instantly implying the two are so to speak equidistant from us and we can and should weigh them in the same balance.' The following year, during the Falklands War, Amis berated the BBC for being similarly 'objective' between the two sides. In November it was the *Observer*'s turn to be berated for implying that the Israelis had carried out the massacres of Palestinians in Beirut and quoting two sources to back their case, 'Yasser Arafat and the PLO's man in New York'. The next year Amis invented a name for this kind of thinking: 'seditious neutrality'. In 1988 he found another example of it in the row over the SAS shooting of three IRA terrorists in Gibraltar. 'When is it all right to shoot at a terrorist? . . . Answer: when he's blown your head off.'

What Amis didn't like about this kind of attitude was its flabbiness, similar in its way to the flabby thinking of literary folk who didn't make up their own minds but followed the trend. 'How I hate all that talk of moderation and reasonableness and flexibility, especially the last, meaning the readiness to give in.' Most of the time, Amis knew exactly where he stood, though he missed having someone like Conquest more expert than himself to discuss politics with. At the Garrick Club, he only saw 'shoot-the-lot reactionaries'. It was 'nice to know they're there,' he wrote, 'but unsatisfying after a bit.'

Amis even tested out ideas about race on Conquest. Though Jews were among his friends from schooldays and he had hated Nashville for its racist climate, he nevertheless wondered on paper if there might not be something in prejudices against both blacks and Jews. In 1984 some 'anon shag' from New York sent him a book by Wilmot Robinson called *The Dispossessed*, which claimed that an alliance of liberals and ethnic minorities were pushing white Anglo-Saxons out of power and influence in America. This set him off speculating. 'Hard on the Jews, natch, but then you pick up the Sunday paper and in the ST [*Sunday Times*] review alone you find Susan Sontag, G Steiner, a Russian-Jewish novelist, Freud, Steven Marcus, a new US Jewish pop novelist and no doubt others – yes, Mel Brooks – all writing or being written about, and you start shamefacedly and reservedly thinking he may have something.' Four years later he speculated that 'the great Jewish vice is glibness, fluency, – also possibly just bullshit, as in Marx, Freud, Marcuse. Pity L. Ron Hubbard not.'

Amis thought that other races had their shortcomings too, which he was equally willing to face. The British philistine, for example, was 'crude, ignorant, aggressively uncultured', while the American variety was 'lettered, art-loving, philosophy-gabbing'. The Americans were 'tremendously anti-British' or rather 'anti-English'. They liked the Scots, as for example Scotty in the television series *Star Trek* and the Campbells, who 'get into that terrible "Tenement Symphony" in "The Big Store" alongside the Cohns and the [fucking] Kellys, the Shagbags and the Vermicellis or whatever, but not the Digby-Featherstonehaughs'. Amis's speculations about race tended to break down like that, into laughter and jokes. There might be something in race prejudices worth thinking about, but they didn't stand up to Amis's scrutiny in the way that women being no good at poetry did, so he did not go further than speculate. For him race, like class, was more about colourfulness than colour, something to raise a laugh from more than to work up a lather about.

Amis was well aware of the vagaries of reputation, not least his own, and the chance ways in which they might be formed against the grain of the truth. Soon after John Betjeman died in 1984, he wrote to Conquest: 'Wherever I go nowadays I hear tales of what a shit Betjeman was to his secretaries, swearing at them, making them cry, having to have a new one every few weeks.' Amis had never made a secretary cry. But Betjeman would 'go down to history as a genial teddy-bear figure beloved of all while a nice old softie like me will be taken as a curmudgeonly old shit.' Who could tell how one might wind up looking in the history books? 'I expect if the truth were known Pope was an absolute charmer encouraging young poets and visiting the sick and Lamb behaved like Dylan Thomas' – the exact opposite of their historical reputation in both cases.

On the whole, though, Amis was indifferent to what people thought of him. He usually spoke up on any issue he was asked about so long as he had something to say, and if the fallout seemed damaging he didn't much care. The received opinion with which his own clashed was usually no more than fashionable and probably wrong. Time would vindicate him. Besides, the older you got, the more immune to criticism you became. Anticipating the imminent publication of his novel *Difficulties with Girls* in 1988, Amis told Conquest he thought the omens were good and added cheerfully: 'you know when you're old enough you can get away with piss, cf G[raham] Greene', of whose work from at least as early as *The Human Factor* of 1978 Amis thought very little indeed, though it continued to be respectfully received by the critics.

There were some things he might think sound enough but perhaps tactless to put in front of the public. Better to confine them to conversation or a letter to Conquest – you could give one hostage to fortune too many. One that he withheld was a poem of which he sent an unfinished draft to Conquest in 1987:

347

Women and queers and children
  Cry when things go wrong:
Not fair! – why me? – can't take it!
  [So] drones/sounds their dismal [eternal] song.

The usual sort of men
  Who hold the world together
Manage to face their front
  In any sort of weather.

With rueful grins and curses
  They push the world along;
But women and queers and children
  Cry when things go wrong.

He polished his lines a bit. The last two lines of the first verse became
'Why me? – not him! you're horrid / Always the same old song.'
But he never got around to publishing them and he has not written
another poem since.

↜ 21 ↝

The Queen's and the Garrick

'My life is just work, family, club (plus Swansea) now and v nice too,' Amis wrote to Robert Conquest in September 1987. He had by then settled firmly into a routine from which he cared less and less to stir. Work, lunch, snooze, more work, television, drinks, supper, a spot of reading, bed. On Saturdays the Queen's pub stood in for the Garrick at lunchtime, on Sundays Odette's restaurant. One night a week there was dinner with Martin and at weekends supper with his neighbours the Houghs, at his place or theirs. Age made an increasingly convenient excuse for refusing all other evening invitations, especially dinner-parties, which Amis had never liked anyway, though he was still available for lunches with the right person at the right place.

In the basement of the house in Regent's Park Road there was his first ex-wife Hilly, and not far away in flats of their own were his son Philip and his daughter Sally. The essential tools – drink, television, books, typewriter – were sufficient and to hand. It was all as comforting and anxiety-suppressing as could be arranged, not perfect but adequate, in its mood as well as much of its detail very like the circumstances of Harry Caldecote, hero of Amis's 1990 novel, *The Folks that Live on the Hill*.

If Harry looked out of his window in Shepherd's Hill he saw a wine emporium called Potandum, while if Amis looked out of his in Primrose Hill he saw one called Bibendum; if Harry wanted a drink he went to the King's public house, as Amis went to the Queen's; to eat locally, Harry would go to Odile's as Amis to Odette's; Harry's club was the Irving, just round the corner from Amis's Garrick;

for newspapers and stamps both Harry and Amis could resort to a sub-post office along the road run by amiable Asians.

Retired from his librarian's job, Harry, though far from rich, has the time and money to do pretty much as he pleases. He has been married twice but considers he has had enough of that state to last him indefinitely. Marriage, he thinks, has 'certain positive advantages' but, he tells his friends at the Irving, 'Many or even most of them . . . seemed to be at least adequately supplied by having one's widowed sister housekeep for one.' Harry spends most of his time taking a benevolent interest in the affairs of a flock of mostly local people with problems that range from drink to money to lesbian bust-ups. They include friends, close relatives and distant ones too: his brother Freddie, his sister Clare, his 'ex-stepdaughter' Bunty, his 'first ex-wife's sister's child', Fiona Carr-Stewart.

'He liked being where he was,' Harry finds himself thinking, 'which these days boiled down to this house, the surrounding quasi-village, the King's, the Irving, the occasional pal'. On the other hand, 'he felt nothing for his house except as the place where he lived.' And the people around him 'are not much of a set of human reasons for wanting to stay in Shepherd's Hill'. So why does he stay, especially when he gets an excellent offer for his line of work from America? His sister Clare comes up with an answer when she tells him why she doesn't want him to go. 'My life is here with you and you're awful . . . but you're family and I'm used to you, and you know I don't think anybody in the world understands as well as I do how important it is to be used to someone.' There is a version in this of Amis's life in Primrose Hill – above all, perhaps, in the importance of having around him someone he is used to.

The one break in his London routine Amis now chose to make was the three weeks he spent in Swansea with his old friends Eve and Stuart Thomas, timed to coincide with the three weeks in August and September when the Garrick shuts for its summer holiday. Here, a loose simulacrum of the London routine was followed: a bit of work in the morning, a bit more in the evening, television, drinks and so forth, with the Bristol Channel Yacht Club on the Mumbles shore standing in for the Garrick. Or at least it stood in until 1993, when Stuart Thomas was expelled after forty years' membership following a row with the committee, whereupon Amis instituted his own boycott, declining to cross the threshold again after being a member for nearly as long as Thomas. Local pubs made perfectly satisfactory substitutes, and the Bees Knees restaurant provided an acceptable lunch at however unconventionally late an hour Amis and party reached it from the pub.

Swansea was a break but not exactly a cure. 'You probably wondered if I was dead,' Amis wrote to Conquest from there in early September 1987; 'well, I bloody nearly am, or at least my demise has been brought appreciably closer by a couple of weeks in Wales (with another yet to come). Not only the raised alcohol intake but the eating of lethal food. Swansea contains the South Wales chapter of the Friends of Cholesterol . . . My Aussie china Jim Durham is here briefly, helping to provide a peak of self-inflicted damage.' Boozing at Swansea was indeed prodigious. It was while Amis was putting his shoes on to go down to the Yacht Club that he got the idea for his 1986 novel, *The Old Devils*, about a bunch of oldsters retired from most serious activities apart from that of drinking their heads off.

Age took its physical toll too: 'had last 6 upper teeth whipped out and denture triumphantly installed – v. comf so far,' Amis told Conquest in July 1986. By then Amis had also become seriously fat, a process which began when he gave up smoking, and drinking for six months too, to compensate for which he went in for comfort eating. Martin remembers his father stuffing himself with chocolates until his face swelled like a football. Amis stopped eating so much when he started drinking again but the newly acquired weight did not disappear, indeed continued to accumulate. Amis's clothes now had to be bought at an outsize shop and he needed size 17½ collars on his shirts. Soon after he gave up smoking he gave up snuff too. This was certainly good for his general demeanour. He had taken to snorting up great pyramids of powder from the back of his hand, triggering gigantic sneezes that had to be staunched with sheets torn from a toilet roll he carried everywhere for the purpose.

Amis's legs began to let him down, suffering from circulation problems which made trips on foot of more than a few yards difficult. The climb up the staircase to the Garrick Club's first-floor bar had, by the early 1990s, to be taken in stages, and the short distance along Regent's Park Road to the Queen's public house had to be made by taxi. Deafness left its mark too. The television volume control now had to be turned up to higher levels for *The Bill* and in a crowded bar or restaurant Amis would lean towards the person he was listening to, sometimes cocking a helpful hand to his ear, so adding a brushstroke or two to his pinkly blimpish image.

In general, though, Amis remained in obstinately robust good health. He might abuse his body with alcohol and lack of exercise and worry in hypochondriacal style about every sign of illness, real or imagined. But illness remained mostly in his imagination. In the letter to Conquest in which he announced the loss of his teeth he also reported that a new doctor had given him a liver test which revealed 'only the faintest murmur of protest'. And so his liver remained – triumphantly resilient against the odds, like his other vital organs. Amis's doctors muttered darkly about his drinking, but what was the point of bullying him about it? Drink was so much part of him by now that telling him to give up or cut down would cause more problems than it would solve.

The passing of years brought with it the inevitable passing of friends. At times death seemed so frequent as to be almost part of Amis's routine. When the television producer and presenter Huw Wheldon died in 1986, Amis told Conquest wearily: 'The [Westminster] Abbey again. My minicab drivers are beginning to know where to drop me for the west door.' Tibor Szamuely had been the first of Amis's greatest friends to die, going suddenly of cancer at the age of forty-seven in 1973. Szamuely was the friend with whom he had grown intimate most quickly and the one he had known least long, having only been introduced to him by Conquest in the mid-1960s, soon after he arrived in Britain, a Hungarian refugee from Communism. In his *Memoirs* Amis had valued him most of all for his 'steady underlying cheerfulness and optimism'. The two of them got into the habit of talking on the telephone most days, for no special reason, just 'like a couple of bloody women', as Szamuely put it.

Philip Larkin, Amis's longest-standing close friend, went next. Amis recorded something of his feelings about Larkin's decline and demise in his letters to Conquest: 'old Philip is back at Newland Park [Larkin's home in Hull],' he wrote in July 1985, 'rather sooner than I thought they gave us to understand [after an operation to remove his oesophagus]. He sounded very back-to-form on the blower the night before last, as regards tone, that is. Content pretty gloomy, though no words were dropped. Doesn't expect to work [at Hull University library] again. Not much one can say, but I intend to keep up a barrage of letters.' Larkin being particularly fond of getting letters, this would be a special comfort to him. In October, Amis wrote:

Have had (perhaps you too) a very gloomy letter indeed from old Philip, talking of his hours of fear 4–8 a.m., continous thoughts of death, dread of possible more surgery, total lack of appetite. But it seems clear from one phrase, about Monica [Jones, Larkin's long-term girlfriend, who was also ill], 'no one knows what's the matter with her *either*' (my underlining), that *he* doesn't know (in the sense in which you know the date) what they found. [Larkin had not been told of the cancerous tumour which had been discovered in his throat, too far advanced for surgery.] I try to write often, avoiding such themes.

Amis thought Larkin was 'hard up for something to do' and wanted to suggest to him that he put his unpublished early poems together. The trouble with that was that it would imply he wouldn't be around to do the job later. Besides, Larkin wouldn't do it anyway, 'the silly old perfectionist'.

Amis had suggested visiting Larkin in Hull, 'but you know how cagey he is.' Larkin replied to Amis asking him to 'leave it for a bit'. He was in no state for a visit of more than an hour or two, and any such visit would be bound to 'centre on LUNCH, and I can't fucking eat fuck all.' On 21 November, Larkin dictated his last letter to Amis. He had been in and out of hospital for tests and that day was there 'for the big one'. The doctors were looking for something and 'I bloody well hope they don't find it.' Larkin's spirits were 'about as low as I can remember'. He thanked Amis's daughter Sally for a letter and photograph, hoping he would be able to write and thank her personally one day. Larkin noticed 'strong resemblances' between Sally and Hilly and added that Hilly was 'the most beautiful woman I have ever seen without being in the least pretty (I am sure you know what I mean, and I hope she will too)'. Ten days later he was dead.

On 9 December Amis went to Larkin's funeral and in a New Year's Day letter described it to Conquest:

It wasn't a bad day as such things go. Up in the train with Chas Monteith [a director of Faber's, Larkin's publisher], Andrew Motion (who seems not a bad chap, though again how one wanted a word from P.) [Motion had been a colleague of Larkin's at Hull] and wife, Blake Morrison [who as the *Observer*'s literary editor had published reviews by Larkin] and Hilly [who was fond of Larkin], natch. Usual BR journey, 40 mins late at Doncaster etc. My first visit to Hull, though we never actually got there, because v. decent solicitor met us at a suburban station, gave us lunch at his house, which commanded wonderful mysterious view of the tops of the [Humber] bridge above the mist . . . drove us to the church in another suburb. Big place, crammed with univ[ersity] people, perhaps 500 strong. Good full service (Oh God our help, Lead kindly light, The day thou gavest). I orated. V. few at graveside: sister and niece, couple from Leicester, some from Hull; not Monica, who couldn't make it on medical advice. No invitation to go to house, so we buggered off rather miserably, in our case without ever having seen it. Back on the train with junk food and Scotch.

Amis never did get to see Larkin's home in Hull before or after the poet's death. In fact he did not see Larkin in Hull for the whole thirty years that Larkin was the university librarian there. Larkin had visited Amis in Swansea, Cambridge and London, and Amis made return visits to him when he worked in libraries at Leicester and Belfast. But Larkin did not seem to want him in Hull. The reasons for this were and remain mysterious. Amis's best guess was that Larkin would have been embarrassed to let his old friend catch a glimpse of the humdrum life he lived, among humdrum people, especially the women he was involved with. Monica Jones had been his steady, though not exclusive, girlfriend since Amis had used parts of her character for parts of Margaret Peel's in *Lucky Jim*. Amis thought she was 'a pest and a bore' and that Larkin probably thought the same and had very likely said so in the diaries that were burned after his death. (Amis at first thought Monica had burned them for that reason, but later discovered it was Larkin's solicitor, Terence Weldon, who was responsible.) But there were other women in Larkin's life, like Maeve Brennan, whom Amis had known nothing about. 'I never heard a word about her,' Amis wrote to Conquest the following November, 'except that solitary "Maeve wants to marry me" blurting; you too, I think. He didn't half keep his life in compartments. But then he may simply have thought she wasn't important enough to go on about. Well, we shall never know.'

A year after Larkin's death, Amis reported to Conquest on an anniversary radio programme, *The Bicycle-Clipped Misanthropist*, which consisted of 'reminiscent chat by various Hull people': 'Honestly, the worst kind of N-country cheapjack homely wisdom you can imagine . . . Everybody talked on the level of "he perhaps had a forbidding exterior but really he was very warm with a word for everyone." . . . The only point of interest was that P had a telescope on his window-sill to get a better view of passing tits.' Amis also went to an exhibition of Larkin's 'stuff' at University College, London. He would have copied out Larkin's poem, 'Letter to a Friend about Girls', from the manuscript on display but that would have taken him all day and it 'would have been too depressing for words'.

Over the years, Amis had found Larkin close in several senses. Larkin stood his round but stopped well short of generosity. The same caution infected the rest of his life too, from his women to his writing. He never married and after two novels never published another. Amis thought he did not dare to take the risk of committing himself to a woman for life or to a novel for the year it might take to write. After all, neither marriage nor novel might work out. He did not, as Amis puts it, 'throw himself at life'. But they got on extremely well together. Theirs was the sort of friendship that could be picked up at once. A gap of days or months made no difference. Conversation would start up again as if they had met only yesterday. Larkin was funny and warm and sociable, and he and Amis spoke to each other on the same wavelength, about writing or politics or just things that were going on in the world. They did not always agree, especially on literary matters, but they knew each other well enough to avoid dangerous territory. Larkin might have wanted to talk literary theory but he knew Amis would not, so he kept off it.

Since Larkin read and advised Amis on the first draft of *Lucky Jim*, he had not had much influence on Amis as a writer, except in the most general sense, as a standard or example, Larkin being the living poet Amis most admired. There was something else about him, though, something more that Amis couldn't quite pin down. He agreed with Conquest that Larkin was 'something over and above friend and poet'. But what? 'I don't know; presence?' Amis was writing a few weeks after Larkin's death. 'I keep forgetting he's dead for a millisecond at a time: I must tell/ask etc. Philip – oh Christ I can't. I didn't realise how often I must have thought of him.' Amis continued to think of him and in his *Memoirs* wrote: 'My sorrow at his death and my abiding sense of loss is tinged with regret. He was my best friend and I never saw enough of him or knew him as well as I wanted.'

Other friends went too: John Betjeman and John Braine, for example. But Amis's greatest personal loss after Larkin was George Gale, the journalist. Because Gale worked in Fleet Street, latterly for the *Daily Mail*, he was more easily available at the end of the telephone for drinks, meals, conversation. When he died in 1990 Amis was cut off in one way or another from all his greatest friends. Szamuely, Larkin and Gale were dead; Conquest and Paul Fussell were in America; his Australian 'china', Jim Durham, the psychologist, was back home down under; Stuart Thomas was in Swansea. There was thus a hole in Amis's daily life. Their absence left him a little lonelier than he would have liked to be.

But there were other compensations. Honours came Amis's way: the CBE in 1981, an honorary fellowship from Swansea in 1985, the Booker Prize for *The Old Devils* in 1986, a knighthood in 1990. Amis enjoyed all these thoroughly, suffering no false modesty or Lefty qualms about the honours system. Since he was profoundly fond of his country and its traditions he was genuinely glad to be recognised through its official system of honours. One omission rankled a bit, though. No university seemed to think him worth an honorary doctorate. Philip Larkin had scored seven, but Amis got none. He wasn't jealous of his old friend who he thought thoroughly deserved all the recognition he got. But why not him too? The scale of his output compared with Larkin's was vast: more than twenty novels, lifetime collections of poetry and short stories, anthologies from science fiction to songs, a couple of collections of criticism, not to mention sundry reviews, polemics and a volume of *Memoirs*. Surely this should count for something at the universities? Probably it was this fecundity that was his undoing. A smallish, well-managed *œuvre* like Larkin's was better calculated to catch the academic eye than Amis's sprawling achievement. Besides, wasn't Amis best known as a comic writer? A chastiser of university standards? And a right-winger of Thatcherite tendency?

To add to his sense of well-being, Amis became quite seriously well off. In 1986 he told Conquest that his accountant had said he could afford to retire when he was seventy-five. Yet well before he reached that age he did not argue if called a rich man. All his novels back to *Lucky Jim* were available in paperback, and bits and pieces of money kept turning up from all over the place – a few quid from South Korea, for instance, when the television version of *The Green Man* was shown there. It all added up to a tidy annual income, enough for him to give up work if he wanted to.

But he did not want to. In his later years, Amis gave up many things: women, parties, travel. He gave up campaigning too – not that he had been more than a reluctant activist since about 1941. In 1984 he turned out as a member of the Writers' Guild to thwart the efforts of the leftish Theatre Writers' Union to have its members made members of the guild as well. 'By God they were cross when I and 40 other specially-alerted righties turned up and voted them down,' Amis told Conquest. 'Great example of democracy in action.' Thereafter his zeal, such as it was, fizzled out. Latterly, only one campaign attracted his attention. It was called Pipedown and its purpose was to stop music from being played in pubs by leaving cards in the bar warning the publican that one at least of his customers would not be returning if the music continued. Amis stopped reading anything new too. Thrillers were admissible because they had plots, characters, action – all the ingredients which seemed to him to have gone missing from so-called serious literature. For reasons like these he told Martin that he would never pick up a novel again that did not begin along the lines, 'A shot rang out . . .' He re-read old favourites – Anthony Powell, Christopher Isherwood, Evelyn Waugh, C.S. Forester. Fiction like theirs could be guaranteed to deliver, so why waste time on new stuff that would almost certainly disappoint?

If Amis lost enthusiasm for reading new novels, however, he did not lose his enthusiasm for writing them. Far from it. After the relatively barren years of the early 1980s his output, if anything, speeded up by the early 1990s. If asked why he carried on writing when he no longer needed the money, he would reply dismissively, 'It gives me something to do in the mornings', or words to that effect. This was a deliberately ambiguous answer, both frivolous and serious. He had to fill in time before lunch, and writing was how he had done that for thirty years, so why change? But being a writer was also how he defined himself: a writer was what he was. If he ceased to write, what would he be? It was better to avoid that question by not ceasing to write. The daily tussle with words was both torture and joy, as necessary to him as the presence of Hilly or the consumption of alcohol. What he feared most was not being able to think of anything more to write. And it is not fanciful to see too the figure of his father standing behind him at his desk and urging him on.

Looking back over the years, Amis could find many reasons for satisfaction with his working life. No other serious novelist and poet of his generation had produced so much in so many forms, created so much controversy and approval, earned more money or public honour and was still so prolifically at work when well past seventy. Father Amis would surely have been proud. But of his private life Amis could be less confident and his doubts about himself in this department show through his latest novel, *You Can't Do Both*.

It is the most biographical Amis has ever written in the sense that its story follows more closely than any other the course of his own life. Its hero, Robin Davies, was, like Amis, born in South London, grew up as the only child in a suburban household, went to school in the City, proceeded from there to Oxford, joined the wartime army, returned to Oxford, made his young girlfriend pregnant, married her and found a lecturer's job at a provincial university. Both Robin and Amis are 'inextinguishably promiscuous by nature' and remain adamantly determined to be so whether single or married. There are smaller point of likeness between Robin and Amis too. Both have mothers called Peggy. In both life and fiction Peggy had to persuade the parents of her son's pregnant girlfriend to come to their wedding. The mood, the atmosphere, of Robin's home life in South London resembles Amis's too in its enforced intimacy between two parents and a single child and in particular the determination of Robin's father to weed out the undesirables from Robin's circle of friends – undesirable because threateningly lower class. But there are differences as well. Robin is Welsh, goes as a schoolboy for holidays with relatives in Wales, takes his girlfriend there on a wartime leave. Amis was, or did, none of those things. No novel of his is ever lifted straight from the life.

Apart from any other reason for this, life is too sloppily arranged to give form to the plot which any novel must have. Amis therefore invented the two incidents that frame the plot and give the story its moral structure. At the beginning of the book, the young Robin has two dates one Saturday when he is free from school. He believes he can quite easily fit in one after the other, but his father tells him, no: he can't do both. At the end of the book, when Robin is married to Nancy and has two daughters, Nancy surprises him at a hotel where he is booked in for a bout of adultery. She too tells him he can't do both. He can have her and the girls or all the other women he wants, but not both.

Just before Nancy arrives, Robin speculates about why he behaves the way he does. He argues himself round in a circle and comes to no more consoling an answer than that he has 'a grim determination not to let a bagatelle like a hurried marriage interfere with what he had always done or at least wanted to do'. When Nancy confronts him he mumbles that he's sorry, 'no excuses, just me being a shit'. But she won't accept that. Even now, she thinks, he doesn't *really* blame himself. If he does something bad he puts it down to how he was brought up, what he has read and been taught, or just being one of the 'products of the age we live in'. When he does something good, though, it's 'full marks to Robin Davies Esquire and his marvellous natural sense of right and wrong'.

The story of the novel may closely reflect the story of Amis's first four decades, but they are not a contemporary diary. The novel was written in 1993, sixty to forty years after the period it describes. The Amis who wrote it was not the Amis who lived it. The writer is an old man reflecting on his youth. And what it seems to be telling him is that his father was not so wrong after all. Arbitrary and absurd though they seemed at the time, there was merit in his codes of behaviour. Even if they were less than perfect, rebelling against such codes by persistent infidelity to his wife was no proper way of readjusting the rules, justified by no fad or fashion of argument, no better, in fact, than selfish self-indulgence. The novel is a novel, taking its place alongside the others in Amis's *œuvre*. But it is also an apology, a message of love and regret. And it is addressed to Hilly.

Epilogue

I wrote Amis's biography in the sequence in which it appears here, beginning with the introduction, 'Portrait of the Artist in Age'. Having drafted that I put it aside and did not re-read it until I had finished the last chapter and begun making revisions to the whole book. Reading the introduction again eighteen months or so after I had written it, the portrait seemed slightly out of focus. Amis was no longer quite the same man I had described. He had aged. He was not so mentally spry. He was deafer, more crotchety. If there was a gap in the conversation, he was no longer certain to fill it. People he did not like or know well were more likely to be made unwelcome in his small circle at the Garrick Club or the Queen's public house. Amis continued his daily routines – work, Garrick, snooze and so on – but somehow the scale and scope of his days seemed to have shrunk. He did not want to go anywhere that could not be reached in half an hour by cab. He was physically slower. His back gave him pain. Seated in the chair from which he watched television, he would constantly shift about to make himself more comfortable, sitting back then leaning forward, propping his feet on a stool then dropping them to the floor. And he fell over. I saw him trip on a rug and fall, toppling to the floor like some great stricken oak or beast, though – a nice Amis touch – without spilling a drop of whisky in his glass.

Change had come by degrees, not through some single setback or series of them, so I did not really notice it was happening until I came to re-read my portrait and saw that it had. But I decided to leave the portrait unchanged. It seemed to give a fuller impression of the man Amis had been at his best, a portrait fairer than would have been one redrawn to show how he had declined. Besides, Amis had not changed out of recognition. Above all he still wrote. If anything, his output of words seemed to accelerate. In the three years it took me to write and publish his biography, he finished and published a book of short stories (*Mr Barrett's Secrets*), started, finished and published two novels (*You Can't Do Both* and *The Biographer's Moustache*) and wrote his own answer to *Fowler's Modern English Usage*, which at his death was still unpublished. As if all this was not enough, in his last months he advanced a hundred or so pages into yet another novel. Amis's energies remained strong, though they had become more narrowly concentrated on fewer things.

It may be that the experience of having his biography written helped shape those last two published novels. Amis always picked up new situations or themes in what was going on around him. *You Can't Do Both* traces the life of a youth from South London who goes to Oxford, becomes a lecturer in a provincial university and is unfaithful to his wife. Did my asking him questions about all these things in his own life put the idea for the book into his head? I don't know for sure because I didn't ask him, but I suspect it must have had at least something to do with it. Certainly *The Biographer's Moustache* was prompted by our biographical encounter. It is about an older literary chap having his biography written by a younger journalist, a description that fits Amis and fits me. The literary chap and the journalist in the novel are not at all like him or me but the basic relationship between them, of subject and enquirer, is the same as it was between us. All Amis ever needed was the right cue – a person, a situation, an idea – and he was off following a new fictional trail wherever it might take him.

We continued to meet after the biography was finished, at the pub and the club and at his home. He suggested that we write a book together, a sort of postscript to the biography, to be based on tape recordings of conversations across his dining table. I don't quite know what he expected of this project, except that it gave us an excuse to renew the Thursday night sessions we had regularly had while I was writing his biography. If nothing else came out of them, they would be at least an agreeable way to fill in time talking and drinking before *The Bill* on television.

In August 1995 Amis went to Swansea for his usual stay over the three weeks of the Garrick Club's annual summer holiday. As I had done in previous years I went there to join him for a few days towards the end of his excursion. He had told me over the phone that he felt as bad as he ever had in his life, but said it insouciantly enough for me not to worry much. But when I saw him it was clear that his banter had covered a very great deal of pain. Eight days earlier he had fallen over after lunch with a friend. The effect was to raise the pain in his back to a far higher intensity than ever in the past. The day before I arrived his speech had started to slur. Two days later I drove him back to the Chelsea and Westminster Hospital in London. His Swansea hostess, Virginia Rush, came too. He was able to come home from there to Primrose Hill but he soon needed intensive care and went back into hospital, this time to University College. On Sunday October 22, he died. There was no struggle at the end. He simply stopped breathing.

Throughout those last two months I always thought that Amis would pull through. He had come to seem somehow indestructible and permanent, still strong in spite of his decline. He told me once that he fully intended to see in the millennium and I believed him. There were moments in those last weeks when he was his old self, as when he said he was busily collecting material from his life in hospital for another novel, to be called *The Last Old Devil* or *The Oldest Devil*. The comings and goings of the medical staff were giving him plenty to think about. But anything less than full recovery would have been most miserable for him. Suppose he had been unable to go to the Garrick, or to write? He would have hated his life to become more circumscribed than it already had. As full recovery became less likely, so death came to seem the more merciful alternative.

30 October 1995

The day after he died, the obituary of Amis I had written several months before appeared in the *Guardian*. I also wrote a tribute to him in that week's *Spectator*. It is with these that I close this most unexpected and unwelcome epilogue to the paperback edition of *Kingsley Amis: A Biography*.

From the *Guardian*

Kingsley Amis has died at the age of seventy-three certain of his place up alongside Wodehouse, Waugh and Powell among the English comic masters of the twentieth century.

His achievement as a novelist tended to obscure his very considerable accomplishments as a poet. Amis was, in fact, a fully subscribed member of that increasingly rare species, the all-round man of letters. He wrote everything, from advertising copy to restaurant reviews, radio plays to television scripts, political polemic – on behalf of both left and right – to literary criticism, science fiction to a James Bond sequel. His style was rich, acute and fastidious and unmistakably his own.

Amis's output was prodigiously large as well as various: twenty-four novels, more than a dozen collections of poetry, short stories and criticism and a large miscellany of other work, including volumes on drink, politics and himself – his *Memoirs*.

It was a perverse indication of his standing that, while he was knighted by the Thatcher government, Amis received no doctorate from a university, unlike his friend the poet Philip Larkin who notched up half a dozen. Amis was not much admired in academic circles where he was seen as having committed several vulgar errors: he was popular, he was politically incorrect, he wrote too much. Furthermore, no donnish or critical exegesis was required to explain the meanings of his fiction; it spoke for itself. Worse still, he came to think not much of the Eng. Lit. trade he had once practised himself and didn't mind saying so.

Time will judge, but it will surely judge that the knighthood was the juster verdict. Amis's wit, shrewdness and verbal dexterity, his ability to turn quite unexceptional characters and everyday situations into something very much larger and more memorable, the sheer pleasure of reading him – with regular breaks in which to bawl with laughter – will ensure that he is read when more fashionable writers are forgotten. Or he will be read so long as ordinary, intelligent people still care about fiction of quality.

In art as in life, Amis was unable to conceal his feelings. If he was bored, he showed it in a frown and if he was amused he laughed hugely. His books, too, reflected his current preoccupations, personal or social, whatever they might be. As a result his fiction tracks closely changes in the morals and manners of his times, from Lucky Jim Dixon's hapless fumblings in the mid-1950s to the permissiveness of the 1960s and the uncertainties beyond. As a moving picture of middle-class relations between the sexes in the second half of the twentieth century, his novels are unmatched.

But they also track change that was going on inside Amis himself. Early Amis heroes may not have thought much of their jobs or their bosses, but they certainly had an appetite for pleasure, especially in women. The later novels are full of a sense of disintegration and disillusion, with women especially. Taken as a whole, the sequence of his novels reflect with remarkable clarity the movement in Amis's own feelings as he moved from youth to age.

In later years the real sensitivity of his work seemed sometimes to get lost sight of beneath his image, that of an arch-reactionary, misogynist, boozer and scourge of modernities in all things from poetry to restaurant food. It was easy to depict him in blimpish red-face, staring apoplectically and uncomprehendingly out of the Garrick Club bar at the fools in the street below.

Amis was by then a fully committed conservative who, for example, admired almost everything about Lady Thatcher except her treatment of the universities, for which he thought Oxford was quite right to withhold the honorary degree it customarily offers any of its graduates who become Prime Minister.

But Amis also liked to be seen in a strong light. If he was to be conservative, let him be thorough-going about it. Should a journalist call to ask his views on some matter of the day, he would get a robust answer or a flea in his ear. Besides, strong opinions were likely to annoy people and were therefore much more fun to hold than half-baked ones.

He believed that, when at the typewriter, absolute integrity should prevail. But when a writer came to sell himself and his wares in the market-place, more or less anything went. He made mischief on his own behalf and in doing so did some mischief to his own reputation.

Amis's prime motive in moving to the right from the Communism of his youth was his detestation of Communist tyranny, something he thought post-war intellectuals far too easily ignored. But there remained a strong thread of continuity in his beliefs and attitudes, deriving from his lower middle-class origins in Norbury, south-west London.

As seen from here, much of the world appeared to be dominated by snobberies and exclusivities which Amis came to detest, in literary no less than social or political affairs. Culture as the possession of a class was just as repellent to him as such a class's hold on other kinds of power. Why should literature be fenced off like a private estate? It should be accessible to all.

Amis was, at the same time, powerfully attracted by popular culture, the films, jazz, detective stories and science fiction that competed for his attention during his childhood in the 1930s. If 'highbrow' art was to keep its primacy over the 'lowbrow' variety, then it, too, must entertain. It was not much fun, he would say, to read a book you had to read another book to understand.

Literature at first seemed to be the possession of the traditional upper class. Later, it appeared to have fallen into the hands of another class, of professional pseuds and trend-hounds and poseurs of all kinds who fenced themselves off behind obscurities and so kept ordinary people at a respectfully admiring distance. These new masters of the literary circus may have been as humble in their origins as Amis, but the exclusiveness they created in their field was not so very different from that contrived by the old upper-class in its.

Latterly Amis rarely read anything new that could be described as 'literary' fiction. Even the much acclaimed novels of his son Martin were more glanced at than read. Thrillers were more acceptable because they had old-fashioned strengths, like plots and characters. They might not promise a great deal but they delivered what they promised. And they were not boring, in Amis's eyes almost the worst thing in literature or in people.

Amis was in favour of élites because he was in favour of excellence. But élites should earn their status by their achievement, not by inheritance or by becoming one of the self-appointed 'experts' who laid down the law for everybody else. The increasing dominance of such 'experts' in literature as in politics does much to explain Amis's steady drift towards conservatism. As someone who owed his own education to scholarships and had then gone on to become a university teacher, he particularly detested what he saw as the vandalism of the Labour Education Secretary, Anthony Crosland, in destroying the grammar schools – in Amis's view a typical piece of both intellectual and social class arrogance.

Amis was born on 16 April 1922, at Clapham, south London, to Peggy and William Amis, a mustard manufacturer's clerk employed at Colman's office in the City. As an only child Amis was both cossetted and ordered about. The domestic atmosphere was warm but smothering. Perhaps as a result Amis in his early teens began to suffer from the anxieties and panic attacks that were to dog him intermittently all his life. He suffered bouts of 'depersonalisation' – a feeling that he was no longer there – and came to fear travelling in lifts or underground trains or being left at night in an empty house. After being treated to a flip around Croydon Aerodrome in 1932 Amis vowed never to fly again and never did.

Partial escape from the restrictions of home came when he went to the City of London School on the Thames at Blackfriars. Here he found the education both rigorous and liberal: rigorous in its concentration on a traditional curriculum but liberal in finding time for art and music, as well as being entirely hospitable to minorities like Catholics and Jews. Among Amis's rightwing attitudes there was never a hint of racism.

Amis won an English scholarship to St John's College, Oxford. Soon after he went up in 1941 he met Philip Larkin with whom he formed the most important friendship of his life. At Oxford they mostly fooled about with other undergraduates: drinking, listening to jazz, behaving badly, like hearties except that they never went near a sports field. Their personal friendship ripened later. Larkin gave Amis detailed advice on his first novel, *Lucky Jim*, but Amis did not ask for his help with a novel again. Nor did they much discuss literary matters. Amis had little taste for this, except when teaching students. He believed the business of a novelist was to write novels, not talk about them. Their influence on each other was mostly indirect: when writing they would have the other as invisible audience whose approval each most wanted.

After four Oxford terms, Amis was summoned to the army, commissioned and assigned as a signalmaster to Second Army headquarters, landing in Normandy at the end of June 1944, his first visit abroad. Being abroad with the army was all right because there were colleagues and a purpose, but Amis never did much warm to being abroad as a civilian on holiday. Though he liked America, where he taught at Princeton and Nashville, very much indeed.

Amis's army job was to organise signals traffic and he saw no action at first hand. But the army enlarged his understanding of people and the world. Idealism in politics was no longer enough since all around him he saw men restlessly pursuing their own interests. Besides, the world was too complicated to be organised according to the easy formulas of leftwing politics. Imagine the Fabian Society, say, or the General Post Office mounting the invasion of Europe! When Amis returned to Oxford on early release in 1945 he was no longer the Communist he had been when he left, though he did not vote for a Tory government until 1970.

In 1946 Amis met the seventeen-year-old Hilary (always Hilly) Bardwell, a model at the Ruskin School of Art, in an Oxford coffee shop. She became pregnant, they married and Philip was born in 1948. In 1949, with a First to his credit, Amis became a lecturer at the university college in Swansea.

His first serious attempt at a novel, *The Legacy*, was turned down by several publishers, but his second, *Lucky Jim*, was welcomed by Gollancz. Its publication in 1954 was a great success, winning a Somerset Maugham Award, and thereafter Amis's literary reputation was rarely in doubt. He became identified with literary movements – 'The Movement' and the 'Angry Young Men' – but he never saw these as much more than labels pinned on him by journalists in search of trends. In 1961 Amis left Swansea for a fellowship at Peterhouse. He came to think of his years in Wales as the happiest of his life and in the 1980s and 1990s always went back to Swansea in the summer.

But Amis stayed in Cambridge only two years, finding university social life unbearably formal and his teaching load enjoyable but, if properly done, too engrossing to leave enough energy over for writing. He resigned and rented a house in Majorca – being abroad an unlikely place, but Robert Graves would be a near-neighbour – planning to try writing there full-time for a year. But his plans were aborted. Amis had started an affair with the novelist Elizabeth Jane Howard, whom he had met while taking part in a seminar on sex in literature at the Cheltenham Literary Festival, of which she was director. This was far from the first affair he had got involved in since marrying Hilly, but it was the first in which he openly took the other woman away with him on holiday.

When he returned he found that Hilly had taken the children – there were now three, Philip, Martin and Sally – off to the house in Majorca. Amis rejoined Howard in London and stayed with her, and when he and Hilly were divorced they married. They were at first romantically happy but after a few years the marriage began to disintegrate. In 1980, Howard 'bolted' – her own word. She offered to come back if Amis would give up drinking completely but he refused. It wasn't that he couldn't – soon afterwards he stopped for six months following an accident and a spell in hospital – but he had always drunk a lot and thought total abstinence an intolerable demand which he would resent for ever if he agreed to it.

Amis very much did not like being left alone in his large house in Hampstead but his sons soon came up with an alternative. Why did he not establish a new home with Hilly and her third husband, Lord Kilmarnock? They were short of money, he of company; it made sense. Philip and Martin did not believe it could last long but it did. The household – like something out of an Iris Murdoch novel, as Amis would say – was not without its tensions. But, in divided quarters, it survived intact until Amis's death.

The 1980s and 1990s were scarcely happy – age and Amis's anxious spirit did not allow for contentment – but they were far from miserable. Amis developed a routine: work, lunch at the Garrick, a sleep, more work, television, supper, books, bed. His divorce from Hilly was always his deepest regret, but she was around and the three children not far away. And he continued to write almost to the end. What more could he hope for?

From *The Spectator*

You would not expect, say, P.D. James or Ruth Rendell to stick a knife in you just because they write about murderers. But ever since *Lucky Jim*, Kingsley Amis has been seen not only to resemble his characters but to actually be them. Over the years he has been variously depicted in the style of his heroes as drunk, philistine, lefty, righty, sex-mad, misogynist, curmudgeon, Blimp and scourge of modernist error in everything from cooking to poetry. There was at least some truth in all these images because Kingsley's heroes always started out from himself even if they finished up very different. And Kingsley, it has to be said, hugely enjoyed hamming it up in the roles he had been assigned to. But it is not in any of those stereotypical guises that I shall remember him. He was above all quick-minded, verbally agile, terribly funny, a vigorous persecutor of bores, pseuds and wankers and a most tremendous mimic. Whenever I think of *The Spectator*'s late editor, George Gale, I can't help seeing Kingsley's growling, shrugging imitation of him, which seemed more like George than George. Only a few weeks ago I happened to rouse him (Kingsley, I mean) from a post-lunch snooze and he turned his waking up into a wonderful imitation of his father being woken in similar circumstances and pretending that he had only been pretending to sleep. Kingsley could be kind, generous, affectionate and charming though he chose not to advertise those sides of himself, probably because they were worthy and therefore dull. The Garrick Club bar already seems a drabber place. I shall miss Kingsley most horribly.

Notes

As indicated in the Preface, the principal sources for this book are Kingsley Amis's published works, which are listed in the Bibliography; Kingsley Amis himself (in conversation with the author); and three sets of unpublished letters from Amis to Philip Larkin, Robert Conquest and Anthony Powell. I have borrowed from Amis's own *Memoirs* throughout. Dale Salwak's *Kingsley Amis, Modern Novelist*, published by Harvester Wheatsheaf in 1992, has also proved a very useful guide. As far as possible I have tried to identify these and other sources as I went along. Where quotation marks are placed around words or phrases without citation of a source it is to indicate that the words are Amis's, not mine, and that his words are usually taken from his *Memoirs* or his conversations with me. Notes on sources by chapter follow.

Introduction: Portrait of the Artist in Age

The lines quoted from Philip Larkin ('the shit in the shuttered chateau . . .') are from 'The Life with a Hole in it', in Larkin, *Collected Poems*, ed. Anthony Thwaite (London: The Marvell Press and Faber & Faber, London, 1988), p. 202.

Chapter 1   Norbury

Essays of Amis quoted are 'From Aspidistra to Juke-Box', 'Dracula, Frankenstein, Sons & Co.' and 'City Ways', collected in *What Became of Jane Austen?* (London: Cape, 1970). For the history of Norbury, see

Barbara Mullen and Graham Harker, *Norbury: The Story of a London Suburb* (privately published). The *Croydon Advertiser* for 24 September 1932 contains the interview with the headmaster of Norbury College, and evidence for the promotion of Arthur Mackness is to be found in the same paper, dated 16 January 1937.

Chapter 2   City of London School

A.E. Douglas-Smith, *The City of London School* (Oxford: Blackwell, 1965), describes the history of the school. C.J. Ellingham, *Essay Writing – Bad and Good*, was published by A. Wheaton & Co., The Paternoster Press (Exeter, 1935). On G.K. Chesterton as an early favourite, see Kingsley Amis's Introduction to the Penguin edition of *The Man Who Was Thursday* (London, 1986). Also quoted are an interview with Amis by Melvyn Bragg, the *Listener*, 20 February 1975, and a letter to the author from C.N. Vokins.

Chapter 3   Evacuation to Marlborough

Amis's reflections on the writing of *The Riverside Villas Murder* are in an unpublished BBC radio broadcast, *The Living Novelist*, 7 March 1973. His views on religion are in 'Godforsaken', in *The Amis Collection*, and 'On Christ's Nature', in *What Became of Jane Austen?*, in which the essay 'City Ways' also contains Amis's thoughts on his schooldays. For CLS's evacuation to Marlborough, see Douglas-Smith, *The City of London School*, and the December 1939 edition of the school magazine. Amis's poem about the icy weather at Marlborough, 'Instead of an Epilogue', is in his *Memoirs*. Some of Amis's youthful feelings towards the military life are in the *Observer Magazine*, 9 June 1968.

Chapter 4   Wartime Oxford

Sources for Amis's first four terms at Oxford include Philip Larkin's Introductions to *Jill* and to *All What Jazz*, collected in *Required Writing* (London: Faber & Faber, 1983); *Selected Letters of Philip Larkin, 1940–85* (London: Faber & Faber, 1992); *Philip Larkin: A Writer's Life*, by Andrew Motion (London: Faber & Faber, 1993); and the contributions of Nick Russel and David H. Whiffen to

Philip Larkin, 1922–1985: A Tribute, ed. George Hartley (London: The Marvell Press, 1988). 'Dockery and Son' is in Philip Larkin's *Collected Poems 1944–1979*. Some of Amis's views on his Oxford education are in 'Pernicious Participation', in *The Amis Collection*, and his later reflections on his Communist youth were broadcast in an unpublished talk, *Brave Causes*, on the BBC Home Service on 6 June 1961. For Homer Lane's story, see *Homer Lane: A Biography* by W. David Wills (London: George Allen & Unwin, 1964). Denis Healey and Roy Jenkins gave their versions of Oxford Labour politics as seen from left and right in their autobiographies: *The Time of my Life* by Denis Healey (London: Michael Joseph, 1989), and *A Life at the Centre* by Roy Jenkins (London: Macmillan, 1991).

Chapter 5   In the Army

Amis's stories of army life, 'My Enemy's Enemy', 'Court of Inquiry' and 'I Spy Strangers', are in his *Collected Short Stories*. Second Army Signals War Diaries are in the Public Record Office, file nos WO 171, 239 and 240. Histories of the Royal Signals and of Second Army are at the Imperial War Museum. The typescript of 'Who Else is Rank' is in the Huntington Library, California. Amis's letter to E. Frank Coles is quoted in Dale Salwak's *Kingsley Amis, Modern Novelist* (Hemel Hempstead: Harvester Wheatsheaf, 1992). Amis's poems are in *Bright November* ('Release') and the *Collected Poems 1944–1979* ('A Reunion').

Chapter 6   Peacetime Oxford

Amis's views on dons are in his essay 'No More Parades', in *What Became of Jane Austen?* The junior common room suggestions book and the kitchen book are in the St John's College archives.

Chapter 7   Swansea

The Senior Commoner by Julian Hall was published by Martin Secker (London, 1933). The manuscript of *The Legacy* is in the Huntington Library and there is a description of it in Dale Salwak, *Kingsley Amis, Modern Novelist*. An account of R.A. Caton is to be found in Andrew Motion's biography of *Philip Larkin*. Doreen Marston's letter to Amis is among Amis's letters to Larkin. Amis's poem 'Their Oxford' is in his *Collected Poems 1944–1979*. *The University College of Swansea: An Illustrated History* by David Dykes (Stroud: Alan Sutton, 1992) gives an account of the college and its staff. 'Moral Fibre' is in Amis's *Collected Short Stories*.

Chapter 8   *Lucky Jim*

For the origins of *Lucky Jim* and Amis's attitude to realism in fiction, see 'Real and Made-up People', in *The Amis Collection*. Amis's encounter with Dylan Thomas is in 'An Evening with Dylan Thomas', in *What Became of Jane Austen?* The Gollancz company files contain Amis's correspondence with the publisher of *Lucky Jim*. Philip Larkin's poem 'Born Yesterday' is in his *Collected Poems*. Reviews of *Lucky Jim* are in Dale Salwak's *Kingsley Amis: A Reference Guide* (Boston, Mass.: G.K. Hall, 1978).

Chapter 9   Movements

The Movement and the Angry Young Men are described in Harry Ritchie's *Success Stories* (London: Faber & Faber, 1988) and Blake Morrison's *The Movement* (Oxford: Oxford University Press, 1980). Amis's unpublished talk, 'Letter to Someone's Son', was broadcast on the BBC Light Programme on 28 February 1963. His poems 'A Bookshop Idyll' and 'Nothing to Fear' are in his *Collected Poems 1944–1979*. 'Communication and the Victorian Poet' is in *Essays in Criticism* (Oxford: Blackwell, 1954). Amis's unpublished thesis is in the Henry Ransom Humanities Research Center of the University of Texas and is discussed in Paul Fussell's *The Anti-Egotist: Kingsley Amis, Man of Letters* (Oxford: Oxford University Press, 1994).

Chapter 10   Politics

For Lord Fulton's views on university expansion, see *The Idea of a New University* (London: Deutsch, 1964). *Socialism and the Intellectuals* was published by the Fabian Society (London, January 1957). *Declaration*, ed. Tom Maschler, was published by MacGibbon and Kee (London, 1957). Amis's letter to the *Daily Worker* was in the issue of 14 February 1957, and his interview with *Life* appeared in its issue of 26 May

1958, while Peter Simple's response came in the *Daily Telegraph*, 3 July 1958.

Chapter 11   Portugal

Articles and letters by Amis, Robert Conquest and others are in the *Spectator*, 1955–6. Philip Larkin's letter is in the *Selected Letters of Philip Larkin*, ed. Anthony Thwaite (London: Faber & Faber, 1992). Other correspondence is in the unpublished files of Victor Gollancz.

Chapter 12   Princeton

Amis's interview after one day in America is in the *Princeton Packet*, 18 September 1958. For his views on teaching, see the *Observer*, reprinted in the *Daily Princetonian*, 6 January 1959. For the Kerouac debate and Amis on literary lecturing, see the *Village Voice*, quoted in the *Daily Princetonian*, 26 November 1958, and *Harper's Magazine*, October 1959. Amis on the Englishman abroad is in the *Observer*, 3 January 1960. See also *Tony Crosland* by Susan Crosland (London: Cape, 1982).

Chapter 13   Leaving Swansea

Amis's letter to Marguerite Cutforth is in the BBC archives. Some of his views on Wales are in 'Where Tawe Flows', in *What Became of Jane Austen?*

Chapter 14   Cambridge

Amis's interviews are in the *Daily Mail*, 23 January 1962, and *The Times*, 21 December 1962. His views on Cambridge and Yevtushenko are in 'No More Parades' and 'Kipling Good', in *What Became of Jane Austen?* F.R. Leavis's comments on Amis are in *English Literature in Our Time and the University* (London: Chatto & Windus, 1969), p. 56, and *Dickens The Novelist* by F.R. and Q.D. Leavis (London: Chatto & Windus, 1970), p. 140. Richard Eyre has written about Amis as tutor in *Utopia and Other Places* (London: Bloomsbury, 1993). Robert Graves's remarks on Amis are in a letter to James Reeves, in *Between Moon and Moon: Selected Letters of Robert Graves, 1946–72*, ed. Paul O'Prey (London: Hutchinson, 1984),

p. 219. Amis's letters to Graves are in transcripts loaned to the author by Richard Perceval Graves.

Chapter 15   Elizabeth Jane Howard

Peter Grosvenor's account of the Cheltenham Literary Festival is in the *Daily Express*, 5 October 1962. For the Amis work routine, see his interview with Michael Barber, *Paris Review*, 64 (Winter 1975).

Chapter 16   Maida Vale

Amis's correspondence with Victor and Livia Gollancz is in the Gollancz company files. See also *Gollancz: The Story of a Publishing House, 1928–78* (London: Gollancz, 1978) and *Victor Gollancz: A Biography* by Ruth Dudley Edwards (London: Gollancz, 1987). Amis's emerging political views are reflected in: *Mademoiselle*, June 1962; the *Observer*, 20 July 1963; the *Morning Star*, 16 May 1966; 'Why Lucky Jim Turned Right' and 'Lone Voices', in *What Became of Jane Austen?* See also *Tony Crosland* by Susan Crosland.

Chapter 17   Prague and Nashville

For Amis in Nashville, see the *Observer*, 3 January 1960. Bruce Cook, *National Observer*, 4 March 1968 (quoted by Dale Salwak). 'Kingsley in Nashville' by Richard Porter is in *Kingsley Amis in Life and Letters*, ed. Dale Salwak (London: Macmillan, 1990). 'South' is in Amis's *Collected Poems 1944–1979*.

Chapter 18   Barnet

The Amises' joint interview appeared in the *Daily Mail*, 15 October 1965. Elizabeth Jane Howard's *Vogue* article appeared in June 1978. Amis wrote about Kipling in *Rudyard Kipling and his World*.

Chapter 19   Hampstead

Amis's views on the role of the artist are in *Contemporary Novelists*, ed. James Vinson (London: St James's Press, 1972). Some of his thoughts on the passing of eras in art, especially science fiction, are in *The Golden Age of Science Fiction*. Amis's piece about his last cruise with

Jane was first published by the *Sunday Times*, 26 July 1981, and is reprinted in *The Amis Collection*. Paul Fussell wrote about Amis in 'Kingsley, as I Knew Him', in *Kingsley Amis in Life and Letters*. 'Senex' is in Amis's *Collected Poems 1944–1979*.

Chapter 20   Primrose Hill

For something of Lord Kilmarnock's life in Spain and before, see *The Road from Ronda* by Alastair Boyd (London: Collins, 1969). Amis's opinions of La Tante Claire restaurant are in *The Amis Collection*, and contrary views are in Egon Ronay's *Cellnet Guide* (Basingstoke: Automobile Association, 1987).

Chapter 21   The Queen's and the Garrick

Philip Larkin's last letters to Amis are in *Selected Letters of Philip Larkin*, ed. Anthony Thwaite (London: Faber & Faber, 1992).

Bibliography

Novels

Lucky Jim. London: Victor Gollancz; New York: Doubleday, 1954.

That Uncertain Feeling. London: Victor Gollancz, 1955; New York: Harcourt Brace, 1956.

I Like It Here. London: Victor Gollancz; New York: Harcourt Brace, 1958.

Take a Girl Like You. London: Victor Gollancz, 1960; New York: Harcourt Brace, 1961.

One Fat Englishman. London: Victor Gollancz, 1963; New York: Harcourt Brace, 1964.

The Egyptologists (with Robert Conquest). London: Jonathan Cape, 1965; New York: Random House, 1966.

The Anti-Death League. London: Victor Gollancz; New York: Harcourt Brace, 1966.

Colonel Sun (as Robert Markham). London: Jonathan Cape; New York: Harper & Row, 1968.

I Want It Now. London: Jonathan Cape, 1968; New York: Harcourt Brace, 1969.

The Green Man. London: Jonathan Cape, 1969; New York: Harcourt Brace, 1970.

Girl, 20. London: Jonathan Cape, 1971; New York: Harcourt Brace, 1972.

The Riverside Villas Murder. London: Jonathan Cape; New York: Harcourt Brace, 1973.

Ending Up. London: Jonathan Cape; New York: Harcourt Brace, 1974.

The Alteration. London: Jonathan Cape, 1976; New York: Viking Press, 1977.

Jake's Thing. London: Hutchinson; New York: Viking Press, 1978.

Russian Hide-and-Seek. London: Hutchinson, 1980.

Stanley and the Women. London: Hutchinson, 1984; New York: Summit, 1985.

The Old Devils. London: Hutchinson, 1986; New York: Summit, 1987.

The Crime of the Century. London: J.M. Dent, 1987. London: Century Hutchinson; New York: Mysterious Press, 1989.

Difficulties with Girls. London: Hutchinson, 1988; New York: Summit, 1989.

The Folks that Live on the Hill. London: Hutchinson; New York: Summit, 1990.

The Russian Girl. London: Hutchinson, 1992; New York: Viking Penguin, 1994

You Can't Do Both. London: Hutchinson, 1994.

Poetry

Bright November. London: Fortune Press, 1947.

A Frame of Mind. Reading: School of Art, University of Reading, 1953.

Kingsley Amis: No. 22. The Fantasy Poets. Oxford: Fantasy Press, 1954.

A Case of Samples: Poems 1946–1956. London: Victor Gollancz, 1956; New York: Harcourt Brace, 1957.

'The Evans Country'. Oxford: Fantasy Press, 1962.

A Look round the Estate: Poems 1957–1967. London: Jonathan Cape, 1967; New York: Harcourt Brace, 1968.

Wasted, Kipling at Bateman's. London: Poem-of-the-Month Club, 1973.

Collected Poems 1944–1979. London: Hutchinson, 1979; New York: Viking Press, 1980.

Short stories

My Enemy's Enemy. London: Victor Gollancz, 1962; New York: Harcourt Brace, 1962.

Penguin Modern Stories II (with others). Harmondsworth: Penguin, 1972.

Dear Illusion. London: Covent Garden Press, 1972.

The Darkwater Hall Mystery. Edinburgh: Tragara Press, 1978.

Collected Short Stories. London: Hutchinson, 1980, 1987 ('Investing in Futures' and 'Affairs of Death' added).

Mr Barrett's Secret and Other Stories. London: Hutchinson, 1993.

Recordings

Kingsley Amis Reading his own Poems. Listen, 1962.

Poems (with Thomas Blackburn). Jupiter, 1962.

Plays

Radio

Touch and Go. BBC, 1957.

Captain Nolan's Chance. BBC, 1994.

Television

A Question about Hell. 1964.

One Fat Englishman. 1967.

I Want It Now. 1968.

The Importance of Being Hairy. 1970.

'See What You've Done'. *Softly, Softly.* 1974.

Dr Watson and the Darkwater Hall Mystery. 1974.

'We Are All Guilty'. *Against the Crowd.* 1975.

Criticism

New Maps of Hell: A Survey of Science Fiction. New York: Harcourt Brace, 1960; London: Victor Gollancz, 1961.

The James Bond Dossier. London: Jonathan Cape; New York: New American Library, 1965.

What Became of Jane Austen? and Other Questions. London: Jonathan Cape, 1970; New York: Harcourt Brace, 1971.

Rudyard Kipling and his World. London: Thames & Hudson, 1975; New York: Charles Scribner's Sons, 1976.

The Amis Collection, with an 'Introduction' by John McDermott. London: Hutchinson, 1990.

Works edited or with contributions by Kingsley Amis

Oxford Poetry 1949 (with James Michie). Oxford: Basil Blackwell, 1949.

Oscar Wilde: Poems and Essays. London: Collins, 1956.

'Afterword' to Samuel Butler's *Erewhon*. New York: New American Library (Signet), 1960.

Penguin Modern Poets 2 (with Dom Moraes and Peter Porter). Harmondsworth: Penguin, 1962.

Spectrum: A Science Fiction Anthology (with Robert Conquest). 5 vols. London: Victor Gollancz, 1961–5; New York: Harcourt Brace, 1962–7.

G.K. Chesterton: Selected Stories. London and Boston: Faber & Faber, 1972.

Tennyson (*Poet to Poet* series). Harmondsworth: Penguin, 1973.

Harold's Years: Impressions from the New Statesman and Spectator. London: Quartet, 1977.

The New Oxford Book of Light Verse. London and New York: Oxford University Press, 1978.

The Faber Popular Reciter. London and Boston: Faber, 1978.

The Golden Age of Science Fiction. London: Hutchinson, 1981.

The Great British Songbook (with James Cochrane). London: Pavilion/ Michael Joseph, 1986.

'Introduction' to G.K. Chesterton's *The Man Who Was Thursday*. London: Penguin, 1986.

The Amis Anthology. London: Hutchinson, 1988.

The Pleasure of Poetry. London: Cassell, 1990.

Miscellaneous writings

Socialism and the Intellectuals. London: Fabian Society, 1957.

The Book of Bond, or Every Man His Own 007 (as Lt-Col. William ('Bill') Tanner). London: Jonathan Cape, 1965.

Lucky Jim's Politics. London: Conservative Political Centre, 1968.

Black Papers on Education. Manchester: Critical Quarterly Society, 1968–75.

On Drink. London: Jonathan Cape, 1972; New York: Harcourt Brace, 1973.

An Arts Policy? London: Centre for Policy Studies, 1979.

Every Day Drinking. London: Hutchinson, 1983.

How's Your Glass? London; Weidenfeld & Nicolson, 1984.

Memoirs. London: Century Hutchinson; New York: Summit, 1991.

Index